Praise for
Blood, Iron, and G

"*Blood, Iron, and Gold* reminds us that the railroads did more than just speed up travel or build up national economies. They literally changed the way human beings experienced, thought about, and lived their lives. Christian Wolmar's book should put all high-speed-rail advocates on notice. Trains can return to the American landscape, traveling twice as fast, reprising the social revolution they set off almost two centuries ago."
—Richard F. Harnish, executive director,
Midwest High Speed Rail Association

"[Wolmar's] work is both a serious history and an adventure story. Highly recommended for anyone interested in the growth and global historical impact of railroads."
—*Library Journal*, STARRED review

"Wolmar explores this fertile subject with a blend of lucid exposition and engaging historical narrative. The result is a fascinating study not just of a transportation system, but of the Promethean spirit of the modern age."   —*Publishers Weekly*

"[Wolmar] covers a great deal of territory in *Blood, Iron, and Gold*, but he keeps the reader engaged by highlighting extraordinary projects like the building of the Trans-Siberian Railway from 1891 to 1904. It connected St. Petersburg to Vladivostok, a distance of almost 6,200 miles. Equally stirring is the saga of Cecil Rhodes and his never-completed Cape-to-Cairo line; and that of Peru's vertiginous Central Railway, which ascends the Andes and passes through the Galera Tunnel, 15,694 feet above sea level. The book also features cameo appearances by such colorful figures as Benito Mussolini, who may or may not have made Italy's trains run on time but who definitely made them run faster and more frequently. Nor does Mr. Wolmar neglect the pop-culture angle: Agatha Christie fans will be sorry to learn that history records no instance of a real-life murder on the Orient Express."
—*Wall Street Journal*

"Timely. . . . A superb new history of the world's railways. . . . In one brilliantly written volume Wolmar relates the story of the first global rail revolution." —Andrew Adonis, *Financial Times*

"It's not clear who first thought of putting carts and carriages on flanged wheels and hauling them over iron rails behind steam engines. But the railroad, writes transportation historian Christian Wolmar, changed everything. And he means everything. . . . It's a vast geopolitical story, but Wolmar manages to tell it without losing sight of the romance and adventure, the triumphs and frequent tragedies that accompanied the advancing rails."
—*Dallas Morning News*

"Most attempts at a generalist approach toward railroad history err on the side of history and slight the rail side. [*Blood, Iron, and Gold*] keeps the two elements in graceful balance. And, thanks to Wolmar's crisp style, it's a pleasure to read."
—*Trains Magazine*

"This authoritative and highly readable book will remain the definitive history for years to come."
—Michael Williams, *Daily Telegraph*

"Wolmar brings great energy to the task of explaining how the railways transformed the world." —*New Statesman*

"Christian Wolmar has long been known as the best-informed expert and deadliest critic of Britain's farcically ill-run railway system. Now, in *Blood, Iron, and Gold* he reveals some of the passions behind all that, and his immense knowledge of how the iron horse has shaped world history since its first invention."
—Stephen Howe, *Independent*

BLOOD, IRON, and GOLD

Also by Christian Wolmar

BLOOD, IRON, and GOLD

How the RAILROADS Transformed the World

CHRISTIAN WOLMAR

PUBLICAFFAIRS
New York

First published in Great Britain in 2009 by Atlantic Books, an imprint of
Grove Atlantic Ltd.

Published in the United States by PublicAffairs™,
a member of the Perseus Books Group.

Printed in the United States of America.

PublicAffairs books are available at special discounts for bulk purchases in the U.S. by
corporations, institutions, and other organizations. For more information, please con-
tact the Special Markets Department at the Perseus Books Group, 2300 Chestnut
Street, Suite 200, Philadelphia, PA 19103, call (800) 810-4145, ext. 5000, or e-mail
special.markets@perseusbooks.com.

Book Design by Jeff Williams

Library of Congress Cataloging-in-Publication Data
Wolmar, Christian.
    Blood, iron, and gold : how the railroads transformed the world / Christian Wolmar
—1st ed.
        p. cm.
    "First published in Great Britain in 2009 by Atlantic Books"—T.p. verso.
    Includes bibliographical references and index.
    ISBN 978-1-58648-834-5 (alk. paper)
    PB ISBN: 978-1-58648-949-6
    1. Railroads—History. 2. Railroads—Social aspects—History. 3. Railroads—
Economic aspects—History. 4. Social change—History. I. Title.

HE1021.W78 2009
385.09—dc22

                                                                    2009038340

10 9 8 7 6 5 4 3 2 1

Dedicated to my wonderful Deborah,
who puts up with my obsessions and foibles,
and inspires me to keep going.

CONTENTS

Photos on text between pages 272–273

LIST OF MAPS
AND ILLUSTRATIONS

Maps

1. Main railway lines of Europe.
2. The main Andean railways at their peak.
3. Transcontinental routes in the United States and Canada.
4. Main railway lines of Australia.
5. The Cape to Cairo railway, Africa.
6. Main railway lines of India.
7. The Trans-Siberian Railway.

These maps are purely indicative and omit many lines and connections for the sake of simplicity. On the maps of Europe and Australia, modern place names and boundaries have been used, but on the maps of India and Africa, place names and borders appear as they were around 1900.

Illustrations

1. Opening of the Liverpool and Manchester Railway. National Rail Museum / Science and Society.
2. Carriages on France's first railway. Photos12.com.
3. *Tom Thumb*. From the Collections of the Baltimore & Ohio Railroad Museum.
4. The Dublin and Kingstown Railway. The British Library / HIP / TopFoto.

5. Drilling machine at the Mont Cenis tunnel. World History Archive / TopFoto.

6. Golzschthal Viaduct. Science Museum Pictorial.

7. Sugar plantation railways in Cuba. The British Library / HIP / TopFoto.

8. The Panama Railway. © The Image Works Archives EIWA0922 / TopFoto.

9. Australian convict railway. National Library of Australia / J. W. Beattie.

10. Commodore Perry and his model train. The Library of Congress.

11. Chinese workers. TopFoto.

12. Meeting of the Union Pacific and Central Pacific railroads. The Granger Collection / TopFoto.

13. American locomotive. Brian Solomon / Milepost 92½.

14. Mountain Creek Trestle Bridge. TopFoto.

15. Sunday worship on the Union Pacific. © North Wind / North Wind Picture Archives. All rights reserved.

16. Traveling by rail in India. Milepost 92½.

17. Gare de l'Est in Paris. Akg-images.

18. Early Japanese railways. TopFoto.

19. A shunting elephant. Milepost 92½.

20. Crossing the Ghats. Hulton Archive / Getty Images.

21. Victoria terminus in Bombay. Dinodia / TopFoto.

22. Railway disaster in India. Illustrated London News Ltd. / Mary Evans Picture Library.

23. Cecil Rhodes. The Granger Collection / TopFoto.

24. Railway bridge over the Victoria Falls. James Burke / Time Life Pictures / Getty Images.

25. Poster for the Orient Express. The Granger Collection / TopFoto.

26. Building the Trans-Siberian Railway. Alinari / TopFoto.

27. Laborers on the Trans-Siberian Railway. Akg-images.

28. Engineers on the Madeira-Mamoré line. South American Pictures.

29. Share certificate for the Brazil Railway Company. Ullsteinbild / TopFoto.

30. The Infernillo Bridge on the Peru Central Railway. South American Pictures.

31. La Paz, Bolivia. South American Pictures.

PREFACE

Trying to draw together the history of the railways across the world and to demonstrate their enormous impact globally in one relatively short volume has been a daunting task but one that is eminently worthwhile. Of course, it means I have made no attempt to be comprehensive and have found enormous difficulties in selecting which stories to tell from the myriad accounts that the 180-year history of the railways has generated. Certain tales, however, had to be included, such as the genesis of various railways, the development of the major European networks, the influence of British technology in so many countries, the creation of the huge systems in India and, much later, China, as well as the building of the great transcontinental lines in Russia and the United States. It was essential, too, to outline the way that the railways progressed, becoming faster, more comfortable, and safer.

I have eschewed nostalgia. Although this book inevitably evokes the past, occasionally even wistfully, it is about the way the railways transformed people's lives and were a catalyst for a whole range of other changes. The impact of the railways is almost impossible to exaggerate. To understand the way they changed the world, put yourself in the position of a person who had never seen a large machine, nor traveled in or witnessed anything faster than a galloping horse. Their horizons were necessarily limited, and the arrival of the iron road changed that forever.

There are many books with titles like *The World's Railways* or *Tracks Around the World*, but most either celebrate the technology of trains or only give cursory accounts of their social impact. I have attempted to show how the railways helped to create the world we live in and stimulated development and change in virtually every country. It has been a gargantuan task, but hopefully this book will at least give a taste of the importance of the iron road and of the very enduring nature of an invention that went completely out of fashion in the second half of the twentieth century but is enjoying a fantastic renaissance.

It is easier to list what the railways did not change than to set out their achievements. Quite simply, between the first quarter and the last quarter of the nineteenth century, the railways transformed the world from one where most people barely traveled beyond their village or nearest market town to one where it became possible to cross continents in days rather than months. Their development created a vast manufacturing industry that ensured that the Industrial Revolution would affect the lives of virtually everyone on the planet. Everything from holidays to suburban sprawl and from fresh milk to mail order was made possible by the coming of the railways.

And this was on a global scale. Between 1830, the opening of the Liverpool & Manchester Railway, the world's first twin-tracked and fully locomotive-hauled railway linking two major towns, and the turn of the century, well over 600,000 miles of railway were built around the world,[1] and few countries were left without at least a section of track. Indeed, as this book shows, the railway penetrated far beyond the obvious places, reaching heights and remote corners of the world that seemed impossible. And everywhere that a spectacular railway was built, there would be an amazing group of men who battled to overcome the obstacles. Of the major schemes covered in this book, virtually every one, except for the Cape to Cairo lines, was completed.

I have focused less on the United Kingdom than on the world as a whole because I have covered Britain's railways in great detail in my previous book, *Fire & Steam*. Britain was a pioneer in many respects because it was not only home to the world's first major railway but also the country that developed the technology and operational practices that spread around the world. Britain's story cannot be omitted entirely, but this book somewhat understates the importance of its railways and its role in the development of rail networks around the world.

I have concentrated particularly on the great railways, the ones that were built in the most onerous and difficult conditions but that transformed their nations, such as the Indian, American,[2] and Russian systems. Just think of the imagination and breadth of ambition that led to the construction of the Trans-Siberian, or, indeed, the network of lines linking India's great cities. The story of these railways deserves to be set out in some detail to celebrate their construction.

There is, for example, far more on the American railroads in this book than on those of many other countries. But there are good reasons for this. At its peak, the U.S. railroad system represented about a third of the world's total mileage. It contributed to the very creation of the globe's most powerful nation, as explained in Chapter 4. Admittedly, too, there is a more comprehensive and accessible literature on U.S. railroad history than on any other country's, and the availability of good studies on various railways has meant that on occasion I have given them disproportionate coverage.

Everyone has a pet railway or favorite journey, and I am bound to have left some out. I have not, for example, mentioned the Indonesian railways, which, built by the Dutch, were reportedly among the best narrow-gauge railways in the world before World War II. Turkey and the rest of the Middle East barely get a mention; neither does the Philippines. Nor have I included the story of many strange and wonderful lines, such as the overhead railway in Wuppertal near Düsseldorf, which has been operating since 1901 and still carries thousands of commuters daily. But most significant railways are given some space, whereas others that may not really be as important have been accorded lots of space because there is a good story to tell and the information is accessible. Indeed, if you are a budding railway author, there are lots of social histories of railways around the world remaining to be written to add to the few that I have managed to track down. Much of railway literature is written for what a fellow author termed "rivet counters," and that is a wasted opportunity. The railways deserve better histories than simple accounts of their construction and technology, and I have listed a few in the bibliography.

I have set out the early history of the railways in Chapter 1, covering in brief the opening of the Liverpool & Manchester and the start of railways in several other European countries. Many early railways were created with freight in mind, but it is remarkable how quickly they attracted passengers, and this, in turn, stimulated the spread of the iron road. Although the early railways used or copied British technology, several different railway traditions soon established themselves. These are examined in the next three chapters. Chapter 2 covers the establishment of the European tradition. These railways were built to more generous proportions, and states tended to be far more involved in their creation on the Continent than in Britain. They were used as a deliberate political tool to unify nations that for the most part still had unstable borders. Chapter 3 looks at the spread of the British railway style, notably in India, where the colonial power designed and built a system that also helped to forge a nation. This chapter also covers Ireland and Australia. Chapter 4 is devoted to the United States. It reminds us of the importance of the railways to the development of the country, which is now largely forgotten. It also covers the crucial role played by the railways in the Civil War and their contribution to the ultimate victory by the North. The American way was different from the European traditions, with heavier locomotives, cheaper track that limited speed, and open-plan carriages rather than the compartments favored in Europe.

In Chapter 5, I focus on the development of the railway network in Europe, which gradually crossed borders, allowing for rapid travel all over the continent. Routes were established through the barriers of the Alps and other mountain ranges, which in turn stimulated the development of industry and tourism. I also examine the role of the state in railways, including the European debate over whether they should be privately or publicly owned, and consider the function of railways in several European wars.

Chapters 6 and 7 tell the amazing story of how and why the transcontinental railways were built. Chapter 6 covers the disastrous Panama Railway, whose construction cost the lives of thousands of men and took far longer than expected, but provided a vital link between the east and west coasts of America, and the creation of the first American transcontinental, probably the most significant of all the early railways. Chapter 7 deals with crossing other continents, notably the Russian part of Asia, where the Trans-Siberian was arguably the most ambitious infrastructure project ever built, and Africa, with the failed but heroic Cape to Cairo line.

In Chapters 8 and 9, I take a breather to look at what traveling on the railways was like in the early days and the social and economic changes that they brought in their wake. Chapter 8 covers a variety of journeys on railways around the world and the technological progress that made train travel gradually more pleasant and, indeed, safer. The chapter also describes the ubiquitous nature of the railways, which by 1900 were established in all significant countries and many others, ranging from small Caribbean islands to obscure African colonies, and were still growing apace. Chapter 9 considers the impact of the railways, the way they affected the lives of virtually everyone on the planet. Their existence had a huge number of unexpected side effects: Famines became less serious, as it was possible to transport food more quickly; urbanization grew, as people were able to commute to work; and the scale of wars was unprecedented, as railways brought in troops and arms far more efficiently than horse-drawn transport had.

In Chapter 10, I describe what is generally considered to be the heyday of the railways, that brief period early in the twentieth century when they ruled unchallenged. Their role in World War I was crucial, and, indeed, possibly decisive, but already the seeds of their decline were being sown with the development of motor transport.

Chapter 11 looks at the interwar years and the first hints that the railways might not rule forever. Passenger numbers were still rising, but the railway companies were struggling in the face of competition and often hindered by governments unwilling to support organizations that had previously exploited their monopoly position. Despite these difficulties, this, too, was to some extent a golden age, with steam locomotive technology becoming far more efficient and the emergence of both diesel and electric trains, which offered unprecedented levels of comfort. Despite the threat of aerial bombardment, the role of the railways in World War II was vital.

Chapter 12 examines the decline of railways in the postwar period when the rise of the motor car seemed to make the iron road redundant. It was not to be, though, as gradually governments and railway companies realized there remained great potential for the railways in the markets where they traditionally dominated: commuting, fast intercity travel, and carrying coal and minerals. And Chapter 13 celebrates their renaissance, in particular the high-speed rail revolution that has given the railways in the twenty-first century a new lease on life, and suggests that the future is rail.

I have been deliberately inconsistent in the use of names of towns and cities. Where there is a common English and American usage, such as Vienna or Turin, I have used it, but elsewhere I have kept the local name. Lyon and Marseille have stayed in the French, without the "s" at the end. Forgive me in advance for omissions and errors. Do please e-mail me via my website, www.christianwolmar.co.uk, with any corrections, errors, and comments for future editions. Readers' comments have proved immensely helpful in the past and I thank you in advance. Above all, enjoy the ride.

CHRISTIAN WOLMAR, LONDON

ACKNOWLEDGMENTS

Writing can be a lonely business, particularly on such a lengthy enterprise as this one. I would therefore like to thank a variety of people who have helped on this project in one way or another, either by offering advice, by helping me source material, or simply by being at the other end of the phone or e-mail. Most of all I would like to thank Tony Telford, who, as with my previous book, has provided expert and detailed critiques of every chapter, spending far more time on the project than he ought to have done. John Fowler, too, read every chapter, offering countless useful suggestions. I would also like to thank Jim Ballantyne, Rupert Brennan-Brown, Roger Ford, Bernard Gambrill, Nigel Harris, Phil Kelly, Gordon Pettit, Fritz Plous, and Jon Shaw, who have all helped in various ways. My apologies to any others whom I have inadvertently left out.

Thanks to all the readers on both sides of the Atlantic who have submitted corrections and suggestions. They include: Geoffrey Bryson, Anthony Burton, Peter Crear, Paul Davies, Tim Fischer, Robert J. Foster, Silvio Gallio, Richard Harrison, John Knewstubb, Keith Koehler, Michael Korn, Norman Macleod, Charles Long, Gerald Peck, Pip Dunn, Joe Rice, Richard T. Salter II, Ian Stewart, Jon Sumner, and Richard Wells. Apologies to any I have omitted.

I would also like to thank my agent, Andrew Lownie, who encouraged me to write this series of railway books, and Toby Mundy, Sarah Norman, and Karen Duffy at my esteemed UK publishers, Atlantic Books, for their help and support. Niki Papadopoulos, Katherine Streckfus, Jeff Williams, and Kay Mariea at PublicAffairs made the American edition possible. The errors, of course, are all mine.

The main railway lines of Europe, showing today's borders and country names.

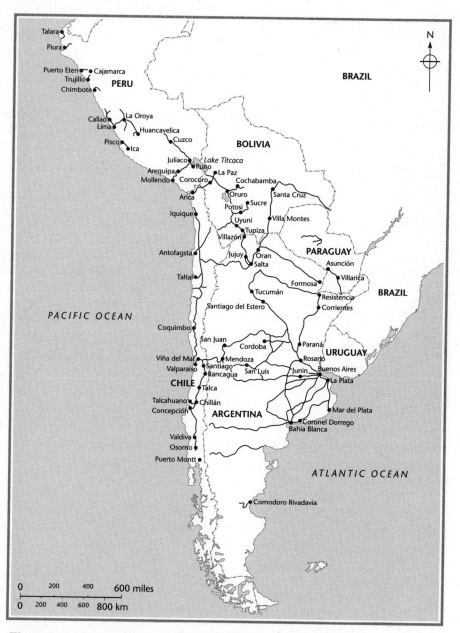

The main Andean railways at their peak—showing connections into Argentina—some of which have now been closed.

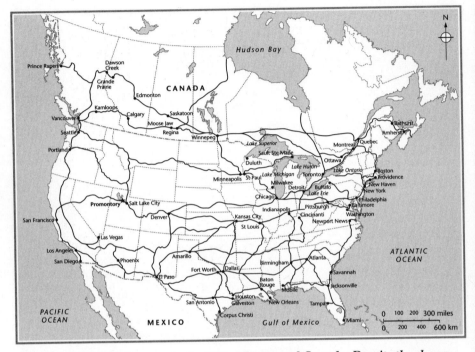

The transcontinental routes in the United States and Canada: Despite the closure of many minor lines, the main transcontinental routes have survived largely unscathed.

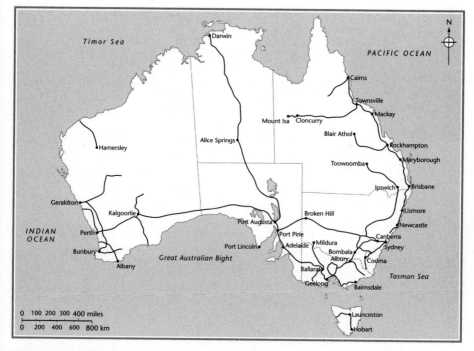

The main railway lines of Australia. The Ghan railway–the north-south transcontinental line through the center of the country linking Adelaide and Darwin–was not completed until 2004.

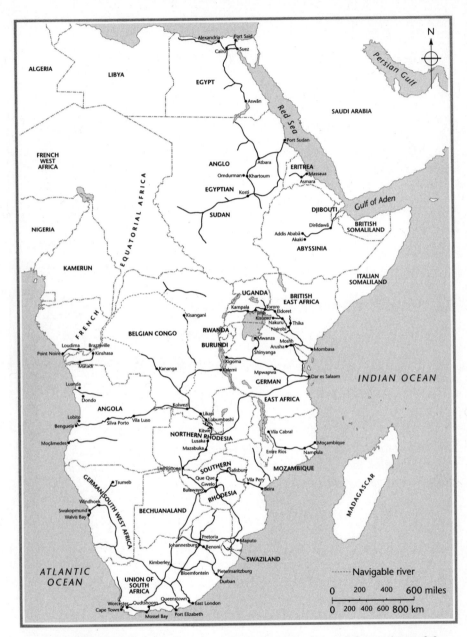

The continent of Africa, showing the colonial boundaries and the sections of the Cape to Cairo railway that were actually completed, plus various feeder railways.

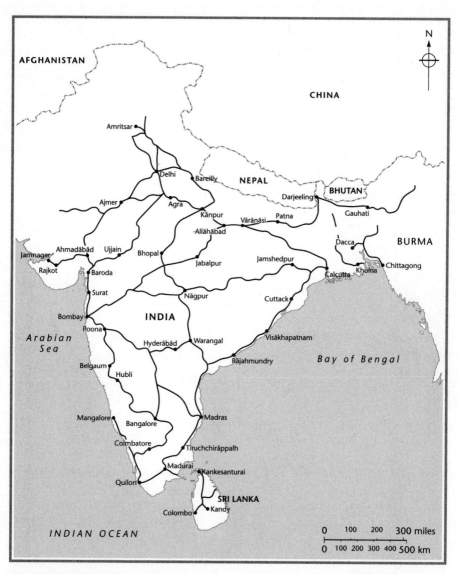

The main railway lines of India, a network conceived by Lord Dalhousie and built in colonial times.

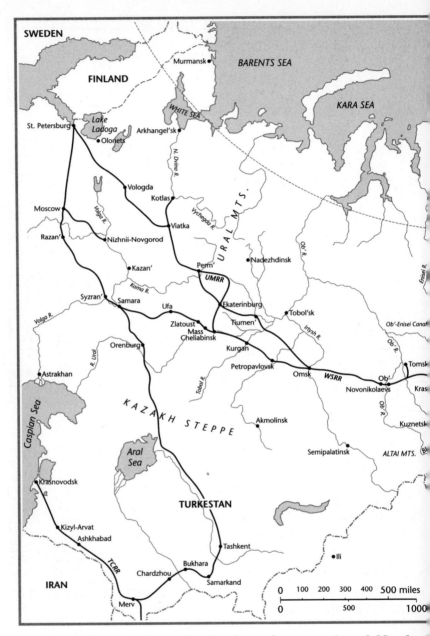

The Trans-Siberian Railway, showing the southern route through Manchuri *territory—which opened in 1916.*

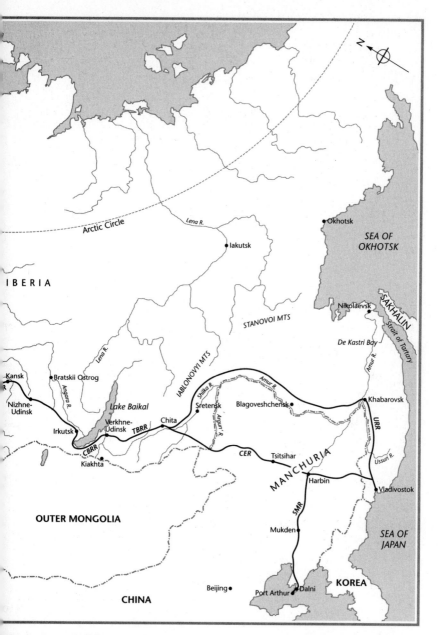

...which was completed in 1904, and the northern route—entirely in Russian

ONE

The First Railways

I T WAS THE WORLD'S FIRST GLOBAL NEWS STORY. IN SEPTEMBER 1830, just fifteen years after the Battle of Waterloo, the inaugural train chugged along the tracks at the opening of the Liverpool & Manchester Railway. This sumptuous event, attended by the victor at Waterloo—the prime minister, the Duke of Wellington—and a host of notables, attracted hundreds of thousands of onlookers. Memorabilia, ranging from penny handkerchiefs and snuffboxes to dinner sets and framed artists' impressions, were on sale, and the whole world seemed to be watching. Newspapers from as far afield as America and India covered the occasion with an awareness that this was an epoch-making event that would change the world. However, not even the most far-sighted and imaginative reporter of the day could possibly have predicted just how fast this transformation would take place or how widely the impact of this new invention would be felt.

The event's significance had not been missed. The Liverpool & Manchester was far more advanced than any of its predecessors or any other line being considered elsewhere in the world. It was double tracked, powered entirely by steam, and connected two of the world's most important cities of the day. It was not, of course, the world's first railway, but while its predecessors had been created principally for the transport of coal or other minerals from a mine to navigable water, the Liverpool & Manchester carried traffic, including passengers, in both directions. Thanks to Britain's place as the cradle of the Industrial Revolution, not only was British technology the most advanced in the world, but its application was far more widespread and developed than elsewhere. Consequently, many foreign dignitaries and, more important, engineers eager to reproduce the technology back home were among the thousands of people who lined the tracks watching the proceedings.

There was, for example, William Archibald Bake,[1] a Dutch artillery officer, who would return home to press for a railway to link Amsterdam with a proposed network of Prussian railways in the Rhineland. Rumors spread through the city that several Americans and Russians were at the opening on fact-finding missions, and xenophobia bubbled under the surface, with dark talk of spies and agents from potentially hostile countries intent on stealing the technology. Indeed, a pair of Americans, Horatio Allen, chief engineer of the Delaware & Hudson Canal Company, and his companion, E. L. Miller, had already dropped in to witness the Rainhill Trials, a competition for locomotives, the previous year. All these people and many more were ready to become proselytizers for railways, taking the message back home that the Iron Horse had arrived and was here to stay.

Without the cheap new method of transportation the railways provided, the economic development stimulated by the Industrial Revolution would have stalled or remained localized for far longer than it did. The railways were the catalyst for the spread of technology and would initiate the process of globalization that culminated with the development of the Internet and the World Wide Web. From its isolation in small communities, the human race was brought together by the railway, for better or worse. Within a decade of the opening of the Liverpool & Manchester, trains pulled by steam locomotives had spread across Europe and started running in North America. Within a quarter of a century, railways had sprung up in the most unlikely places, ranging from Cuba and Peru to Egypt and India. While these new opportunities to travel had huge beneficial effects, they also facilitated the fighting of wars and hastened the decline of many industries.

Britain's role in this process was seminal. Although jingoistic British writers sometimes exaggerate the country's importance in world history, with regard to the history of the railways it is almost impossible to do so. British technology formed the basis of so many different railways that the British tradition was dominant for decades, and its capital helped to fund projects not only in the United Kingdom but also in the rest of Europe and in Latin America. The locomotives of George Stephenson, who was largely responsible for the engineering of the Liverpool & Manchester, for example, would provide the basic design for many railways. A prominent part of the British legacy is the gauge of 4 feet, 8.5 inches—the distance between the rails that Stephenson chose for the Liverpool & Manchester—which would rightly become known as "standard" because it is the most widely used gauge around the world.

Arguments about gauge cannot, unfortunately, be dismissed as a mere technical matter that is outside the scope of this book. Quite the opposite. Gauge plays an all-too-important role in this story because disputes over that crucial distance between the rails encompass a diverse range of other issues, such as cost and speed, and making the wrong choice has often resulted not only in massive sums of money being wasted but also in jeopardizing the profitability of whole railway networks. Gauge was a compromise between cost and practicality, and Stephenson got it about right, which explains the popularity of his choice. Wider railways obviously cost far more to build and take up much more land, but could offer greater standards of comfort. Narrower railways were cheaper to construct, but they were slower and could not accommodate as many people. The width between the rails is not, however, the only aspect of gauge. There is the "loading gauge"— the size of the "envelope" required to accommodate trains, which determines the size of tunnels, the location of platforms, and the placement of lineside equipment—and this is normally larger on standard-gauge lines on the Continent than in Britain. Stephenson did not always succeed in persuading the various foreign railways he advised to adopt his gauge, and the legacy of that failure still proves costly today. In Spain, for example, which the aging Stephenson visited in the 1840s, the nascent Red Nacional de los Ferrocarriles Españoles (RENFE)[2] rejected his pleas to adopt the standard gauge and instead chose 5 feet, 6 inches,[3] which was later used in several other countries, notably India and parts of Latin America.

The debate over gauge occurred in every country with a railway, even in Britain, where the standard gauge was adopted relatively early following a Royal Commission on this vexed issue in 1845. That was already too late for the Great Western Railway, ultimately linking London to western England and Wales, which by then had more than 200 miles of line using the 7-foot, 0.25-inch gauge favored by Isambard Kingdom Brunel, the line's engineer. The Great Western would not fully convert until the end of the century, causing great inconvenience, not least to Queen Victoria, who was forced to change trains on her journeys from Windsor to Scotland, and enormous expense. This brief reference to gauge, a subject that comes up all too often, demonstrates why it is necessary to start this brief international history with an account of the prehistory and early history of the railways in Britain. Though that story has been widely covered elsewhere,[4] a short recap is essential for an understanding of the full account of the global spread of the railways.

The railways were made possible by a series of technical inventions over the space of a couple of centuries involving the development of steam engines, locomotives, and rails. Railways were the answer to the long-established problem of how to transport heavy loads of coal and other minerals to rivers or the sea, and later to canals, where they could be transported for far greater distances. There is some evidence that putting goods in wagons to be hauled by people or animals along tracks predates Jesus Christ, and the earliest surviving representation of such a scene, dating from 1350, can be found in the cathedral at Freiburg im Breisgau in Germany. There were enough such lines to be discussed in a book published in 1556, and certainly by the sixteenth century in Britain there were numerous wagonways[5] using crude wooden rails to help haul heavy wagons out of mines. Horses had begun to replace manpower to boost efficiency, and combining the two ideas, horses and rails, which allowed for far greater loads to be pulled, was the obvious next step. By the early eighteenth century in the principal German coal-producing area of the Ruhr, rather more sophisticated wooden wagonways were developed that used a type of flange—an extra lip on the wheels to keep them on the track—to guide the trucks and prevent them from becoming derailed. These precursors of the railway had an important economic impact in the early days of the Industrial Revolution as coal consumption in Britain increased (there was a tenfold increase between 1700 and the early 1800s[6]), serving both industrial and domestic needs.

The network of wagonways that emerged in the northeast of Great Britain was so extensive that they became known as "Newcastle Roads." By 1660, in the Tyneside region of northeastern England,[7] there were nine such wagonways. The rails became increasingly necessary as the more accessible coal near the surface was extracted and the pits extended deeper. In 1726, a group of coal owners, the Grand Allies, developed the idea further by agreeing to use a shared wagonway to link up their mines, which allowed them to rationalize coal movements. They even created a "main line," a joint route, much of it double tracked, from several mines to the water. This line included the Causey Arch, a bridge with a 100-foot span that lays claim to being the world's first railway bridge and survives today. These railways made extensive use of gravity, since most of them led down to a waterway, and therefore the horses had the relatively easy task of hauling the empty wagons back up the hills. As the putative railways increased in sophistication and length, wagons were coupled together to improve efficiency, and by the 1750s, iron rails were introduced that proved far more durable than the wooden ones.

The other major technical development required for the establishment of the railways was, of course, the steam engine and, later, the development of self-propelled locomotives, a far more complex and difficult process. Again, the idea of steam power dated to classical times, but the first working steam engines were probably those of Thomas Newcomen, an ironmaster from Devon who built them in the early years of the eighteenth century. Applying principles that had been observed by a French scientist, Denis Papin, who had noticed that a piston contained within a cylinder was a potential way of exploiting the power of steam, Newcomen produced engines to pump water from the mines. He created something of a cottage industry, making sixty engines himself, and after his patents ran out another three hundred were built by other engineers over the next half-century, many for export to countries such as the United States, the German states, and the Austrian Empire, where one was even used to drive the fountains for Prinz von Schwarzenberg's palace in Vienna.

Toward the end of the eighteenth century, it was James Watt who made steam power commercially viable. He did this by improving the efficiency of steam engines and adapting them for a wide variety of

purposes. The engines manufactured by the company he formed with
Matthew Boulton were used to provide power for everything from ships
and looms to the sugar mills of the West Indies and the cotton mills of
the United States, but they were not used for steam locomotives. Other
inventors, however, did try to put steam engines on wheels. The first to
do so was the Frenchman Nicholas Cugnot, whose *fardier* was intended
to be used as an artillery tractor. On a test run in Paris, it reached a
speed of 2.5 miles per hour but hit a wall, overturned, and was declared
a public danger by the city authorities. It would never have run far anyway, since there was no way of replenishing the steam once it ran out.
Various other inventors in England, Scotland, and the United States
built similar steam road locomotives, but a historian of the railways dismissed these early efforts, writing that "none of these pioneers made
any contribution to the design or development of the steam locomotive."[8] Their problem, which explains why railways were developed
more than fifty years before road vehicles, was that the roads, poorly
built and little-maintained, were simply too bad to support their weight.

It was when Richard Trevithick, who had a short but crucial role in
the history of the railways, hit upon the idea of putting steam engines
on rails that a workable form of transport was developed. Trevithick, a
Cornishman, has the best claim to the much disputed title of "father of
the locomotive." Whereas Watt and Boulton had insisted on only building low-powered engines, Trevithick developed the concept of using
high-pressure steam, enabling him to obtain more power for a given
weight. In 1801 he produced the world's first successful steam "road
carriage," which drove into a ditch because there was a crude steering
mechanism and then exploded because he and his colleagues went off to
the pub, forgetting to extinguish the fire under the boiler. When Trevithick developed an improved model the following year at Coalbrook-
dale, an ironworks in Shropshire, he had the brainwave of putting it on
rails,[9] which not only dispensed with the need for steering but also gave
it a firmer base than the muddy lanes that, at the time, passed for roads.
In 1804, a Trevithick engine hauled wagons weighing 9 tons at a speed
of 5 miles per hour at Pen-y-Darren in Wales, another ironworks, which
was certainly a world first. However, the primitive rails were not up to
the task, as the locomotive was too heavy, and consequently it was soon
converted into a stationary engine powering cables to haul the wagons.

Although steam engines proliferated, with 30,000 being in use in Manchester alone by 1830, the development of locomotives was slow, not only because of the technical difficulties but also as a result of doubts about whether they would ever justify the large amount of investment required to perfect them. When Trevithick built a locomotive with the playful name *Catch Me Who Can* and demonstrated it successfully on a circular track near the present site of Euston Station in London, there was no interest in producing it commercially. Poor Trevithick gave up and went to South America to develop stationary steam engines for use in the gold and silver mines of Peru.

Other engineers attempted to build locomotives with little success, and it was not until 1814, when George Stephenson, a self-taught engineer from Northumberland, produced his first one, that the idea began to be seen as viable. Stephenson is often wrongly referred to as the inventor of the steam locomotive, but he has the best claim to being the "father of the railway" because it was his drive and energy, together with his skill in making use of available technology, and, indeed, improving on it, that ensured that railways came into being. Born into a poor family in 1781, he received no formal education but learned on the job from the age of twelve, when he started working as an assistant to his father, a fireman on mining-related steam-pumping engines. His talents were soon recognized and he became an enginewright in charge of all the stationary engines at Killingworth, a large mine in Northumberland. Stephenson soon realized that engines that could run on rails and haul loads would be far more flexible than the traditional stationary ones. His first one, *Blücher*, proved reasonably successful (though not very reliable), and over the next seven years he built sixteen more, both for Killingworth and for the Kilmarnock & Troon, the first Scottish railway. The tracks at Kilmarnock & Troon were not strong enough to support the weight of the locomotives, so they were quickly converted into stationary engines. He then formed a locomotive construction company with his son, Robert, who would also have a career as a great railwayman, notably as the engineer for the London & Birmingham Railway, and who, at the remarkable age of nineteen, became responsible for locomotive development.

In the early 1820s, George Stephenson was asked to advise on the construction of the Stockton & Darlington Railway, and he later became both its surveyor and engineer. He was thus able to ensure that his engines would be used on the railway. Though it has a claim to be the first steam-hauled public railway, the Stockton & Darlington was really only a logical extension of the mining tramways that had been developed over the previous couple of centuries. As mentioned above, the Stockton & Darlington was a single-track local line whose principal purpose was hauling coal, and it carried few passengers. Most of the trains on the line, including all the early passenger services, were horse-hauled, and the single track not only meant that speeds had to be kept low but also resulted in fierce arguments about who should give way and reverse to the nearest crossing point when trains met. Moreover, when the line opened in 1825, only one of Stephenson's locomotives was available for use, and even when more were delivered, they were notoriously unreliable. Indeed, at one point the directors contemplated turning the railway back to horse-haulage, and they only relented after hearing desperate pleas from Stephenson.

Stephenson was to play an equally vital role in the construction and development of the Liverpool & Manchester, and his company's locomotives were to become far more reliable thanks to the efforts of his son. Stephenson *père* was again the surveyor and later the engineer, and, crucially, when the promoters hesitated over whether to use horse or locomotive power, he was able to convince them that steam engines were the future. In 1829, the year that public mass transport started in London with the introduction of the horse omnibus, the remarkable Rainhill Trials were organized by the directors of the Liverpool & Manchester to find the best locomotive for the line. The event proved to be a virtual walkover, as Robert Stephenson's *Rocket* was the only entrant to complete the course without mishap. His locomotive covered the 1.5-mile course repeatedly at an average speed of 14 miles per hour without any problems, while the three other entrants all suffered breakdowns. On the final run Stephenson opened the valve to let the engine go faster, and it reached 30 miles per hour, to the amazement of the assorted crowd of the great, good, and merely curious. Stephenson not only won the prize of £500 and the contract to build four more locomotives for the line over the next three months, but was also, on the basis of the *Rocket*'s performance, able to persuade the promoters to use only steam locomotives for traction (except on a section of the line where stationary engines had been mandated by the Parliamentary Act that had given permission for its construction).

Obtaining the right to build the line was not the only difficulty facing the directors. Technically, the railway was far more sophisticated than any of its predecessors, and George Stephenson, who for a period had been shunned by the directors in favor of another engineer, Charles Vignoles, was recalled to ascertain the most suitable alignment. This involved the crossing of Chat Moss, a damp marshy plain, which required another innovation: "floating" the railway embankment on a bed of brushwood and heather. It was the first example of how railway engineers would learn to develop new engineering methods on the hoof in order to surmount the ever more daunting range of obstacles that slowed the progress of so many railways.

Development of the 31-mile-long Liverpool & Manchester Railway had been stimulated, like its predecessors, by the need to carry freight. Liverpool was the main arrival point for raw cotton, which had to be processed in Manchester's mills. It was a booming port, with rum and sugar imported from the West Indies and tobacco from Virginia, and movements of coal in the whole northwestern region of the United Kingdom were rising sharply as factories using steam equipment sprang up. By the 1820s, the road between Liverpool and Manchester was completely inadequate to the task, and canal transport, the only alternative, was expensive and slow. Given the booming economy in Lancashire, then the most industrialized region in the world, it was not simply happenstance that led to the world's first major railway being constructed between these two growing towns.[10] Although freight had been the main motivation for its construction, the railway opened initially for passengers, as the freight wagons were not yet ready, and the railway's owners quickly discovered that there was enormous demand for travel.

Despite the notorious tragedy on the opening day, when the former government minister William Huskisson was killed by a passing train, people were immediately attracted to this new and fast means of travel, turning the Liverpool & Manchester into an instant success. In 1831, the first full year of operation, it carried nearly 500,000 passengers, enabling the company to pay generous dividends to its investors. Additional services for the journey between the two towns were quickly introduced, and goods also began to be carried. At first, the freight consisted principally of cotton and coal, but soon live animals were being taken. In May 1831, a consignment of forty-nine squealing Irish pigs was carried to Manchester, each costing the owners 1s 6d (7.5p) to transport, and later sheep were carried at half the price, just 9d. This was far cheaper than carrying them to market on horse-drawn carts and ensured the animals arrived at market in far better condition. Farmers and fishermen soon realized that the railway opened up a huge market for their produce. The transport of fresh dairy products, vegetables, meat, and fish helped to revolutionize the diet of ordinary people, in particular the urban masses, who previously had rarely seen fresh food. Passengers, though, remained the mainstay of the company's income, a lesson that was not lost on railway promoters around the world.

The railway in Britain spread rapidly. Within a decade there was a main line through the spine of England, linking London with Birmingham, Liverpool, and Manchester, and lines from the capital to Bristol and Southampton. Even though the trains were slow, averaging barely 20 miles per hour in the early days, journeys that had taken days by stagecoach could be undertaken in a few hours. Goods that previously had to be transshipped from carts to canal boats and back again could be taken rapidly right into the centers of towns, saving shippers and industrialists vast amounts in transport costs.

Britain's more advanced industrialized position allowed it to keep ahead of its European counterparts, and as a result it was the first nation to exploit fully the boundless potential of this new technology. It would retain that lead for some time, experiencing a series of railway manias, most notably in the early 1840s, the opening years of Queen Victoria's reign. These periods of growth would result in the construction of more than 7,000 miles of railway within two decades of the opening of the Liverpool & Manchester. Although other countries would also undergo periods of railway mania, the British version was the first and one of the most fruitful.[11] Britain was, too, the first country to experience a railway scandal. George Hudson, who melded together a vast railway empire controlled by the Midland Railway, for a time Britain's largest railway company, was exposed as a fraudster who had cheated investors out of their money. The railways were also beginning to be recognized in arts and literature. In 1844, J. M. W. Turner painted the first major artwork to feature the railways—*Rain, Steam and Speed*, a rather romantic view of a train speeding out of swirling mist over a bridge.

In the early 1830s, Britain may have been the pioneer of railway development but it was not alone in taking an interest in this new technology. As we have seen, there had been wagonways on the Ruhr for hundreds of years and similar systems had been developed elsewhere in Europe and in the United States. Indeed, even before the opening of the Liverpool & Manchester, promoters of railway lines were emerging in several other European countries, although, apart from France, they would not see their first lines opened until the mid-1830s.

Mostly, as with the Liverpool & Manchester, the purpose of these early lines had to do with economics, but there were other reasons for their construction as well, including one that I will refer to many times in the pages that follow: nation-building. In France, as in Britain, it was a coal-producing region that became the site of the country's first public railway. In 1823, the reinstated monarch Louis XVIII signed the act permitting the construction of the country's first railway, the 11-mile line between St-Etienne and Andrézieux in the Massif Central, a mountainous area in the center of the country. It was intended to carry coal from the mines to the Loire River. France's answer to George Stephenson was Marc Seguin, though, as a scientist and inventor, he was a very different man from the self-educated Northumbrian. Seguin, in fact, advised Stephenson prior to the Rainhill Trials to use the Frenchman's concept of the tubular steam boiler for the design of the *Rocket*, and this was a crucial part of the locomotive's design, greatly improving its efficiency and consequently contributing to its victory.[12] It was to prove a useful Anglo-French cooperation, as Robert Stephenson went on to build locomotives designed by Seguin.

Since work started on the St-Etienne line in 1825, the same year as the opening of the Stockton & Darlington, the line bore far more resemblance to that railway than to the more sophisticated Liverpool & Manchester. When the French line opened in 1828, the trains were hauled by horses, usually with two in tandem, and it was principally used for carrying coal. When, in 1832, the horses were put out to grass and replaced by locomotives, with the line being extended all the way from St-Etienne to Lyon, a total of 36 miles, the railway began to attract considerable numbers of passengers as well as freight. The facilities for passengers, however, were rather more comfortable than those endured by travelers on the Stockton & Darlington and even on the early trains of the Liverpool & Manchester, whose carriages consisted of little more than stagecoach bodies mounted on wheels. The French coaches, though also based on a stagecoach design, were much more elaborate, and the accommodation was divided into separate compartments, a design that became the blueprint for early railways around the world. Moreover, some were designed with two decks; the lower ones had open sides for all three compartments, but with curtains that could be drawn to give some limited protection against the weather, while passengers in the upper deck, who were effectively in second class, found themselves exposed entirely to the elements.

Apart from France, the other country that began to develop railways early was the United States, and, typically for America, it was on a large scale. The population of the country was growing. In 1831, it had 13 million people, 800,000 more than Britain, concentrated largely on the East Coast. Not for the Americans a little coal railway of a few miles, but instead a long double-track railway different in scale from its predecessors. The United States already had various small lines serving mines or wharves, using mostly either standing engines or horse power, but they were small in scale and mostly limited to carrying freight. The equivalent of the groundbreaking Liverpool & Manchester was undoubtedly the Baltimore & Ohio Railroad, the first American railway[13] to carry both passengers and freight on a regular schedule using steam locomotives. The railroad was conceived by two far-sighted Baltimore citizens, Philip E. Thomas and George Brown, who had visited the Stockton & Darlington and other railway projects in Britain in 1826.

When they returned to Baltimore, they organized a meeting of local merchants, an echo of a similar gathering that had been the genesis of the Liverpool & Manchester. At this meeting the idea for a 380-mile double-track line linking Baltimore with the Ohio River at Wheeling, West Virginia, was broached. There was a clear need for better communications between the four major seaports of the Northeast—New York, Philadelphia, Washington, and Baltimore—and the rapidly developing hinterland west of the Allegheny Mountains. All four were jostling for primacy, and their leading citizens realized that communications were key to their development. Except for Baltimore, the cities opted for canals, and in the case of Philadelphia, a grandiose, ill-fated project called the Main Line, which ran nearly 400 miles to the Ohio through a combination of canals involving passage through no fewer than 174 locks, inclined planes using stationary engines, and sections of railroad. The journey necessitated frequent inconvenient changes between canal and rail. The obvious disadvantages of this convoluted system pushed Baltimore into choosing a railroad. Rivalry was the key stimulus. Local businessmen were particularly concerned that the 364-mile-long Erie Canal between Buffalo and New York City across upstate New York, completed in 1825, would give New York a huge competitive advantage. Proposals to construct a Chesapeake & Ohio canal that would parallel the Potomac River also threatened the viability of Baltimore's port, and consequently there was widespread support for the railroad project, which quickly obtained permission from the Maryland state legislature. It was realized that the railroad would provide a far faster alternative for goods going between the East Coast and the Midwest than the putative Erie Canal with its eighty-three locks and horse-drawn barges.

It was a courageous decision, as the economics of building such a long line were totally unknown and the project would require the enormous sum of $5 million in capital investment. It was brave, too, from a technological point of view, since it was unclear whether the British know-how, developed for a milder climate and shorter distances, could be adapted for the United States. Indeed, as we will see in Chapter 4, American technology soon deviated from its British origins.

Construction started in 1828 with a ceremony that suggested the importance of the Baltimore & Ohio as a national event. With typical American exuberance, the first sod of earth was turned with a silver spade by the only surviving signatory of the 1776 Declaration of Independence, the ninety-year-old Charles Carroll, who, at the ceremony, said: "I consider this among the most important acts of my life, second only to my signing the Declaration of Independence, if even it be second to that."[14] Given the role of the railways in uniting the United States of America and creating a nation out of separate states spread over a distance from east to west of nearly 3,000 miles, his statement was prescient indeed.

The railway largely followed the line of the rivers to make the gradients as gentle as possible; it was feared—wrongly—that the resulting sharp curves would make it difficult for steam locomotives to be used because the wheels would not be able to negotiate the bends. Initially, just as in Britain, in the United States there was a debate about what form of traction to use. Peter Cooper, an industrialist and inventor from New York, had invested heavily in land in Baltimore and was keen to ensure the success of the railway, which he realized could only be guaranteed with the use of steam locomotives. Cooper, who also later patented gelatin, the principal ingredient of Jell-O, built an engine called *Tom Thumb*. As the name suggests, it was a relatively small locomotive, described by one onlooker as having a boiler "not as large as the kitchen boiler attached to many a modern mansion." Nevertheless, on a test run along the first 13 miles of track, it reached an exhilarating 18 miles per hour, impressing the assorted investors and VIPs who had come along for the ride.

On the way back, Cooper exceeded himself and agreed to race his locomotive against a horse to prove that it was superior. The powerful gray horse was initially in the lead, thanks to its faster acceleration, but once the locomotive engine gained purchase on the track and its safety valve opened up to supply extra power, *Tom Thumb* glided past the galloping steed. Cooper's machine was a quarter of a mile ahead when disaster struck. Just as the horse's rider was ready to give up, the belt that drove the pulley on the locomotive snapped and the engine gradually eased to a halt. Cooper struggled to replace the belt, and he did manage to finish the course, but with his hands lacerated by burns caused by the heat of the engine, he was by then hopelessly behind.

The equine victory, though, proved pyrrhic. The investors had been sufficiently impressed with the performance of the little engine to realize that using steam haulage, rather than horse power, was the only way to make the line viable. Although some horse traction was used early on, locomotives were dominant, and by the time the line was extended westward, they were used exclusively. In fact, progress inland was slow as a result of opposition, not least from some legislators, and it would actually take a quarter of a century for the line to reach Wheeling, the original planned destination. By that time the East Coast was peppered with railways (see Chapter 4) and the great project to link the two coasts had been started. By then, too, the Baltimore & Ohio was linked with various other railways, including a branch to Washington opened in 1835 that was partially funded by the Maryland state government.

In Europe, meanwhile, railways were progressing rather more slowly and on a scale that was far less grand. It was not only the more advanced state of the economy in Britain, the original seat of the Industrial Revolution, that gave it the early advantage. It was also the greater political stability, particularly after the fall of the Duke of Wellington's antireform government in 1830. The political upheavals, revolutions, and wars being experienced by various European countries did not create the right climate for the long-term investment required by railways, and by the mid-1830s the railways had established little more than a toehold in some countries. However, their value, often as a tool of nation-building or for helping to move military forces around to quell protests, was beginning to be widely recognized by even the most reactionary governments. Already a distinction was emerging between types of railways: Very broadly and with inevitable exceptions, they could be defined as British, continental, and North American.[15] These are covered in the next three chapters.

TWO

Europe Makes a Start

THE FIRST RAILWAYS IN MOST EUROPEAN COUNTRIES EITHER USED British technology or copied it. Moreover, many employed British train drivers—engineers, as they were often rather misleadingly called—who were sought because of their experience and who were prepared to travel abroad, thanks to the higher wages and the esteem they could attract. This spread of British technology and know-how was hardly surprising given that Britain had a decade-long start on its rivals and was, for a time, the only country producing locomotives. Britain also had the Stephensons, whose unique skills and experience meant that putative railway promoters would "send for the Stephensons" to obtain advice in designing and constructing railways.

Even the men who built these European railways were sometimes British. The British "navvies"—so called because they were a particularly tough group of workers like the navigators who built the canals—had rightly earned a fierce reputation for both hard work and hard drinking, and they were recognized as superior to local workers, who were often recruited from agriculture and did not have the experience and strength required for railroad work. The navvies were brought over by British contractors—for example, by Thomas Brassey for the Paris-Rouen railway—and might then travel around the Continent finding work on different projects. They impressed the locals, as one of Brassey's timekeepers on the Paris-Rouen reported with admiration after looking "on as fine a spectacle as any man could witness . . . every man with his shirt open, working in the heat of the day." "Such an exhibition of physical power," he continued, "attracted many French gentlemen . . . at Paris and Rouen," who said, *"Mon Dieu, les Anglais, comme ils travaillent."*[1]

Europe's early railways were constructed on a continent that was markedly different from its geopolitical situation today. Although France's boundaries remain similar, neither Italy nor Germany existed as we know them now. Much of the French Riviera belonged to an Italian-speaking state called Piemonte-Sardinia, one of many on the Italian peninsula, and though there was a "German people," they resided throughout central Europe. Prussia was the biggest country in the north and the Austrian Empire dominated the south, and thirty-seven smaller Germanic states were sandwiched between them.

Other than France, the countries that made a start in the 1830s in Europe were Belgium and "Germany" (Bavaria and later Saxony) (1835), the Austrian Empire (1838), and Italy and Holland (1839). The success of the Liverpool & Manchester's locomotives did not immediately convince every railway promoter that steam was definitely the future. In the Austrian Empire, lengthy railways using horse power continued to be developed well into the 1830s. Indeed, Austria boasted the world's longest horse railway, connecting Linz in Upper Austria with Budweis in Bohemia (home of the famous beer, now in the Czech Republic), a distance of 90 miles, and it would soon be extended even further, to the salt works at the health resort of Gmunden. By 1836, just before steam began to replace the horses, the network of interconnected public horse railways in Austria covered an impressive 170 miles.

Nevertheless, for the most part locomotive traction was seen as the favored option, although horses were used for some services, especially transportation of goods. Although the various European railways retained elements of British practice, such as running on the left, and used similar semaphore signals, they soon established their own traditions and practices. The continental railways diverged quickly from the British model in several important respects, notably the size of the trains. While for the most part they used the Stephenson gauge of 4 feet, 8.5 inches, the loading gauge—the limits above and outside the rails that determine the size of the rolling stock[2]—were more generous, allowing for wider and thus more comfortable and spacious carriages.[3]

There was another crucial difference, too: the role of the state. In this respect, Belgium probably represents the opposite extreme from the British model. The development of railways in Britain rather took the government by surprise, and they were operating before any policy toward them could be formulated. In Britain the railways were laid out according to the requirements—whims, even—of the private companies that developed them, whereas on the Continent the state's involvement was invariably more prominent. British railway promoters did have to obtain a bill from Parliament for the construction of a line; this did not involve the government, however, but rather the elected members of Parliament (MPs), whose decision-making process was not based on any strategic view of the overall needs of the country. Quite the opposite, in fact: Obtaining the legislation was a haphazard process that was often dependent on whether the MPs concerned had a vested interest in favor—or indeed against—the line.

This was not the case on the Continent where the government of each country, or, in instances where there was still an absolute monarchy, the king or emperor, would decide whether to support a railway proposal. In several countries, the state itself would draw up plans for a line and designate the route. The railway historian Michael Robbins has characterized the continental approach as more "authoritarian,"[4] and certainly it illustrates two contrasting styles of governance that extend far beyond railway policy. These styles, which could be broadly characterized as Anglo-Saxon and European, represent differences that still pertain today. The British method, which was also adopted in the United States, was more organic, a bottom-up process driven largely by the obvious economic benefits to local towns and cities of better transportation connections, which reduced the price of coal or dramatically cut the cost of carrying goods or minerals for use in nearby factories. Awareness of these benefits was translated into pressure from local entrepreneurs to build a line, and it was often they who drew up the proposal, drafted the bill so that the parliamentary process could begin, and sought financing for the project.

Right from the beginning, European governments on the Continent were aware that the railways were such an important part of their country's infrastructure, and would play such a vital role in economic development, that the state had to be involved. Whereas in Britain the government stood aloof, allowing Parliament, with its various factions and vested interests, to determine what railways should be built, European governments on the mainland took an active role in planning their railway networks. As a result, they avoided the duplication that was a feature of the British model and the unnecessary expense of many towns being served by two or more rival companies.

As mentioned in the previous chapter, it was not only the obvious economic advantage that stimulated the building of railways in Europe, but also the realization that they could help to unify nations, especially the capital cities, out of which the network invariably radiated. Nor did it escape the attention of many governments that having a fast method of transportation with outlying regions of their country would give them a way to respond quickly to unrest and rebellion. The advantages of a developed railway network appeared so obvious that the British state, with its casual disregard for the way the railways spread out higgledy-piggledy across the land, became the exception rather than the norm.

The continental state-sponsored approach was taken up most strongly by the Belgian government, and for good reason. Belgium had only recently, in 1830, carved itself out of the Netherlands. Building a railway system was perceived as a way of stimulating a sense of nationhood in a country that, to this day, remains fractured between two distinct groups, the Flemish speakers of the north and the largely French-speaking Walloons in the south. The Belgian railway was both planned and owned by the state, and there was a military aspect to its development, which, as we shall see, is a recurring theme. The separation of Holland and Belgium had left the latter without direct access to the main waterways of northern Europe, particularly the Rhine, leaving the country vulnerable to a blockade in the event of a dispute. Since these waterways provided the only efficient route for goods to and from western Germany, an alternative was needed quickly.

The country's first king, Leopold I, readily approved the design of a railway network, and in 1834 work started on the first line, which was intended to link Antwerp with Brussels and Mons in the southeast and eventually to continue right into Prussia via Aachen to Cologne, a total of 154 miles. Together with an Ostend-Liège line, broadly on an east-west axis, the plan was to provide Belgium with a national rail network right from the outset. The first section, from Brussels to Malines, was opened within a year, and inevitably George Stephenson played a role. Not only were the three locomotives produced by his company, but he rode incognito on the first train, which was carrying the royal party; when it broke down, he went to the engine to help fix the problem. He was soon knighted by the king, and then he continued his travels to Switzerland and Spain to advise on their inaugural railways.

The advantages of government support and a coherent plan were many. The Belgian system was able to expand quickly, not least because the government could brush aside reluctant landowners, thus conveniently dispensing with a problem that had caused much difficulty in Britain and elsewhere. In 1836 the line was extended to Antwerp, providing a route between the port and the capital by land instead of by use of the inland waterways, and by 1843 most of the basic cross that forms the heart of the Belgian railway system had been completed. The heavily industrialized areas of Belgium had the densest railway network in the world in terms of mileage relative to surface area. There was, however, still much ambivalence about state dominance, and this later led to the development of several lines by private enterprise, the first of which, the Liège-Namur railway, was opened in 1851 by a British company. The state, though, remained in control, as private railways were only granted twenty-year licenses before their lines reverted to government control, and its aim of using the railway to unify the country succeeded: "Without the revolution [of 1830]," an anonymous politician is quoted as saying in the official history of Belgium, "the railway could never have existed; without the railway the revolution would have been compromised."[5]

Contrast this with the situation for Belgium's neighbor Holland, from which, as mentioned above, it had separated in 1830, where railways developed much more slowly, hampered not only by the difficulties of building lines across the heavily used waterways but also by the competition from those canals and by the lack of industrialization. The Dutch railways only emerged after several stillborn efforts. William Bake, who, as mentioned in Chapter 1, had attended the inauguration of the Liverpool & Manchester Railway, tried to raise money for the Rhenish Railway between Amsterdam and the Rhineland, but despite the support of King William, who backed the project with his own money, opposition from landowners and lack of sufficient funding prevented the scheme from going ahead.

Another line, the Holland Iron Railway, thus became the first Dutch railway, opening in 1839 between Amsterdam and Haarlem and reaching Rotterdam eight years later. Unlike most early railways, the Holland Iron Railway was built specifically for passenger traffic, an emphasis that would remain in much of Dutch railway development, and its promoters had not even ensured there would be a connection with the port of Amsterdam. Passenger traffic was high, however, and the railway quickly killed off competing coach and barge traffic along the same route, but it provided a strong stimulus for such services to link in with the railway stations, an early example of integrated transport.

The Rhenish Railway was eventually started in 1838 but took eighteen years to reach the Rhine. Its funding ostensibly came from the private sector, but in fact bonds had been issued on the basis of King William's personal guarantee, which ensured that they were quickly taken up. Unlike the Holland Iron Railway, the Rhenish was designed for freight, which from the outset contributed about half its revenue. On the orders of the Dutch government, both lines were initially laid out to a unique broad gauge, 1,945 millimeters (6 feet, 4.5 inches), which had to be changed in 1866 to standard gauge when the owners realized the importance of establishing easy connections with the rest of Europe's rail network.

The different emphasis on rail development between the two low countries, which are of similar size, would have long-term implications. By 1860, there were still only 211 miles of railway in Holland, only half the mileage built in Belgium by 1848, and ultimately Belgium would end up with a network three times the size of its neighbor. The different levels of progress can certainly be attributed to the lack of interest shown by the Dutch state, which left the railways to be developed by private interests until 1860. The first state railways were then authorized in parts of the country too sparsely populated for profitability or for lines that were essential to creating a viable network (see Chapter 5). Holland was also difficult terrain for railways, with low, marshy land, often reclaimed from the sea, large rivers, and canals that had to be crossed while allowing for navigation. As it was impossibly expensive to raise bridges high enough, they had to be movable to enable the shipping to get through, and a great variety of alternative designs were used, ranging from simple lifting devices (like a castle drawbridge) to swinging spans that moved across when a train was due. No fewer than fifty-eight such bridges were required between Amsterdam and The Hague, a remarkable number that amounted to more than one for every mile of railway. There was a risk, of course, of leaving the bridges open when a train was due, but this was minimal in the early days when there was very little rail traffic, with a mere four trains per day in each direction on the Holland line, and just two on the Rhenish. Later on, as traffic increased, there were several accidents, and eventually tunnels and high bridges replaced the swinging spans on major routes.

The builders of the Rhenish Railway encountered surprising difficulties, given that Holland is, in the main, famously flat. The railway passed through country that was hillier than elsewhere in Holland, and at one point when bad weather stopped work, the unpaid navvies rioted. The army had to be called in to quell the protest. Moreover, an 1810 law enabled landowners to exact outrageous concessions for the loss of even small parcels of land, and to avoid litigation the railway would normally acquiesce to these demands. Most notably, at one stage the Holland railway was forced to divert around an enormous loop simply to avoid a little lane near Delft where the proprietor was asking an exorbitant price, but after being lampooned in the press, he relented and simply gave away his rights. On the Rhenish, a landowner demanded that all trains be required to stop at a station near his home, a lonely farm between Amsterdam and Utrecht, and he later had to be bought out at great cost.[6]

THE FRENCH AND "GERMAN" railways developed at a very different pace, too. Intuitively, France, with its unified, well-organized state and its early start, would have been expected to build a network of railways far faster than its neighbor, which, in the 1830s, still consisted of numerous small and squabbling states. Yet the converse was true. Growth of the railways in France was slow. The system was laid down very early on paper but remained fragmentary until the 1860s, while in Bavaria the opening in 1835 of the first line between Nuremburg (Nürnberg) and Fürth was quickly followed by a proliferation of services in several German states.

Part of the explanation lies in the French suspicion about the value of this new technology. As Nicholas Faith pointed out rather baldly, "the French typically started by theorizing."[7] Parisian intellectuals debated furiously over whether the railway would ruin the peace and quiet of the countryside or represent progress and development. The poet Théophile Gautier complained about the noise of this mad invention, while Edmond de Goncourt, who had actually traveled on a train, warned that "on the railway itself, one was so jolted about that it was quite impossible to collect one's thoughts."[8] Similar though more prosaic doubts had been expressed in Britain, too, where opponents of the railway had warned about cows being so terrified that they would stop producing milk, of sheep being discolored, and of passengers being unable to breathe when speeding along at 30 miles per hour.

However, as in Britain, the doubters lost out. Pierre Larousse, the encyclopedist and developer of the eponymous dictionary, eulogized: "Railway! A magical aura already surrounds the word; it is a synonym for civilization, progress and fraternity. Up to now man had gazed at the denizens of the air and of the sea with envy and a certain feeling of inferiority; thanks to the railway, the birds and fish no longer have an advantage over him."[9] The battle of minds was eventually won. A few years later, the French historian Jules Michelet took the train to Versailles and wrote: "The *château* represents pleasure, the caprice of one man; the railway is for everyone's use, bringing France together, bringing Lyon and Paris into communion with one another."[10]

The intellectual doubts could be overcome, but there were also practical and economic barriers to the growth of the railway in France. Thanks to Napoleon, the French had a far better network of roads and wide canals than Britain did, which made investment in railways less attractive and remunerative. There was a delay of several years before the inauguration of France's first main line aimed principally at carrying passengers rather than coal, the 12-mile-long Paris-Le Pecq, which was finally completed in 1837. Financed largely by the Rothschild banking family, the railway offered a pleasant scenic journey, crossing twice over the winding Seine. The original intention had been for the line to connect Paris with the old and picturesque market town of St. Germain, but the hill on which it sits was too steep for the primitive locomotives to climb, and passengers had to take a ferry to reach it. While at first attracting only a few curiosity seekers, the railway quickly became very well used by Parisians seeking fresh air in what was then the country-side. Soon, with the addition of a second track to boost capacity, it also became a very early commuter route for workers, and within two years over a million passengers had been carried on its trains. The masses traveled in open wagons, while the better off were provided with *berlines fermées*, stagecoach bodies transferred from their road chassis onto rail bogeys with a huge winch like those used by large container cranes in ports today. A 2.5-mile-long branch was soon built to Versailles following the Seine and tunneling under the park of St. Cloud. The Le Pecq line was built to Stephenson's gauge, 4 feet, 8.5 inches, which was adopted in most European countries apart from Spain and Russia, which deliberately opted for alternative sizes for military or economic reasons.

French progress was hampered, too, by a serious accident that would remain for many years the world's worst rail disaster. A competitor to the original line from Paris to Versailles via Le Pecq had been built in 1840 but proved to be less successful than its rival, even though it was shorter. A train returning from Versailles on this line was packed with day trippers who had been to the palace to watch royal celebrations. The train was so heavy that two locomotives were being used, and the leading engine suffered a broken axle, a relatively commonplace incident in the early days of the railways. Both locomotives derailed, along with three carriages; that would not have been so disastrous except that they quickly burst into flames triggered by the hot coals. Over fifty people died, though the precise number was never ascertained because the bodies were incinerated in the conflagration. The casualty list was certainly lengthened as a result of the railway's practice of locking in passengers in order to prevent them from jumping off between stations. Among the victims was the explorer and adventurer Jules Dumont d'Urville, famous for bringing the *Venus de Milo* to France, who was burned to death along with his whole family.

The accident prompted further hand-wringing about the future of the railways in France. The poet-politician Alphonse de Lamartine took a pragmatic line shortly after the accident during a debate on the future of the railways in the Assemblée Nationale, saying: "Gentlemen, we know that civilization is a battlefield where many succumb in the cause of the advancement of all. Pity them, pity them . . . and let us go forward."[11] There would be many such spectacular accidents over the ensuing years; indeed, safety improved only through the experience of such disasters. The death toll among those who built the railways and those who worked on them, though far more numerous, would attract much less attention (see Chapter 9).

Given these difficulties, the French state, aware of the importance of the railways, was called upon to take a more active role than its British counterpart had. As early as 1837, Louis Legrand, the public works minister, produced a program for the development of the railways, setting out a pattern of main lines radiating from Paris. He envisioned a kind of public-private partnership, with the state being responsible for creating the track bed and constructing the necessary tunnels and bridges, and the private companies laying the track and providing the trains. The government granted concessions to various companies, which fused quickly into six large concerns,[12] each enjoying a monopoly over its area. Despite this plan, the doubts, the disaster, and the well-developed road system all meant that initially the French network grew slowly. Only 350 miles of various short unconnected lines had been built by 1840, compared with 2,000 miles on the other side of the channel.

More surprisingly, Germany, too, would soon overtake France in terms of railway mileage. As Michael Robbins pointed out: "In 1850 the traveller arriving in France from overseas at Calais, Havre, Brest, or Bordeaux could not have got through at any point on the French northeastern, eastern or southern frontiers; but a passenger from Bremen, Hamburg, or Stettin could cross Germany to Cracow or Prague, and could get within easy distance of the western frontier near Cologne and of the southern near Munich."[13] The Germans had a good reason to develop their railways, spurred by that perennial driver of railway construction: nation-building. A federal Germany was a dream for many intellectuals who were dismayed at the disparate nature of the various German states and saw the development of railways as a way of bringing about a unified state.

As early as 1817, Friedrich List, an economics professor at Tübingen, had argued that poverty and political fragmentation were inseparably related, suggesting that a nation could only prosper through trade and industry. At the time, German industry was little developed, and List argued that famines like those that occurred in the late 1810s could be prevented if fast, efficient steam railways, ignoring arbitrary frontiers,[14] could carry food around the country. The railways, he argued, would bring about prosperity as well as the political unification of the federal states.

List was a kind of freelance railway promoter both in Germany and the United States, where he helped to build the 21-mile horse-drawn railway line in the Blue Mountains to carry Little Schuylkill coal from the mine to the nearby canal of the same name. When completed, in 1831, it was the longest railway line with iron-capped wooden rails in America. His enthusiasm for the railways, born of his support for free trade and international communication, did not always do him much good. He was thrown in jail for a time for political reasons and then forced to emigrate to the United States, but he returned in the early 1830s still proselytizing. He was not alone. Johann Wolfgang von Goethe, too, was an early supporter of the idea of developing efficient transportation systems as a way of uniting Germany. In 1824 he wrote: "Railways, express mails, steamboats and all possible means of communication are what the educated world seeks."

Before his departure to the United States, List had published a pamphlet setting out a plan for a railway system for the whole of Germany, with six lines radiating out of Berlin to towns on the German borders such as Munich (München), Basle in Switzerland, and Cologne (Köln). On his return, List decided to focus on Saxony, which had the most developed network of horse-drawn railways in Germany, as wagonways had been used there to help extract coal and ore since the sixteenth century. The exploitation of the silver and tin mines was helped enormously by the use of the *hund*, a four-wheeled vehicle that ran on tracks, initially wooden rails but later iron ones. Indeed, there is some debate about whether the first use of wheeled vehicles on rails had been there or in the northeast of England, which, as we have seen, was also littered with primitive wagonways.

List saw Saxony as the most obvious site for a major railway. Soon after coming back from the United States, he developed the idea of a line between Leipzig and Dresden, pressing the state government to fund and build it. By then, Germany's first steam-hauled railway had already

opened—the short stretch between Nuremburg and Fürth, which had
been given the royal assent in 1834 by the Bavarian king. The king's
doubts must have been assuaged by the fact that the promoters were going to name the line after him, calling it the "Ludwigsbahn." The 4-mile
line, which opened in December the following year, was built by Jo-
hannes Scharrer of Nuremburg, and it proved to be an immediate success. It paralleled the busiest highway in Bavaria, helping to relieve
congestion, just as the Liverpool & Manchester had done a mere four
years previously. Scharrer had realized that a railway largely catering to
passengers could be a paying proposition because of the peculiar local
circumstances. Since medieval times, the civic authorities of Nuremburg
had refused to allow laborers or foreigners to stay overnight in the his-
toric center, and therefore neighboring Fürth turned into what would
now be called a dormitory town for merchants and artisans. Within a
year, passenger numbers on the railway, which ran both steam and
horse-drawn trains, had built up to 9,000 per week,[15] and the shareholders were being paid handsome annual dividends of 20 percent.

In support of his idea for railways, List, who is now recognized as an
important early political economist, wrote a pamphlet in which he set
out their advantages. It could be taken as the manifesto for railways
anywhere in the world:

> Railways would carry wood, turf and coal at less than half the present
> costs. Bavaria, where flour, meat and other foodstuffs are 50–100 per
> cent cheaper than in Leipzig, could export its surplus to the Erzgebirge,
> the Elbe and the Hanseatic cities. Cheaper food and fuel would partly en-
> hance the well-being of the working classes, and partly lower money
> wages, increase population and increase the extent of industry. Cheap
> building materials and low money wages would encourage building and
> lower the rents in the new and more distant parts of the city.[16]

He also recognized how value would be added more widely to an
area by the advent of the railway:

> Increased population and industry would increase the rents, and by that
> the value of the houses, in the centre of the city, well placed for trade and
> industry. In one word: population, the number of buildings, industry,

trade, and the value of land and houses in Leipzig would be doubled and in a short time . . . this increase in value in Leipzig alone would in a few years exceed the total capital costs of the new railways.[17]

Too true, but a point that the promoters of railways would find difficult to prove to the satisfaction of investors and governments skeptical of their value, particularly in countries with a laissez-faire tradition such as Great Britain.

Even so, the Saxon government was convinced and was prevailed upon to build Germany's first major railway, linking Leipzig with Dresden.[18] List was nothing if not far-seeing, since he saw this line as part of a European network. Moreover, for List, the railway was part of a grander scheme: the reunification of the German states.

That was to be his undoing. Although the Saxon aristocrats who ruled the state were prepared to countenance building a railway, they were not interested in any wider ambitions to link the system with places beyond their borders in order to create a united Germany. Saxony was a middle-ranking state, albeit in an important geographical and strategic position between Prussia and Austria, the two great powers. Moreover, as List had recognized, there was an obvious need for a railway. In the previous thirty years, more than two hundred factories had sprung up in Saxony, and many industrialists realized that a railway would be invaluable in transporting coal and ore. At a meeting in May 1835, the scheme obtained the support of local businesspeople, who were quick to raise the share capital (£210,000, the equivalent of about £21 million today, or roughly $35 million in today's U.S. dollars), and work started shortly afterward. However, List, who was, by all accounts, not an easy man to deal with, did not get a major role in the construction of the railway. He was merely paid a small sum for his efforts and, disillusioned and penurious, shot himself a few years later.

The line, though, was built remarkably quickly thanks partly to the use of British techniques and personnel. The initial survey of the line was carried out by James Walker, a prominent Scottish engineer who had worked on the Liverpool & Manchester. A route had been suggested, but Walker, spending two weeks surveying the area, chose a flatter line that involved less earth moving, with the idea of reducing the cost of construction. He handed the detail work to a younger man, John

Hawkshaw, who was later to have a successful career on various projects, including the Severn Tunnel (the longest and dampest tunnel on
the British rail network), the abortive attempt to build a tunnel under
the English Channel, and railways in India, Russia, and Egypt. Like all
these early lines, the railway between Leipzig and Dresden was built essentially by hand with the help of nothing more than picks and shovels
wielded by men from nearby villages.

Since no locomotives were being built in Germany at that time, an experienced British manufacturer, Peter Rothwell and Company, supplied
three modest engines. The first locomotive, the *Komet*, which cost
£1,383 (about £112,000 in British currency today, or roughly $185,000
in U.S. dollars), was shipped to Germany in fifteen boxes and assembled
with the help of John Robson, a former driver on the Liverpool & Manchester, who was also called upon to drive the first train. The first sixteen
locomotives came from Britain, and there was also an American engine,
the *Columbus*, manufactured in Baltimore. The engines were crated up
at the factory and shipped along with skilled personnel to Leipzig, where
they were reassembled with the technical assistance of the foreign locomotive drivers and mechanics. The first train, with Robson at the controls, ran on the inaugural 7-mile section between Leipzig and Althen in
April 1837. Using British drivers had become something of a tradition
since the outset of the spread of the railways in Europe, as another former Liverpool & Manchester engineman, William Wilson, had taken the
controls of the first train on the short Nuremburg-Fürth line. Wilson had
been so impressed with the whole experience that he had settled in the
country.

The railway created an elaborate set of rules for passengers that
seemed designed to make life difficult for them. The bureaucratic mien
of the railways had first been exhibited when the Liverpool & Manchester opened: Passengers had to buy tickets a day in advance and provide their age, place of birth, and occupation. But the railway
companies soon realized that these requirements were unnecessary and
dropped them. The rules on the Leipzig-Dresden line were even more
authoritarian, a trend that Michael Robbins reported was the norm
across Europe: "The passenger was checked and supervised at all stages
to a degree that astonished the British traveller."[19] When the section of
line opened to Althen, there were no advance ticket sales, and access to

the station was only allowed a quarter of an hour before the train left, establishing from the outset that irritating practice, widespread in Europe, of opening the ticket office only shortly before the departure of the train. Children under twelve were banned from traveling, and there were no discounts for older ones. Passengers were required to purchase a return ticket on a train that would depart for Leipzig one hour after the arrival in Althen, which suggests that the line was being used principally as a tourist attraction, but they could apply for a refund from the Althen ticket office should they not want to use the return. The railway immediately asserted its quasi-judicial status by imposing fines of four groschen (the German currency until 1872; four groschen would have been equivalent to about 20 U.S. cents of the time) for walking on the track—and double that for riding a horse on it.

Driver Robson was not the only British worker on the Leipzig-Dresden railway line. The second locomotive used on the opening day was driven by a fellow Briton, John Greener, and the coaches were designed by Thomas Wordsell, who had built the tender for the *Rocket* at the Rainhill Trials. Wordsell, in fact, had been headhunted from the Liverpool & Manchester Railway by the Germans and had set up a coachworks, which grew quickly to employ fifty workers building carriages for the Leipzig-Dresden line. Greener, who came from a railway family, with two brothers who worked on the Stockton & Darlington Railway, stayed on, and several other British men arrived over the next few years to apply their skills to the running of the line. Not surprisingly, this caused some friction with the Saxon workers, who were worried that work was being taken away from their people. According to one historian of the railway, "the local newspaper reported several times on these disputes over the employment of foreigners on the railway."[20]

The traffic in skills did not go only one way. A Saxon, Carl Beyer, was to play an important part in the development of British locomotive design. Beyer, a self-made young man, had traveled to England in 1834 to study textile machinery and despite offers of employment in Germany had returned to work in the drawing office of the Manchester-based Sharp, Roberts & Company, which had built a locomotive for the Liverpool & Manchester. Although the initial locomotive had been rather unsuccessful, the company established itself as a pioneering manufacturer, and Beyer spent nearly twenty years there before founding his own firm, Beyer Peacock and Company, which would later build loco-motives for the early London Underground lines and would ultimately construct 8,000 locomotives, many for the British Empire, before the firm's demise in 1966.

Like the early railways in Britain and France, the train service on the completed stub of a line between Leipzig and Althen was an immediate success. Within a couple of months, there were six daily trains, each ca-pable of carrying 150 people in each direction. Sunday brought more customers than any other day, suggesting not only that the tourist mar-ket was dominant, but also that, unlike in the United Kingdom, there were no religious objections to Sunday running. Indeed, a new railway tradition was born, since on the Continent the norm is for the same timetable to operate on Sundays and weekdays, whereas in Britain a curtailed and slower service was provided on Sundays. By the end of the first year of operation, 400,000 people had taken a trip on the line, some 1,000 per day, bringing in useful revenue to finance the continued works and confirming that there was tremendous latent demand. The passenger trains were all hauled by locomotive, but horses were used for the freight trains, which traveled at night.

Over the next three years, work was completed in stages on the rail-way, which included a tunnel that was a third of a mile long and two bridges over the River Elbe, and the grand opening took place on April 7, 1839. The first train consisted of fifteen coaches, with three classes of seating, hauled by two locomotives, one of which was named the *Robert Stephenson*, driven by an Englishman, a Lieutenant Peters.

By the time of the full opening of the line, railways across Germany were expanding and a railway mania was taking hold. The German plains were filling up with lines. Speculators began to realize that there was enormous potential in the railways for making money, and the state governments were becoming aware of the powerful economic impact of the iron road. Prussia even passed a railway law regulating the industry in anticipation of this bonanza before any lines were built. In Bavaria, the king had granted a concession in 1836 to the München-Augsburger Eisenbahn Gesellschaft to build a 40-mile line between Munich and Augsburg, but the start of construction, as in so many early railway schemes, was delayed because of difficulties encountered in raising funds—in this case, for two years. By August 1840, a line had been built from Dresden to Magdeburg that went through Leipzig. It proved so profitable that it quickly became double tracked. Saxony was clearly benefiting from being a pioneer, and it was regaining some of its former economic strength thanks to the railways.

The Leipzig-Dresden railway encouraged economic interdependence among the German states and stimulated Saxony's industry. Moreover, List's vision of a cross-Germany railway network was being realized, with all the benefits he had predicted. The increase in cross-state trade stimulated by the railways built up pressure on the states to reduce customs fees and bureaucracy, pushing Germany toward unification. In effect, the railways were the biggest ally in the nationalist cause, and the economic development they brought in their wake was a key catalyst for the unification of Germany. The railway companies were not passive bystanders in this process. As early as 1846, they created the Union of German Railway Administrations, which established uniformity in tariffs and made state boundaries all but redundant. Although building a railway confined to the relatively small geographic area defined by a state's boundaries was unlikely to be profitable, once the line extended over the border into other parts of Germany its viability was guaranteed. This was a crucial economic point. The governments of the states were slow to understand this, but once they did, the expansion of the railway system was assured, and so was the creation of a federal German nation.

Unification had been the dream of a growing intellectual movement in nineteenth-century Germany, but the extension of railways across borders made it far easier to turn this dream into a reality. As a consequence, by 1871 the thirty-nine states that made up Germany, each with its own border controls and customs dues, were brought together by the railway. The railway also stimulated the development of Germany's industry. At the time of the creation of the first railway, 90 percent of Germany's iron was smelted using charcoal, a method long since abandoned in Britain. By the beginning of the 1850s, coke-using blast furnaces were being built in the Ruhr. This would not have been economically viable without the cheap transport of coal afforded by the railways.[21]

Even international borders were being breached. With its vast size, and the absence of mountains or rivers to protect its borders, Germany was becoming the heart of a wider European network. Fortunately, the far-sighted promoters of the early schemes had stuck to the standard gauge developed by George Stephenson, which meant that cross-border routes were feasible. The first, completed by the Rheinische Eisenbahn, linked Cologne (in Prussia) with Antwerp (in Belgium) in 1843 and became a rival to the Dutch Rhenish Railway. Bigger projects, such as crossing the Alps (see Chapter 5) were being discussed. Germany soon had the second most extensive network of railways in Europe, with nearly 4,000 miles of railway in 1850. This was still some way behind Britain, which after its period of railway mania in the 1840s had 6,000 miles. It would take until the mid-1870s for Germany to overtake Great Britain and become the country with the most extensive railway network in Europe.

GROWTH IN THE other principal European countries was more gradual. Italy, which was not unified until the creation of the Kingdom of Italy in 1861, began on a very small scale and in a rather strange place, Naples, in the impoverished south, rather than any city in the more industrialized north. Naples, though historically poor, was the capital of the Kingdom of the Two Sicilies, which was ruled by one of the last absolute monarchs in Europe, the portly Ferdinand II. Although politically backward-looking, he liked to portray himself as a modernizer and had responded favorably to an approach from a Parisian, Armand Bayard del la Vingtrie, who proposed building a 22-mile railway starting from the shore of the Bay of Naples to serve the pretty little towns under Mount Vesuvius. Ferdinand's enthusiasm for the project may have had something to do with the fact that the railway would allow him quick access from his main palace to his other residence, at Portici, which was the initial terminus when the first 5-mile section of line was inaugurated. The railway was intended primarily for passengers, though transportation of goods was also an objective, as one of the first three locomotives on the line was dedicated to that purpose.

The line opened in October 1839 with the king at the opening ceremony, though he did not venture onto the first train—the royal party would allow lesser mortals to take that risk. Instead, he traveled on the second train, thereafter becoming a regular passenger. Like new railways in other countries, the line attracted many curiosity seekers.

People flocked to the area to ride on the line, and soon more than 1,000 were traveling on it every day.[22] To attract even more passengers, discounts were offered to "ladies without hats, servants in livery and noncommissioned officers."[23] Although the promoter was French and the builders Italian, the British influence was still apparent: The first three locomotives were made by Longridge and Company, a pioneering manufacturer based in Northumberland.

The progress of the iron road in Italy was speeded up by the modernizing Pope Pius IX, who was elected to the papacy in 1846 and controlled a large chunk of central Italy through the sovereignty of the Papal States. His predecessor, Gregory XVI, had been adamantly opposed to new-fangled devices—not only trains, but also other inventions such as gas lighting in the streets of Rome—and the route through central Italy was blocked until Gregory's death. In fact he was remembered long afterward through a joke told by Romans. On his way to the Pearly Gates, Gregory asked St. Peter if there was a long way to go because his legs were getting tired: "Ah," said St. Peter, "if only you had built a railway, you would have been in paradise by now!"

Pope Pius IX supported railway development, and once he was inaugurated, with papal blessing now assured, lines in Italy quickly emerged. This was despite a topography that is not easy for railways in Italy, given the mountain ranges in the north and down the spine of the country, along with a series of large, fast-flowing rivers. Indeed, it would take until 1864 for an iron bridge to replace the wooden structure across the River Po, allowing express trains to at last travel uninterrupted from Berlin to Rome. There was an economic barrier, too. While the north of Italy was relatively affluent, though nowhere near as industrialized as northern European countries, the south was—and would remain—an economic backwater that could barely sustain a few major lines, let alone a network of minor ones. Despite this, an extensive network of smaller, state-supported railways were built in the south, few of which were ever profitable.

In the kingdom of Lombardy-Venetia, which was still ruled by the Austrians as part of their empire, progress was rapid once the first line, from Milan to Monza 8 miles away, had been completed. By the mid-1840s, lines had spread out from Milan; crucially, by 1857 they had reached Venice, with a branch line to Mantua, cementing the power of

Austria over its Italian provinces. The line to Venice from Mestre, on the coast, was remarkable in that it crossed the lagoon on a 2-mile-long viaduct with 222 arches, an expensive structure that cost £190,000 at the time (equivalent to about $30 million in today's U.S. dollars). However, the most impressive of the early Italian railways was the 100-mile line from Turin to Genoa, in Piedmont, which has a reasonable claim to being the first mountain railway. Heading inland from Genoa, it crossed the Apennines, reaching a maximum height of 1,180 feet just 14 miles out of the port.

Piedmont boasted the most developed economy in Italy, so it was no surprise that its railway ambitions exceeded those in the rest of the peninsula. The heavy traffic between the bustling port of Genoa and the state's thriving industrial capital of Turin made it an obvious route for a railway. The Italians had hoped to obtain advice from Robert Stephenson, who passed through Genoa during the early stages of construction, but, according to one chronicler, "the great man merely remarked enigmatically that he would not care to be responsible for the operation of the incline and went on his way."[24] The railway was opened by Camillo di Cavour, the prime minister of Piedmont, in December 1853, and the king, Victor Emmanuel II, gave his blessing by traveling on the whole route the following February.

The progress of the railway in Piedmont was helped, too, by the active part played by the government. Just as in Germany, the various states had contrasting policies on the involvement of government in the building of the railways. In Italy, the private sector was left to develop the railways in all the states except Piedmont, where Cavour, one of the architects of the Risorgimento (the movement for Italian unification), favored state development of the railways on the Belgian model. Cavour's rationale resonates today in debates about the roles of the public and private sectors. He argued that it was always easier for governments to obtain credit on the capital markets than it was for private firms, something that remains true today. Only when the government could not obtain funds, he said, should private companies become involved, and even then they should be prepared to offer guarantees of profit levels.[25] Twenty-first-century politicians still use this argument to justify resorting to the more expensive but widely available capital raised by the private sector to fund major new railway schemes. In

Piedmont, the main lines were developed directly by the government, but strictly regulated private companies were allowed to build the minor ones.

Even in other parts of Italy, where the lines were built by private capital, the state was always there to pick up ailing companies, as happened with the Monza line in 1851. But the government was disinclined to retain ownership and sold the line to the Paris house of the Rothschild bankers, which created the Société IR Privilégiée des Chemins de Fer Lombards-Vénitiens et de l'Italie Centrale, quite a mouthful (Lombardy-Venetian and Central Italian Railways Limited).

Just as in Germany and Great Britain, once the railways in Italy reached a critical mass and began to prove their worth, railway mania developed along with a rush to build lines. There had already been much speculation in railway shares. Schemes were promoted as soon as the early 1840s, and several other waves of speculation followed, especially because the absence of clear legislation on joint-stock companies allowed all kinds of fraudulent practices to thrive.

In Lombardy, the drive to build more railways was led by the Rothschild company, which obtained the concession to build two major railway systems: the completion of the main east-west artery from Trieste across to the Piedmont border beyond Milan, and the construction of the line in central Italy heading south from Piacenza to Bologna in the Papal States and to Pistoia in Tuscany, a total of more than 650 miles. The Rothschilds were to be the first railway moguls, helping to join up a European network that would change the face of the continent.

The role of the railways in unifying the Italian state, while just as important as their role in unifying Germany, came later, when the Risorgimento was on the point of victory. According to one historian of the Italian railways, investing in infrastructure, especially railways, was seen as a way of bringing the country together: "More than half the spending on infrastructure in the 1860s and 1870s was on the railways and for the whole period between 1861 and 1913, the railways absorbed around 13 per cent of the total budget and 75 per cent of the amount spent on public works."[26] Interestingly, Italy, an undeveloped country at that time, did not start industrializing until the final decade of the nineteenth century, and it benefited less from the arrival of the railways than its more economically advanced counterparts. Most of the early capital for Italian railways came from abroad, and because of the lack of coherent objectives for the network, other than in Piedmont, passenger numbers were lower than elsewhere, limiting both the usefulness of the system and its ability to make profits to fund expansion.

The effort to connect the Italian network with foreign railways was helped by its universal use of the standard 4-foot, 8.5-inch gauge on its main lines, as Stephenson's gauge had triumphed almost everywhere in Europe. The Great Western Railway in Britain, for which Brunel had chosen the generous width of 7 feet, 0.25 inches, was one of the main exceptions. As we have seen, the Great Western was eventually forced to change this because of a governmental commission in 1845 that decreed that all future railways should use the standard gauge. Other exceptions were Ireland, which used a gauge of 5 feet, 3 inches, on its main lines, and, as we have seen, Holland, at least initially, with its 6-foot, 4.5-inch, gauge. Many branch lines and minor railways across Europe would be built with narrow gauges of varying sizes to save money, but Stephenson's standard gauge would prove to be an invaluable asset to Europe's railways when they started to provide services across frontiers.

There was to be another important exception, the rather isolated Iberian system. Of Europe's most powerful nations, Spain was the most reluctant to join the railway age, and when it did so, the government chose a 5-foot, 6-inch, gauge. This would prove to be a major obstacle in uniting its system with that of the rest of Europe. Even a visit by Stephenson to attempt to persuade the government to adopt his gauge before construction of its first line began proved to be of no avail. This was not a case, as it was elsewhere, of the gauge being selected by railway promoters on the grounds of comfort or expense. Instead, the isolationist Spanish government had made a decision on the basis of military considerations: It was felt that a change of gauge at the frontier would hamper an invading army.

Spain had stuttered its way into the railway age. The government had rejected private promoters' attempts to build a railway three times during the 1830s, and by the time the Spanish government was more amenable to the idea of a railway, the political situation in Europe had deteriorated, making Spain's rulers even more eager to make sure there would be a change in the gauge at its frontier. They wanted to slow down any putative French attack. In the event, France never showed more than a cursory interest in invading its Iberian neighbor, being rather preoccupied over the next century in trying (and failing) to defend itself against its eastern one (Germany). For its part, Spain would come to rue its decision, which not only stymied its economic development but reinforced its geographic and political isolation from the rest of Europe. Moreover, the wider gauge made Spain's railways more expensive to build, as it required extra land and pushed up construction costs, which in turn caused railway companies to favor the use of a narrower gauge for many minor and branch lines, creating yet more problems of integration. The first Spanish railway was actually built in one of its colonies, Cuba. This was the 46-mile-long line between Havana and Güines, which was also the first railway in Latin America. The first 16 miles of this line, from the capital to Bejucal, were completed in November 1837, making Cuba the seventh country in the world to obtain a railroad. Cuba thus predated its colonial power in developing a railway by a decade.

Spain's decision on gauge left Portugal, whose only land boundary is with its Iberian neighbor, with no alternative but to choose the same gauge for its first railway, completed in 1856.[27] Russia, too, chose a gauge different from Stephenson's standard; it was also different from those adopted in Spain and Ireland: 5 feet.[28] As in Spain, the principal reason for this decision in Russia was military. This provides insight into the tsarist regime's isolationist thinking; ironically, however, the different gauge would also later hamper Russia in its military ambitions. Russia was a fertile place for railway growth, given its vast territory and the difficulty of creating good transportation and communication systems there. The roads were mostly mud and sand at that time; they became swampy when it rained, and broken bridges frequently blocked any progress. The rivers, which broadly run south to north, were the main means of transport, but they were frozen for half the year and sometimes impassable in summer as well due to drought. It could take two years for goods to reach St. Petersburg from Astrakhan on the Caspian Sea, though, in the 1830s, Tsar Nicholas I belatedly began to attempt to improve the transportation links.

As in Italy, in Russia the first railway connected two palaces, linking the main one at St. Petersburg, at the time the capital, with Tsarskoe Selo, the enormous royal residence favored by Catherine the Great. It took a foreign promoter, however, to persuade the Russians, who were always divided between isolationists and those who looked to the West, to accept the need for a railway. There had been a number of industrial wagonways in late eighteenth-century Russia, and the first horse-worked railway had been built early in the nineteenth century to link the Zmeinogorsk mines in Siberia with factories nearby. Nevertheless, Russia, with few factories and an agricultural economy, was still very much in its pre-industrial stage when, in 1836, Tsar Nicholas I was persuaded by an Austrian engineer, Franz Anton von Gerstner, to build a railway between St. Petersburg and the largest city, Moscow, which were 400 miles apart, about twice the distance between New York and Washington.

Gerstner suggested the construction first of a test-bed railway of 16 miles from the center of St. Petersburg to the royal palace of Tsarskoe Selo to assess whether the concept was viable. Nicholas was persuaded of the railway's huge commercial and military advantages. Gerstner stressed how the Liverpool & Manchester Railway had improved the economy of the local area and, cannily, also mentioned how, through its link with Liverpool Docks, "it had provided troop movements to Ireland to quell disorder."[29] Gerstner, too, promised that railways could operate in the harsh conditions of the Russian climate, pointing to lines in Austria and the United States. Nicholas was relatively open to the idea, but he had to use his power as an absolute monarch to push the project through in the face of opposition from the conservative faction of the Russian nobility, which argued that any available resources should be channeled toward improving agriculture. There was also the usual scaremongering in newspaper articles warning of the damage the railway would do to the forest and how it would affect the horse industry.

Russia was painfully lacking in capital, and therefore Gerstner, who would have to find investors mostly outside the country, demanded a series of concessions before he would proceed with the construction—such as a twenty-year monopoly on railway construction and the grant of large swathes of land for development around the railways. The line was built to a high standard using technology and locomotives that Gerstner imported from western Europe, as no Russian manufacturers were able to provide the equipment. The project proved to be a useful test for future development, as the railway was virtually entirely straight, passing through swampy areas, on which embankments were laid, as well as through forests, which had to be cleared. In the autumn of 1837, the same year that the writer Alexander Pushkin was killed in a duel, the railway was opened with the usual fanfare and posse of dignitaries, including many of the government ministers who had strenuously opposed its construction. Within five years, the government approved the whole line between Moscow and St. Petersburg, and the line was completed in 1851 (see Chapter 7).

The railways also arrived later than in most of Europe in neighboring Scandinavia. Railway development on the Continent tended to move northward and eastward from the center, reaching Denmark in 1847, Norway in 1854, Sweden in 1856, and finally, Finland in 1862. The reason for the tardiness of the Scandinavia was obvious. These were poor and sparsely populated countries, with a harsh climate that was not conducive to easy railway construction. Norway, for example, had just 2 million people in the middle of the nineteenth century. Spread over a vast country, most eked out a living from agriculture, timber, and fishing. The first short line ran between Christiania (now Oslo) and Eidsvoll, and although plans to build a network connecting all the major towns quickly emerged, the scheme took fifty years to complete because of the distances involved and the lack of capital. Sweden, by contrast, built a long line linking its two main cities, Stockholm and Gothenburg, 260 miles apart, remarkably, completing the project in 1862. Sweden, through a combination of state and private railways, implemented a plan that would eventually lead to the creation of a system of 10,000 miles, giving its citizens more track per inhabitant than any other country in the world.[30] Finland, which was even more sparsely populated, and a fiefdom of Russia until World War I, struggled to make a start on its railways and then built them to the Russian 5-foot gauge rather than the European standard used by its Scandinavian neighbors, connecting Helsinki with St. Petersburg in 1870.

THE EARLY LINES of Europe were, for the most part, relatively modest affairs, going nowhere in particular, but that was because they were conceived from the outset as part of a wider network rather than an end in themselves. Although several of these first railways were designed primarily as passenger railways, even those built with freight in mind found themselves catering to large numbers of people who were

keen to use this new invention. These early railways proved more popular than many had expected them to be, partly because people were so
curious about them; but the fact that they offered such a fast and relatively cheap form of transportation, immeasurably better than any of
the alternatives, was also crucial to their success. In almost every country, particularly those with developed banking and finance systems,
within a decade or so of the inauguration of the first railway a rush to
build new lines ensued. These were often so frenetic that they could be
termed manias.

The promoters of these railways were a varied bunch, but for the
most part they were forward-looking entrepreneurs, often visionaries,
who knew that railways would change the world, but possibly not just
how much. There were a few crooks and unprincipled speculators out
to make a fast buck at the expense of gullible investors, too, with varying degrees of legality. But, as Michael Robbins said, the results of the
development of the railways were virtually the same everywhere: "In all
countries where the national railway system, whether private or state-
owned, has succeeded in creating a network which permits, even en-
courages, a free flow of passengers and freight . . . from end to end of
the territory, the railway has at once become a source of economic
strength to the whole nation, not merely to a locality."[31] The crucial
point there is that the successful railways united countries and brought
disparate regions together. In some cases, such as when different gauges
were used, or when the insuperable physical barrier of a mountain
range or wide river prevented the development of a network, the economic value of the railway was sharply reduced.

In terms of patterns of ownership, the countries of mainland Europe
were reluctant to allow the kind of free-market experimental structure
that was allowed to emerge in Britain. The European states on the Continent kept a much tighter rein on their railways than Britain did, sub-
jecting them not only to the need for approval but also to far greater
control. One key reason was that their leaders recognized the military
importance of the railways. There was also less concern about government interference in enterprise in these countries than in Britain, something which remains true to this day. Though Britain had a part in
stimulating the creation of railways across Europe, its huge initial influence waned somewhat as other countries developed their own manufac-

turing bases and local workers developed the necessary skills. Several European countries were quick to import ideas from across the Atlantic. British locomotives, however, remained a major export for the cradle of the Industrial Revolution right into the twentieth century. And the British influence would remain strong in its colonies, as we shall see in the next chapter, and later in many parts of the world, from South America to the Far East.

THREE

The British Influence

BRITAIN HAD A GREAT ADVANTAGE IN ITS ATTEMPT TO REMAIN AT THE forefront of railway development. Not only did it serve a large domestic network, which grew rapidly, albeit in fits and starts as the various railway manias flared up and ran their course, but it also had the advantage of a captive market in its expanding colonial empire where it could export locomotives, rails, and other railway supplies. Consequently, as Michael Robbins put it, "Until about 1870 . . . Britain was the heart and centre of railway activity throughout the world."[1] Britain's own network suffered from unnecessary duplication, with a proliferation of lines that could never be viable because the government deliberately eschewed planning or any attempts to control the private companies building the network. Its lead, however, was such that its technology, expertise, and finance were exported to many countries, including very unlikely ones, such as several in Latin America and Asia with little previous connection with the British Empire. British technology was therefore widely imitated, and its financing, in the last quarter of the nineteenth century, became vitally important for many systems, but the British style of laissez-faire planning for the railways, characterized by lack of interest from the state, was rarely imitated.

Ireland, then part of the United Kingdom, was the most obvious country to be influenced by British technology, but oddly, this did not extend to the choice of gauge. Ireland was an early starter, and its first railway, the 6-mile-long Dublin & Kingstown Railway, was built to standard gauge, but it was to prove the exception. The Dublin & Kingstown was intended to provide improved communications between Dublin and the harbor at Kingstown (now called Dun Laoghaire), where ships from Liverpool and Holyhead docked. Rather uniquely for these early projects, its promoters had carried out detailed surveys of traffic on the existing road to assess potential demand before seeking investment for the line. They had to allay the very vociferous opposition to the railway, which was based on suggestions that the line's builders were crooked and on fears that the line would affect existing commerce. It took much determination on the part of the contractor, William Dargan, to take the project to completion, as he had to appease two large landowners who sought massive compensation. One of them, Lord Cloncurry, demanded not only a private bridge over the line, so that he could retain access to his own secluded bathing area, but also a short Romanesque tunnel through an embankment in order to maintain his privacy. Dargan, who came from a humble farming background and had learned his trade building highway projects under Thomas Telford in England, went on to construct 800 miles of railway in Ireland. He also funded the Great Exhibition in Dublin in 1853, helping to establish the country's national gallery. Like his fellow contractor Thomas Brassey, who built more of the English railway system than anyone else, he had a reputation for paying good wages and treating his men fairly, a rare quality among railway contractors.

The Dublin & Kingstown line was built by 2,000 navvies in the space of two years and opened, slightly later than expected, in December 1834, with locomotives imported from England. The railway's opponents had certainly been right in predicting that the patterns of commerce would change, since the crowded road between Dublin and the harbor did lose much of its traffic to the railway, despite the latter's early teething problems, which included a collapsed bridge that was hastily replaced. The railway had been intended for the transport of both goods and people, but it was the passengers, some of the very earliest daily rail commuters in the world, who became its mainstay.

Ireland, a poor country with little mining and no heavy industry at that time, was not very fertile territory for railways. Although it was initially quite densely populated, the country would lose 20 percent of its population in the famine of the 1840s, and even more over the ensuing decades as a result of emigration, which left it a sparsely populated island with income levels well below those of the rest of the United Kingdom. Moreover, the terrain was not easy, with many difficult geological features for railways to cross. Therefore, the railways developed more slowly and in a very different way from those in England, not least in terms of financing. Even before the famine, the British government, recognizing Ireland's peculiar circumstances, provided both loans and grants for the construction of railways, a policy that ministers had always refused to adopt in England itself. This financial support from the state provides a rare clue into Great Britain's early recognition of the economic development that railways could stimulate, but ultimately the subsidies would cause Ireland to have too many railways[2] in relation to its size and its economy. The railways were developed in Ireland for political and social purposes, and yet, in England, such measures would have been seen as totally unnecessary—scandalous, even—interference with market forces. Even today, Northern Ireland's railway remains publicly owned, operating as an integrated business, while in the rest of the United Kingdom the network has been split up and privatized.

The British government's attitude in relation to the gauge of the tracks on the island was strangely incoherent. Given that the first Irish railway had used the standard gauge, and that, by 1845, the Gauge Commission in England had ruled in its favor, it was surprising that ministers did not insist that Ireland, which, after all, was part of the United Kingdom, should follow suit. Not so. The second railway to be built, the Ulster, which ran between Belfast and Portadown in Northern Ireland and was completed in 1842, used the generously wide gauge of 6 feet, 2 inches, following the recommendation of an 1836 Royal Commission. Then came the Dublin & Drogheda, which planned to use a gauge of 5 feet, 2 inches, despite the fact that eventually it was expected to connect with the Ulster Railway.

At that point, the Board of Trade stepped in. The board's inspector, a Major General Pasley, sought the advice of the Stephensons, who, strangely, rather than pressing the case for the standard gauge they had created, bizarrely suggested something between 5 feet and 5 feet, 6 inches. As a historian of the Irish railway, H. C. Casserley, sarcastically put it: "The major general came up with the discovery that the average between the two figures was exactly 5ft 3ins and this was the figure which was decided upon."[3] Casserley noted, quite rightly, that the bit of extra room available in most Irish coaches makes an appreciable difference in comfort to passengers compared with the standard-gauge trains in England. Although the choice of gauge was curious, at least the mainline railways were all built to it, which ensured that they could be connected, although many minor lines used smaller gauges, mostly 3 feet, in order to save on construction costs.

This unique gauge, which would later spring up on railways in diverse places around the world, including Brazil, three Australian states, and New Zealand, undoubtedly hampered the development and economics of the railways in Ireland. Much to the inconvenience of passengers traveling across the Irish Sea, the break in gauge precluded the establishment of ferry boat trains, which became widely used across the Channel between Britain and France and several other places in the world after first being developed in Denmark in the 1870s. When, between the world wars, the Southern Railway introduced the *Golden Arrow* Pullman and the *Night Ferry*, whose sleeping cars went right through from London across the Channel to the Gare du Nord in Paris, nothing of the sort could be developed for the Irish Sea. Instead, overnight passengers on the *Northern Irishman* were unceremoniously awakened at an ungodly hour at the small station of Stranraer on the West Coast of Scotland and had to walk, carrying their luggage, onto a ferry for Larne in Ireland, all because of the daft decision over gauge made at the outset. Moreover, the 5-foot, 3-inch, gauge necessitated the construction of specially adapted rolling stock and prevented the shipment of rail wagons from the United Kingdom, all of which added to the expense of operating the railway.

Ireland's railways developed quickly, thanks to the combination of state aid and entrepreneurial activity. Although the railway mania was less intense there than in England, there was rapid growth in the 1840s, and by 1853, the main towns—Dublin, Belfast, Cork, and Galway—had all been connected. There were 830 miles of railway on the island by 1853, 130 of which had opened that year. Given the economic difficulties and the low population density of Ireland, this was a remarkable achievement so early in railway history. Casserley suggested that occasionally the rural nature of the country and its famously relaxed ways sometimes compromised safety: "The casual attitude of the southern

Irish is well illustrated by the story of the level crossing keeper who kept one gate over his single line open to the railway[,] and the other to the road, because he was 'half expecting a train.'"[4]

Ireland may have been legally part of the United Kingdom, but the British government treated it very much as a colony rather than an integral part of the country. Although India was much further away, the British influence over its railways was just as powerful. The railways were developed on the subcontinent rather later than in Ireland, but they were to have a dramatic effect on the country, beyond what even the most far-sighted of the railway promoters might have imagined. As we have seen, the British influence on the railways was enduring and worldwide, but India was the jewel in its crown. The British Raj created a system designed and built by British engineers—helped, of course, by hundreds of thousands of Indian laborers—using equipment exported from Blighty in thousands of British ships. This system was to form the backbone of the nation, and, indeed, it remains today a vital part of the infrastructure for both passengers and freight. Moreover, the breadth of conception of the railways in India, laid out as a network from a very early stage, made railway development in India the biggest public works project worldwide since the construction of the Pyramids.

The railway was brought to India by the British East India Company, effectively the commercial arm of the British government, which ruled India from 1757 until the late 1850s. The initial motivating force behind the introduction of the railway system in Bombay[5] was the need to transport the cotton on which the city's economy depended, the same product that had stimulated the creation of the Liverpool & Manchester. The failure of the American cotton crop in 1846 pushed the cloth manufacturers in Manchester and Glasgow to seek new sources of the material, such as India. However, in order to guarantee a steady supply, transportation to the port at Bombay needed to be improved, and therefore the cotton magnates pressed the British government to build a railway there.

In fact, there had been talk of railways in India since the early 1830s, but bureaucracy, and the long distance between the country and the motherland, ensured there was much procrastination. It took months to receive a reply to a letter, which slowed down the decision-making process, and there were strong doubts in Britain about whether a huge investment in a railway in a distant colony—and huge it would have to be—would reap dividends. There were doubts, too, among the British rulers in India about the value of a railway to a country where most people lived in desperate poverty and where society was divided by a caste system that created a huge number of "Untouchables," with whom people of higher castes would not want to travel: "Will they be able to pay for the fare of a ticket? Would they feel at all for the necessity to increase the pace of their life? Would the masses ever travel by train, if it were introduced? These were the doubts that assailed the authorities in those days," wrote a historian of the Indian railways.[6] The answers to all three questions would all be yes, but it is understandable that such serious reservations were expressed. The need for a stable cotton supply appears to have brought about the turning point in the decision to proceed: "The increasing British demand for alternative supplies of cotton from India possibly tipped the balance in favour of quickly providing this rapid means of transport," according to another history of the Indian railways.[7]

The first scheme for an Indian railway was set out in 1844 by Rowland MacDonald Stephenson (no relation to George and Robert), who argued that a railway could operate safely and profitably even to remote corners of the country despite the harsh weather and challenging topography. He stressed the benefits it would bring to a land whose population was then around 100 million—a vast number, and yet fewer than one-tenth of today's total: "The commercial benefits that are likely to arise from railways are unquestionable and almost incalculable."[8] He went on to list the potential uses of a railway in India for both freight and passengers, such as increasing the trade between Calcutta,[9] the capital at the time, and the Upper Provinces, which would be "sure to increase with the facilities of railway communication." While conceding that most Indians would not be able to afford railway travel, he concluded that "the growing classes of intelligent natives would freely resort to it." He even, very presciently, suggested that there would be a big take-up among pilgrims going to holy cities such as Benares and Allahabad, but he warned that separate carriages would have to be provided for "Mahomedans, and high and low caste Hindoos." And if there were any female passengers, they, too, would have to be provided with separate accommodation. With 11 million passengers using the railways of India daily today, Stephenson's predictions have proved correct.

Confusingly, another Stephenson, George's son Robert, now comes into the story. He supervised a survey of the line after his appointment as the consulting engineer of the first Indian railway company, the Great Indian Peninsular. However, he remained in England, leaving the actual work of chief engineer to James Berkley, his pupil.

Unlike the rail network in the United Kingdom, which developed through a haphazard process, the rail network in India was created through centralized planning with the deliberate intention of linking all the major cities as soon as possible. The plan was laid out by Lord Dalhousie, the governor general of India from 1848 to 1856, who supported the idea of building two experimental lines, one from Bombay, near the present Victoria Station, to a small town, Thana, 21 miles away, and a second that would be a far greater undertaking, a 121-mile-long line in Bengal, stretching between Howrah—on the western side of the Hooghly River across from Calcutta, which could then be reached only by boat—and the small town of Raniganj, chosen because it was situated conveniently among the coalfields of Burdwan, where previously, according to Robert Stephenson, it had taken "two seasons . . . to bring the coal down a direct distance of seventy five miles"[10] by a very circuitous route. Both of these railways were clearly intended to be the start of a network linking Bombay with Pune and eventually Madras, and linking Calcutta with Delhi and later, through the newly conquered Punjab, extending right through to Lahore in what is now Pakistan.

Dalhousie had been pressing for the creation of a strategic rail network since his appointment as governor general. He was the sort of dynamic modernizer that the British Empire occasionally produced and later claimed he had unleashed in India the "great engines of social improvement, which the sagacity and science of recent times had previously given to Western nations—I mean railways, uniform postage and the electric telegraph."[11] Indeed, he had a past interest in the railways, having chaired a parliamentary committee in 1844–1845 that attempted to put some order into the chaotic situation when, at the height of the railway mania, Parliament had been literally inundated with bills petitioning to build lines. Under pressure from the railway companies, the committee had been abolished within a few months of its creation, but Dalhousie had clearly understood that the haphazard development of railways was a wasteful process. Dalhousie set out the basic framework for the railways, not only devising the outline of the network, but deciding on the gauge, which he argued needed to be wider than the British standard gauge because narrow trains might be blown off the tracks in tropical storms.[12] He wanted 6 feet but under pressure from London settled on 5 feet, 6 inches, 9.5 inches wider than the standard gauge. His vision was far-sighted. Bridges and tunnels were to be built to a scale that could accommodate a double track, even though all the initial lines, apart from a few sections near major stations, were single.

Just before work started on the first line, Dalhousie wrote a famous "minute"[13] expressing the hope that "this great instrument of improvement may be extended over all the land bringing with it the rich and numerous benefits it is calculated to produce," and arguing for the need to learn from "the errors which we have committed at home in legislation for the regulation of railway works." In other words, he argued, the railways should be viewed not merely as "private undertakings but national works, over which the Government may justly exercise a stringent and salutary control." Dalhousie recommended the adoption of a form of compulsory purchase order, ensuring that landowners on the path of the railway could not hold the railway to ransom by demanding an exorbitant price for their land.

There was no attempt to disguise the fact that this was a nakedly imperial project. It later would attract the wrath of nationalists, including Mahatma Gandhi. The British rather than the Indians would decide on the location of lines and the timetable for building them. If necessary, parts of India not controlled by the British would be crossed irrespective of the views of the local maharajahs. Dalhousie's "minute" was effectively a blueprint for how to build a railway network in an underdeveloped country for the benefit of the imperial power. Although the British implicitly accepted that the Indians would benefit in countless ways from the railways through their impact on the development of the economy, the network was designed to serve imperial interests above all and was seen as a way for Britain to establish itself in its most lucrative colony. As Ian Kerr, a prolific writer on the Indian railways, wrote: "The interests of the Indians were incidental although, as represented in the writings of Dalhousie and many Britons, the progressive consequences for India of the railroads was a self-evident truth."[14]

Dalhousie recognized, too, from the outset the tremendous military advantages of railways for a small colonial administration stretched over a vast nation (which at the time included present-day Pakistan, Bangladesh, Sri Lanka, and later, even Burma), since they would allow for the rapid deployment of troops anywhere throughout the country in days rather than months. Indeed, these advantages were shown early in the history of the railways when, in 1857, they were used extensively to transport forces to put down the Indian Mutiny. By then the railway network was still in its infancy, with only a few major lines, but it nevertheless greatly facilitated the movement of troops, and the authorities recognized that had a more extensive network been available, the Mutiny might have been quelled more quickly. Given these military and economic benefits, the British viewed the cost of the railways as a small price to pay if it allowed them to maintain their rule over their most precious colony.

Britain would also benefit from the railways through the easier access they provided to India's produce, especially cotton, and through the facility gained in exporting manufactured goods back to the subcontinent. There was no doubting the scale of Dalhousie's ambition: "The complete permeation of these climes of the sun by a magnificent system of railway communication," Dalhousie wrote, "would present a series of public movements vastly surpassing in real grandeur the aqueducts of Rome, the pyramids of Egypt, the Great Wall of China, the temples, palaces and mausoleums of the great Moghul monuments."[15]

The first completed railway in India, as in many other countries, was a seemingly modest affair, the experimental line between Bombay and Thana, but the name of the company that built it, the Great Indian Peninsular Railway, rather gave the game away as to the railway's ambitions. Work started in 1850. The Thana line, now part of Bombay's extensive bustling suburban network, went through attractive countryside that presented several difficult obstacles to the railway, including a marsh that, like Chat Moss on the pioneering Liverpool & Manchester, required a complex embankment to be built, and a hill, which meant cutting through a mile and a half of hard rock. These major undertakings were typical of those that would be needed for virtually every future railway on the subcontinent. After the line left the urban areas, it crossed attractive countryside with views of the Thana River and the Ghats, the mountains that would prove to be a daunting barrier to the railway's further progress out of Bombay.

The line was the first to be built in the Far East, and its opening in April 1853 was a momentous event, not least because India, unlike the European countries where railways were springing up, had hardly seen any industrial development by this time. Steam power was still a novelty. The *Illustrated London News*,[16] which reported on the opening two months later, when the dispatch reached London by boat, was effusive, stressing that the event would be remembered far longer than the recent battles that had brought India into the British Empire. Those military victories, the paper wrote, "seem tame and commonplace" in comparison to such an event, whereas England's power "was never so nobly exemplified as . . . when the long line of carriages conveying nearly 500 persons glided smoothly and easily away amidst the shouts of assembled thousands."

The crowds, including many people who had perched on walls or climbed trees to get a better view, were all the more impressed because of the sheer scale of the first train. Since trials of the locomotives and rolling stock had been undertaken for several months prior to the opening, the authorities took the risk of running an enormous train, carrying all the VIPs in a train of fourteen carriages, which was sent off with a twenty-one-gun salute. As had become traditional at railway openings, a public holiday was declared, and a banquet was provided for the VIPs, who reached Thana after a trip of fifty-five minutes in which they averaged a speed of more than 20 miles per hour. The sightseers included visitors from as far afield as the East Coast of Africa, the Persian Gulf, and Afghanistan, and, according to the *Illustrated London News*, they were a mile thick at Thana, where they slowed the progress of the train by spilling onto the tracks, a habit that has survived to this day on the subcontinent.

The railway was presented by the British as the embodiment of civilization, which they were graciously offering India. One speaker at the banquet, at which the Indians were seated on separate tables from the British, summed it up this way: "A well desired system of railways, ably and prudently executed, would be the most powerful of all worldly instruments of the advancement of civilization in every respect."[17] At the start there were just two trains per day, and no freight, since the wagons were not ready, but traffic built up rapidly as more equipment, all imported from Britain, arrived.

A separate company, the East Indian Railway Company, had been incorporated to build the more ambitious Bengal line and started work in 1851. Various mishaps held back progress: A ship carrying the first railway coaches sank at Sandheads on the Bengali coast, for example, and a vessel carrying locomotives from Britain was erroneously routed to Australia instead of Calcutta. The locomotives eventually reached their destination at Calcutta a year late, for which, presumably, some poor clerk must have received a royal rollicking, although perhaps the mistake was understandable given that Australia, too, was joining the railway age (as described below). Despite these setbacks, progress on the line was fast. The topography was easy and in September 1854 the first section opened; the entire 121 miles between Howrah, across the river from Calcutta, and Raniganj were completed by February 1855.

Down in the south of India more railways were being built. The Madras Railway Company opened its first 64 miles of line in 1856. It ran from Madras to Arcot on easy terrain and stretched from coast to coast by 1861. The pattern was the same throughout the country, with sections of line being opened almost annually, focusing on the trunk routes between major cities.

Other large companies sprang up to build lines in the Punjab (the Sind, Punjab & Delhi Railway) and northwards from Bombay (the Bombay, Baroda & Central India Railway). Dalhousie's "minute" had expressed a vision for a railway that was both strategically important but also profitable, stating that once opened, these railways would, "as a commercial undertaking, offer a fair remunerative return on the money which has been expended in their construction."[18] That was an ambitious aim, which he realized would only be possible with initial state aid. Therefore, most Indian railways were built through an arrangement combining the public and private sectors. Conventional joint-stock companies, based in the United Kingdom, would raise capital, mostly from British investors, to fund the construction, but, to ensure that the money could be raised, the government of India guaranteed a healthy 5 percent rate of return. This was essential, as it took many years, sometimes decades, for the companies to achieve profitability, and the government had to be the financial backstop to pick up the shortfall.

The two first companies were soon expanding, and the initial extension of the line out of Bombay was to be the most ambitious project yet attempted by railway builders anywhere in the world. It involved crossing the Western Ghats, the mountain range that rises up from the coast. This was arguably an even more daunting task than building the Semmering crossing of the Alps (see Chapter 5), since the conditions the workers had to face, including extreme heat and monsoon rains, were more challenging than those found in Europe.

The Western Ghats seemed like an insuperable barrier. They may only rise to 2,500 feet, but they present steep faces and rugged terrain, and the few roads that had been cut into them were narrow and full of perilous hairpins, not at all a suitable alignment for a railway. The route through the Ghats was decided by the engineer, Berkley, with advice from Robert Stephenson, but it would take eight years to build the 15-mile pass, and poor Berkley would not live to see it. He died young, like so many of his fellow engineers, back in England, his health broken by the tropical climate. He had produced 3,000 maps and drawings to work out the precise line up and down the mountain, creating the railway equivalent of a hairpin bend in order to keep to the 1 to 37 gradient, the maximum that the locomotives of the day could handle.[19] There were two major inclines, the Bhore and the Thal, as well as several tunnels, and Berkley devised a novel way to build the line up the hill. Instead of constructing a continuous line, he carved a reversing section out at a bend near the summit, obviating the need for a stationary engine that would have provided power through a cable. It was a neat solution that survived well into the twentieth century, despite the obvious inconvenience of having to reverse all trains, and the same method was later used on railways in Brazil and the Andes.

The raw statistics of the Ghat project were staggering. It involved constructing twenty-five tunnels and eight substantial viaducts and cutting 54 million cubic feet of rock. The cost was £70,000 per mile (about £5 million in today's money, or roughly $8 million in today's U.S. dollars), about one and a half times the amount spent on an average line in Britain, where wages were much higher. Except in the tunnels, work on the railways in India stopped during the four-month period of the monsoon, which meant that for most of the time construction took place in intolerable heat. The working conditions were perilous in the extreme. On several cliff faces, there were no footholds, and consequently workers had to be let down on ropes to drill and blast the face. It only took a break in the rope, a stumble, or the careless use of blasting powder to result in a cruel death down the ravine, from which the bodies were never recovered.

Accidents were, however, not the main source of death and injury. Diseases were rife and, in those days before antibiotics, frequently deadly. Cholera was the most prevalent, but other killers included malaria, typhoid, smallpox, and blackwater fever, which together wiped out a large proportion of the malnourished and overworked coolies. One historian estimated that 25,000[20] lives were lost in the construction of the Ghat inclines, mostly to disease. As, even during the busiest part of the project, there were at most 42,000 people working on the project, such a death toll suggests that the average life expectancy of the workers was remarkably short. The colonial attitude toward the suffering that the building of the railways engendered is neatly encapsulated in the chilling language of an official report: "The fine season of eight months is favourable for Indian railway operations, but on the other hand, fatal epidemics, such as cholera and fever, often break out and the labourers are generally of such feeble constitution, and so badly provided with shelter and clothing, that they speedily succumb to those diseases and the benefits of the fine weather are, thereby, temporarily lost."[21]

One could ask exactly whose responsibility it was to provide that food and shelter. As Anthony Burton, the author of the book that quotes this memorandum, put it, "the notion that lives—and the inconvenient loss of working time—could be saved by providing proper shelter and decent conditions does not seem to have been considered."[22] The death toll was simply startling, reaching numbers normally seen only on the battlefield. Deaths of navvies during the construction of the railways in Britain were a common occurrence,[23] but the rate at which they occurred was nothing like that in India. No wonder, at times, the railway's agents had to scour far and wide to find willing labor, especially as there were frequent delays in paying them.

It was, of course, not only the Indians who suffered. The poor engineer Solomon Tredwell, sent from England to work out how the railway could scale the Ghats, perished from dysentery just four weeks after his arrival. Later, when cholera struck, in April 1860, 25 percent of the British contingent working on the Bhor Ghat Incline succumbed.[24] These workers were a precious source of skills and experience, and such a depletion of their numbers posed a threat to the progress of the project. Overall, there were an estimated 500 British engineers working on the various Indian railways during the hectic decade of the 1860s, when at any one time some 1,000 miles of line were under construction, involving perhaps 250,000 Indian laborers and craftsmen. It was a remarkably small number of engineers to take responsibility for so much work, and the quality of these men was variable, to say the least. Many were rough fortune-seekers who treated the workers lamentably, but even they were seen as having a vital management role to ensure projects were completed. The quality of the British contractors also varied considerably, as they ranged from undercapitalized men of little experience to the likes of the ubiquitous and conscientious Thomas Brassey.

The worksites for the scaling of the Ghats are reputed to have been the worst in the history of the railways, though those on the Panama Railway (see Chapter 6) must have run them a close second. When, in January 1859, the Indian workers turned violent and attacked their bosses with sticks and stones, because their delayed wages had been paid out at half the agreed rate for the job, a British official commented wryly that it was surprising that they had not revolted sooner: "It is evident that labourers have been most grossly abused in the matter of their wages," he said.[25]

The working methods, too, were completely different from those that were common in Britain, and it took the British a while to realize that they could not just impose Western ideas on the workforce. Bridge construction, for example, required erecting the primitive but effective bamboo scaffolding used in India, "a forest of jungle wood,"[26] as a British engineer called it, although it is far safer than its flimsy appearance suggests. Sometimes trying to force British practices on the Indians resulted in low comedy, such as when Indian coolies, ordered to use wheelbarrows, simply carried them on their heads, a sight that finally persuaded the British to leave well enough alone. The British also tried to stop women and children from working on the sites, but eventually they had to acquiesce to the employment of whole families. The men would break up stones, and the women and children would carry them away, which might have seemed primitive and labor-intensive, but these methods were tried and true, proven by time: They had, after all, proved good enough for the construction of the Taj Mahal.

Religious sensibilities could not, however, be compromised. Henry Fowler, one of the main contractors to build the line from Bombay to Thana, made the mistake of taking water from a workman's pot only to find that the man immediately broke the vessel on the ground because it had been defiled by a person outside his caste. The British did not repeat the mistakes that led to the chaotic and lawless scenes on the Ghat; instead, they began to try to blend in with the existing culture, rather than riding roughshod over the local workforce. In the words of Ian Kerr, a historian of the Indian railways, "The lessons the British learned from the Bhore Ghat extended well beyond civil engineering and they were applied well beyond the Ghat construction."[27] Conditions improved on later railways, though they were still awful even by Victorian standards, with perilous working practices and disease ravaging the work camps at regular intervals.

The British, too, realized that techniques imported from back home were not always suitable for tropical conditions, notably when building bridges. The Mhok-ki-Mullee viaduct, part of the Bhore incline project, was an eight-arch bridge designed in the classical style to cross a 135-foot ravine. It had been the subject of concern because of rumbling noises reported from the river below and the development of small cracks, but upon examination nothing untoward was found. However, suddenly, in July 1867, a platelayer tightening bolts on the track on the bridge noticed the ground start to give way soon after the passage of a train. He managed to run to the end of the bridge as it was collapsing and then warn an oncoming train. Miraculously, no one perished in the incident. The precise cause of the bridge's failure was never ascertained, though there were suspicions that "scamping," the deliberate use of inferior materials, may have contributed to its weakness. More significantly, the builders of future bridges had to adapt to water flows caused by monsoons and snow melt from the Himalayas far in excess of those encountered back home. As Kerr suggested, "it took some decades of bitter experience before the engineers understood how deeply below the riverbed they had to place bridge foundations in order to secure them from the scouring effects of the extremely high flow levels carried by many rivers."[28] The railway builders had to become hydraulic engineers, shifting water courses and guiding flows away from the piers, in order to protect their bridges. Overall, the learning process was slow but steady, ensuring that railway building did eventually become routine: "Those who managed the construction process slowly came to know much better how railways should be built in India. A process only poorly understood in the 1850s and 1860s—whose first model was railway construction in Britain—had by the 1890s become familiar, routinized to a certain extent, and better adapted to the natural and social conditions of India."[29]

The problems encountered by the two inaugural railways would be replicated throughout the massive expansion that took place over the rest of the century. Despite the array of difficulties and obstacles, both human and physical, the railways were built, and built rapidly. Within ten years of the opening of the Thana line, there were 2,500 miles of railway in the country, and double that a decade later. This progress was not quite as fast as in the United Kingdom, where the network reached 7,000 miles in just over twenty years, but it was excellent going, given the conditions. By the 1890s, the basic network of trunk lines set out by Dalhousie covering the whole country had been completed.

The railways were a great engine of economic growth, not so much for India as for Britain. For the first couple of decades, virtually every item of investment in the equipment was shipped from the United Kingdom. By the end of 1863, barely a decade into India's railway age, 3 million tons of railway material had been sent to India in 3,751 ships.[30] It was not only the manufacturing and shipping companies that benefited: The city of London, too, enjoyed the Indian railway boom, as British money financed the construction, and every shipment had to be insured—rightly so, since there were considerable losses—further boosting business in the Square Mile, London's financial district.

And if the economic benefit was largely going to the British, so was the best of the railway service. First class, which provided a completely different order of comfort compared with the third-class carriages used by the Indian hoi polloi, was largely the preserve of the whites. It provided a standard of accommodation that was far better than anything in Europe. One traveler enthused: "Soda water is offered to you just as you are conceiving a wish for it. Tea comes punctually at 6 A.M. No sooner have you passed your hand over your stubby beard, a barber appears to shave you in the carriage. You get a 'little breakfast of eggs and bacon with bananas and orange at eight, a delightful tiffin [light lunch or snack in India] in the heat of noon, and a good dinner at sunset.'"[31] Even with these luxurious facilities, rail journeys, which were often measured in days rather than hours, were tedious affairs, and constant vigilance had to be kept for skilled thieves who developed special tools, such as long shears, to cut off jewelry through windows. Women traveling on their own were put into Ladies Only compartments, but "these were a favourite target of the thieves."[32] On the long hot trips up to the cool hill stations, blocks of ice would be placed on the floor of compartments every morning, but the passengers still stewed, and they were engulfed in dust once the sun rose.

Even affluent and educated Indians were not made welcome by the white sahibs in these relatively comfortable surroundings, and there are numerous accounts of petty discrimination. No Indian would attempt to enter a compartment occupied by a European, and, while the converse was true, problems could arise when there were no vacant compartments available. When Justice Nanabhai Haridas, the first Indian judge at the Bombay High Court, was returning to Bombay from Surat in June 1885 with his son, he found only one first-class compartment with two spare spaces, but it was already occupied by a Captain Loch, who objected to the judge and his son being seated there because his sick wife was traveling with him and was in a state of partial undress (which probably meant her shoulders were bare). The stationmaster at first tried to persuade the captain to accommodate the illustrious Indian, pointing out that the compartment had not been reserved, whereupon the officer simply paid for the reservation. The judge and his lad ended up in second class.

The Indians did, though, occasionally get revenge. In one incident, Sir Ashutosh Mukherjee, another renowned lawyer, who became the first Indian chief justice in Calcutta, dozed off in a first-class carriage and found, upon awakening, that his sandals were missing. A white plantation owner had taken objection to them and thrown them out the window. But then the white man made the mistake of going to sleep himself. When he woke up, he found his jacket was missing, and when he complained to Sir Ashutosh, he was informed that "your coat has gone to fetch my slippers."

The Australian railway system developed under quite different circumstances. Whereas in India, detailed planning and strong direction from the state led to the creation of a coherent and integrated system that became the backbone of the country's infrastructure, in Australia[33] the failure to impose a uniform set of standards, notably through an agreement on gauge, was to prove deeply damaging to the railways. As in India, the Australian railways made an early start, and they also came up against natural barriers. These barriers were not as difficult to breach as the Ghats, but they nevertheless posed a formidable obstacle for railway pioneers and delayed progress.

Astonishingly, the first railway in Tasmania did not use steam or even horse traction but was human-powered, and since this was Australia, the muscle inevitably belonged to convicts. The inaugural railway was a 5-mile line across the Tasman peninsula in what was then called Van Diemen's Land. It was built to enable passengers from the capital, Hobart, to avoid a lengthy sea trip around Cape Raoul and through the appropriately named Storm Bay, a perilous and rough journey. Using the railway, boats would be able to anchor in the quiet Norfolk Bay, and passengers and goods could then be taken across land on the railway to a bay that provided easy access to the convict settlement at Port Arthur. The idea for the railway came from the commandant of the settlement, Captain O'Hara Booth, who used the prisoners to build the crude railway, which had wooden rails about a foot apart and followed the contours of the land in order to minimize heavy earthworks.

More controversially, though, he then organized the prisoners to pull the trains, which was by no means an easy task, as the gradient leaving the prison was quite steep. An early visitor, a Captain Stonor, described the carriages as "very rude construction, very low, double seated with four very small cast iron wheels. On either side projected two long handles which the prisoners lean on to propel the carriage."[34] Stonor was appalled at the sight of the poor prisoners, who were dressed in saffron prison garb complete with arrows, having to "puff and blow pushing on the carriage" up the hill, and he noted that, "when descending they jump alongside of you and away you go, dashing, crashing, tearing on."[35] At the midpoint of the line, there was a rest station where the crew changed, and a maintenance gang was housed there. The fresh team then had to struggle further uphill for a mile, after which there was a similar length of descent to the jetty which was, according to Stonor, "quite a nervous affair and as the speed increases each moment you expect to be dashed over into some precipice or deep jungle alongside the tram."[36] Braking was crude in the extreme: There was a guard on the train who could "check the wheels with a drag." Later visitors claimed that the speed on the final descent toward the jetty was 40 miles an hour, but this seems unlikely.

There was no shortage of critics shocked at the use of such labor in torrid conditions, especially as sometimes the men did three return journeys in a day, covering 30 miles (though they did ride the down sections on the train): "It jars harshly against the feeling to behold man as it were lowered to the standard of the brute, to mark the unhappy guilty creatures toiling and struggling along, their muscular powers exerted to the utmost and the perspiration bursting from every pore," wrote one.[37] The writer did later point out that many of the "free" navvies building Britain's railways toiled in similar conditions for days on end. For the passengers, however, who had paid a shilling for the journey, it was actually a lovely ride through heavily forested country that in springtime was "thick with pink, red and cream heath and stunted little golden wattles."[38]

Tasmania persevered without recourse to steam engines until 1871, but even after that date built the longest horse-drawn railway in the British Empire. The line was occasioned by a rush to exploit the richest tin mine in the world in a mountain originally known as Mount Bischoff, but later unimaginatively, though accurately, described as "the mountain of tin." The railway would quickly help to reduce it to a hill of debris. The Van Diemen's Land Company, a London-based concern, laid down a railway from Emu Bay to the mountain, a distance of 45 miles, and operated the line with horse-drawn trucks. It was a crude affair built entirely with local materials, including railroad ties made of myrtle and stringybark, which tended to expand in winter and shrink in summer, and it made for a bumpy journey. Navvies imported from Victoria built the 3-foot-gauge line, cutting through the rugged countryside where few people had even traveled, and which was full of wide gullies that had to be crossed using crude bridges.

Irritatingly, the railway ended short of the mountain because the company could not get the right of way through the adjoining farm, which meant the freight had to be manhandled on a muddy track for the final two miles. The service, though, proved amazingly popular, with a train scheduled to travel in each direction on alternate days, except Sundays, so that on many occasions freight and passengers had to wait a couple of days for the next train. Horses were used on the line for half a century. They were specially bred coach animals with strong legs that were selected for their relatively modest appetites, since every ounce of food had to be brought in on the railway. Four horses, which were rested every 10 miles, pulled the trains, and the trip took seven or eight hours, though it could take twice as long when snow covered the ground on the high point, 2,000 feet above sea level, or when one of the horses acted up. The Tasmanian tradition of building horse-drawn lines survived into the twentieth century. Early in the 1900s a 4-mile timber-tracked line was laid between Tullah and Boko Siding in the west of the state, where a mine was being developed. It quickly proved inadequate and was replaced within a decade by a narrow-gauge steam railway.

Horses were used on the first iron road on the mainland, too—on the 7-mile-long line in South Australia between Goolwa on the lower Murray River and Port Elliot on the coast, which opened in May 1854. It was built to give access to the seaport for goods taken down the Murray River, especially wool, and was built with the ambitious intention of providing an important artery. But like so many such early schemes, it was not properly thought out, as Port Elliot proved too small to be a viable seaport.

Meanwhile, the first steam-hauled line was being built in Victoria using the Irish 5-foot, 3-inch, gauge, and the inaugural train ran the 2.5 miles between Flinders Street in Melbourne to Sandbridge, now Port Melbourne, in September 1855. There had been talk of building a canal instead, but as Patsy Adam Smith pointed out, "railways won the argument against canals and a 'navigation canal' has never been dug in Australia."[39] New South Wales entered the railway age barely a month later, only missing out on being the first because its inaugural railway was delayed by the insolvency of its promoter, the Sydney Railway Company, which went bankrupt. The project was taken over by the state government, thus becoming the first in the British Empire to be government owned. The 13-mile line between Sydney and the inelegantly named Dog Trap[40] Road, Parramatta, was built using standard gauge and included one remarkable structure, the stone-arched Lewisham Viaduct.

Along with using controversial labor for its first railway, and persevering with horse-drawn trains far later than in Europe, Australia can also be regarded as the country where the textbook of how not to build a railway network should be written. The almost simultaneous opening of these two railways in neighboring states sowed the seeds of the disastrous failure to agree to a common gauge that was to dog the development of Australia's iron roads. The failure to ensure a uniform gauge proved to be an insuperable obstacle to the establishment of a coherent system, ensuring that coastal shipping freight would remain competitive in a country whose sheer size should have favored railways. Indeed, the railways in Australia, though in many ways as important to the economy as elsewhere in the world because of their impact on the mining industry, did not have the same nation-building effect as in the world's other large countries, such as the United States, Canada, or Russia.

How on earth did this gauge chaos occur? The reason that the Australian states, which inevitably blame each other for the situation, adopted different gauges is a confusing tale with roots in the obduracy of a couple of railway engineers in New South Wales. In 1846, the British colonial secretary had recommended that the Australian railways be built to standard gauge, but F. W. Shields, the chief engineer of the Sydney Railway Company, was an Irishman, and he was familiar with the 5-foot, 3-inch, gauge and persuaded the New South Wales government to adopt it. The Victorian and South Australian governments decided to fall in line and altered their gauges to conform with that of New South Wales, and several railway companies that had sprung up in Victoria during 1852 ordered rolling stock on that basis. Then, suddenly, Shields resigned over a salary cut, and his replacement, James Wallace, was an Englishman with an attachment to the standard gauge. Despite protests from the other two states, the change went through. New South Wales built its railways to the standard gauge, while Victoria and South Australia held firm, arguing that it would be too costly to alter their stock. As the chroniclers of the history of the Australian railways put it, "the glorious bungle began."[41]

In fact, it was not very glorious, but it was to prove expensive, and it became a permanent handicap to the railways. Millions of dollars are still being spent to remedy the situation in the twenty-first century. When the railways finally met at Albury in June 1883, it "showed the broken gauge in all its limitations and folly." Moreover, two other mainland states, Queensland and Western Australia, along with Tasmania, adopted a narrow 3-foot, 6-inch, gauge, which meant that a transcontinental journey could require two train changes. South Australia has the dubious distinction of having three different gauges—narrow, broad, and standard, and there are innumerable problems related to the gauge changes.

As elsewhere in the world, each of these inaugurations was marked by ceremonies and trips on the new line by the great and good, but in Australia there seemed to be an unspoken competition among the states over who could hold the grandest gastronomic display. At the opening of the 45-mile-long Geelong-Melbourne line in June 1857, the town of Geelong organized a repast consisting of "two and three quarter tons of poultry, two and three quarter tons of meats, three quarters of a ton of fish, three quarters of a ton of pastries, half a ton of jellies and ices, half a ton of fruit, a ton of bread and unlimited wines, spirits and ales."[42] This seems far-fetched, but despite this cornucopia, the VIPs never got their tucker. After various hold-ups caused by problems with the locomotive and lack of coordination, the guests from Melbourne only arrived at Geelong when the dinner was over; the local populace had taken advantage of the delay to have their fill as never before. The opening, incidentally, like the inauguration of the Liverpool & Manchester Railway, was marred by a fatality: Henry Walters, the locomotive superintendent of the Geelong & Melbourne Railway Company, was killed when he was knocked from the train while passing under a bridge 2 miles from Melbourne.[43]

It was on the line heading westward from Sydney toward the interior of New South Wales that Australian engineers faced their most difficult challenge, when, within a few miles, they found the Blue Mountains, which extended up to 3,000 feet, barring their path. The local population was insufficient to provide the muscle required to carve the railway through the mountains, so in 1858 a labor force was recruited by Thomas Brassey, the British contractor, who found 2,000 men, mainly Scottish, who were willing to travel to the other end of the world to build a railway. According to Brassey's Victorian biographer,[44] many of these men decided, upon arrival, that they would rather work in an easier job, and therefore wages, and consequently costs, were notably higher than back home. The engineer, John Whitton, originally wanted to use a gradient of 1 in 20 to scale the highest ridge, which was 3,336 feet above sea level, but no locomotive of the time could have managed such a climb. He eventually settled for a series of zigzags, similar to those on the railway up the Ghats in India, and a mile-long tunnel at Glenbrook through the mountain. There was a considerable amount of blasting required, including the need to dislodge a 40,000-ton chunk of rock, an event so spectacular that the Countess of Belmore, the wife of the governor general, was called upon to push the button setting off the explosion. Australia was on the way to getting a cross-continental railway, though it would not be until 2004 that a line would link the north with the south through Alice Springs, located in the middle of the country.

Railways came late to another major part of the British Empire, Africa, where in the last decade of the century, the great transcontinental project, the Cape to Cairo railway, was begun. Although it was never completed, large sections were built (see Chapter 7). The British tradition would prevail in its colonies apart from Canada, as well as throughout Latin America (covered in Chapter 8), but the other great style of railway—the American railroad—was beginning to assert itself. Although that type of railway did not extend much beyond the American continent, it would have the most profound influence, not least on the country where it originated.

FOUR

The American Way

THE BURGEONING AND VAST UNITED STATES EMBRACED THE RAILWAY. Many of the myths and images of railways that are popular worldwide are American: those perilous-looking trestle bridges, the massive locomotives with domed chimneys, the hobos jumping freight trains, carriages under fire from bandits, and the small Western town waiting for the Iron Horse to arrive. It is the United States where the arrival of the railway in the nineteenth century had the greatest influence, both in creating the nation and in stimulating its economic development. Quite simply, without the railway, the United States would not be the United States of America. The railway's role has been largely written out of the nation's history, however, because of the American love affair with the car and the airplane, and even today the railway's importance in moving freight across the vast country goes unrecognized.

Given its size and economic growth, the United States was obviously fertile ground for the railway, and it was inevitable that the new nation would soon boast more mileage in rails than the rest of the world put together. The continued importance of the iron road can be shown by the fact that, by the beginning of World War II, the U.S. network still represented a third of all the globe's railways and was by far its largest network.[1]

American railways contrast sharply with those in Britain and Europe, and not just because they are called "railroads" on the western side of the Atlantic. As A. F. Garnett put it in his book *Steel Wheels*, "the difference can be summed up in one word: scale."[2] This does not simply refer to the huge distances covered by American railroads, which link the two oceans with three different routes, as well as connecting Canada and the American South and Southwest through several lines. Nor does it describe only the amazing feat of crossing major natural barriers, including mountain ranges such as the Appalachians and the Rockies and countless large rivers, such as the Mississippi and the Ohio. It also encompasses the simple fact that American trains were bigger, taller, longer, and heavier than those in Europe. They look massive even in photographs, by comparison.

Though American trains run on the same 4-foot, 8.5-inch, standard gauge as the trains of Europe, their maximum height is 15 feet, 6 inches, nearly 3 feet higher than the 12-foot, 8-inch, maximum in Great Britain. Moreover, as there are very few overhead crossings on the American railroads, double-decked passenger trains can easily be accommodated; more importantly, this added headroom allows road trucks to drive straight onto flat-bottomed wagons, and containers can be stacked on top of each other. Finally, there is another sense in which scale is of a different order, as this chapter will show. The commercial drive to build the railroads, and the wheeling and dealing that process involved, which quite frequently lapsed into sheer criminality and skulduggery, was also on a much grander scale in America than anywhere else. There is, though, a need for some caution, too, in assessing America's contribution to railway development. While there is no doubt that the American railroads were pioneering in many respects, and that their growth, particularly in the thirty years after the Civil War, was quite extraordinary, there is a tendency among some U.S. railroad historians to claim just about every technical improvement and development as their own. The truth is more nuanced, with inventions often being made more or less simultaneously on both sides of the Atlantic and innovations being adapted to particular local circumstances.

In many respects, the U.S. railroads bear a greater resemblance to their British counterparts than to continental railways. A complete rejection of central planning—except when the transcontinental route was being debated—together with the same emphasis on competition as in the United Kingdom, rather than on coordination, resulted in many similar problems, such as unnecessary duplication of lines; inconvenience for passengers, who often had to change trains and buy separate tickets for a single journey; and a proliferation of unconnected stations in major cities. There was another similarity between America and the whole of Europe. There was no question that the railroads would be open to all comers on the turnpike principle—that is, that anyone prepared to pay a toll would be allowed to use the line. On the contrary, the decision was quickly taken by the railroad companies that they would run the whole business, operating all aspects of the service from running the trains and determining the timetable to providing the trains and maintaining the track. As in Europe, a system of "running rights" and access charges when companies traveled on tracks belonging to another railroad soon developed. The U.S. railroads, too, were saddled with the obligations of being common carriers—required to transport anyone or anything offered to them, often at rates determined by regulation—and again, as in the United Kingdom, one of the first organizations in the states to spot the enormous advantages of carriage by rail was the government postal service. If the railroads were sometimes the victims of regulation, however, more often they benefited from the lack of it. Railroads in the United States were true monopolists and, on occasion, were able to exploit that situation to the fullest.

Interestingly, right from the start, in distinct opposition to the style of their European counterparts, the U.S. railroads had cars—always called "cars"—which were open plan on the inside (though fortunately, they were closed to the elements), without compartments or any class distinctions—what one writer described as "a democratic palace instead of a nest of aristocratic closets."[3] More likely, though, it was not the democratic mien of the Americans that dictated the design, for this was a country where even the supposedly free black people of the North were soon forced into their own separate accommodation, but rather an early understanding that traveling long distances in a cramped compartment in very hot or cold weather, without access to a toilet, would be unbearable. Nor were open-plan carriages necessarily always a pleasure, since many early passengers complained of the habits of their fellow travelers, such as spitting and chewing tobacco, and there was nowhere to escape. A more serious danger, particularly before those bulbous chimneys became commonplace, was the risk of clothes being set on fire by sparks coming in through windows. That left passengers with a cruel choice in summer: opening the windows and risking wrecking their clothes—one woman, Harriet Martineau, complained in 1836 of thirteen holes in her gown after just one short journey on a southern railroad—or stifling in the oppressive summer heat. Most chose the latter. People's lungs, too, suffered from the smoke that came from the locomotives, especially those that were wood burning.

Passengers soon had to face the rigors of night travel. On the first long line to be completed, between Albany and Buffalo, the trains were supposed to take fourteen hours, but often the journey was twice as long. The nights, with only a couple of candles to illuminate the carriages, were the worst part of the journey for the passengers, and only partially alleviated by the introduction of crude sleeping cars with accommodation, which were curtained off on one side of the carriage. These were quite different from the individual compartments that became standard in Europe. The common fear of passengers in these early night trips was the prospect of being stuck for a lengthy period on one of the perilous trestle bridges, not, oddly, because of concerns about falling into the abyss below, but rather because of the fear of catching a chill or some disease from the "bad air."

In fact, the passengers on these night journeys, which became increasingly necessary as the railroads extended their routes, faced far greater hazards than lack of privacy or malodorous swamps. The enginemen could not see the track, which might be blocked or have suffered a broken rail, and therefore at first some companies sent down locomotives on their own[4] in advance of the passenger trains to ensure the line was clear and complete. Consequently, unlike in the United Kingdom, where headlights were considered an unnecessary extravagance, in the United States lighting was seen as essential. Working out how to provide lighting, however, was not easy in those pre-electricity days. One railroad, the South Carolina Canal & Railroad Company, hit upon the rather crude idea of hitching a flat wagon with a bed of sand and a blazing pine log fire to the front of the locomotive. It was not a successful concept, and once night travel became frequent, in the 1840s, the rather more secure system of using kerosene lamps with mirrors to provide a powerful beam was universally adopted.

We are indebted to Fanny Kemble, a famous British actress of the nineteenth century, who loved the railways and wrote elegantly about her trip on the Liverpool & Manchester just before its opening[5] for one of the earliest descriptions of a railroad journey in the United States. The American trains must have seemed strange to her, as she was used to the compartment design used in Britain. She wrote with the same acuity and wit in America as she did about her experiences on the tracks in Britain, evoking the hurly-burly that, oddly, is rather evocative of train travel in India today: "The windows . . . form the walls on each side of the carriage which looks like a long greenhouse upon wheels; the seats which each contain two persons (a pretty tight fit, too), are placed down the whole length of the vehicle, one behind the other, leaving a species of aisle in the middle for the uneasy (a large proportion of the travelling community here) to fidget up and down."[6] She took a dislike to the tobacco chewers, who spat in the aisle, and the "whole tribe of itinerant fruit and cake-sellers to rush through distributing their wares at every stop." These included water boys, who had only a long spouted can and two grubby glasses, but so clammy was the atmosphere in the trains during the summer that they did a good trade. There was only one class on the railroads, but in keeping with the American spirit, the affluent and the posers could avoid all these discomforts by hiring their own car for between $5 and $25 per journey. It could be just a flat wagon onto which their buggy or stagecoach was lifted and secured, or it could be a well-furnished private carriage for the exclusive use of their party.

The southern railroads were the most uncomfortable, providing far fewer facilities than their northern counterparts. The contrast was emblematic of the economic and social gulf between the two halves of America, who were soon to wage war against each other. The illustrious architect of New York's Central Park, Frederick Law Olmsted, who traveled widely in the South before the Civil War, complained that southerners felt that the need of the Yankees for a fresh clean bed, digestible food, and trains that made their advertised connections at least half the time was illustrative of northerners' weak-bellied nature. For his part, Olmsted was critical of the lax safety practices of the southern railroads and could not understand why basic precautionary principles were not followed, such as ensuring that lubricants were on hand when axles overheated, a common occurrence on early railroads.[7] He warned presciently that the inadequate railroads would be to the detriment of the South: "There is nothing that is more closely connected, both as cause and effect, with the prosperity and wealth of a country than its means and modes of traveling."[8]

Food, in fact, was rarely wholesome anywhere on the railroads, North or South, mainly consisting of quickly prepared meals at lunch stops that caused passengers to complain of indigestion. But that may have been because they tended to bolt their food down in order to reclaim their seats, which generally were not reserved. Indeed, unlike the system in Germany and Britain, where pre-booking was mandatory, it was possible to board most American trains without a ticket, a practice that proved lucrative for many conductors, who would sell tickets and pocket all or part of the proceeds, a practice known as "knocking down." Passengers who embarked on longer journeys that required traveling on routes run by different railroad companies, and frequently entailing transfers between steamboats and trains, became prey to "runners" working for forwarding houses who had their own ways of extracting money illicitly. The agents of the forwarding houses bought large numbers of tickets in bulk cheaply from the railroads, steamboat lines, and stagecoaches and then sold them on to passengers. The runners would make all kinds of promises about connections, the inclusion of meals, and timing, but the poor passenger would often arrive at a change-over point to find that the runners' ticket, which was usually beautifully produced with pictures of modern, fast-moving trains and ships, was not valid for the rest of the journey.

Once on the train, passengers were still not guaranteed that they would reach their destinations. Derailments and accidents were relatively common in this early period (though their consequences tended to be less serious than they are today because of the slow speeds of the trains, which averaged below 20 miles per hour until the 1840s). Trains did not even follow a set schedule. They were generally only guaranteed to leave at a particular hour—an arrival time was not specified—and therefore there was none of the imperative for speed that was to prove so deadly later in the century in a series of train wrecks. Where there were timetables, they were so unreliable that the expression "to lie like a timetable"[9] became common parlance.

However, there was one notable early casualty. In January 1853, Franklin Pierce, who had just been elected president but had not yet taken office, was traveling with his wife and eight-year-old son, Bennie, from Boston to their home in Concord, New Hampshire, on the Boston & Lowell Railroad. Although the carriage may have been better furnished than early ones, its structure and safety features were little different from the basic coaches used on the inaugural railroads twenty years previously. Wooden throughout, and resting on two bogies (sets of wheels), each consisting of four wheels—in contrast to contemporary British carriages, which were usually fitted only with four wheels—they only had one amenity: a small potbellied coal stove. It was big enough to prevent passengers from freezing in the winter, but not enough to keep them warm. The seats, which were cushioned and sprung, unlike the wooden benches of their predecessors, were an improvement. The technical equipment, though, was still primitive, with brakes operated by a wheel on the outside platform at the end of each car and crude link-and-pin couplings connecting the three or four carriages, which is all the wood-burning locomotive could manage. It was the faults in those couplings that were to prove fatal, as the one holding the Pierces' coach snapped when the train lurched, flinging the car off the rails down an embankment. Surprisingly, the parents were unscathed, but poor Bennie was crushed under a seat and died instantly, a tragedy that cast a pall over Pierce's undistinguished four years in the White House, which were marked by his failure to prevent the Union from slowly disintegrating.

The biggest danger was head-on collisions, known, appropriately, as "cornfield meets." Virtually all the railroads were single track, and early signaling was not so much primitive as nonexistent. Outside of major towns and junctions, U.S. railroads used a system of train orders, which authorized a train crew to proceed to a particular point, which could be dozens of miles away, such as a station or a loop, to pass a train coming in the other direction. Until the advent of the telegraph, which was first used in the 1850s, there was no means of communication between drivers and dispatchers, who occasionally made mistakes by sending two trains on a collision course. And when a train broke down, it was in great peril, as the dispatcher may well have sent out others in its wake on a time-interval basis. The rear of broken-down trains had to be protected quickly by the train crew with warning flares. These all-too-frequent early accidents were often later portrayed in silent films as deliberate, with villains trying to arrange cornfield meets by nobbling the dispatcher and issuing false orders so they could steal from the wreckage. A less perilous but still irritating aspect of early train travel in the United States was delay, which was caused by the paucity of passing loops. The single sections of tracks had centerboards halfway between passing loops, and the railroad rule was that the first train to get to that midpoint would have priority; the other would have to backtrack in reverse to make way. This led to disputes between enginemen over which train should go back, and even passengers were known to join in.[10]

Once the first line, the Baltimore & Ohio, had been completed, there was some delay before a concerted rush toward building railroads occurred in the United States, replicating an almost universal pattern in the major economies. The nation was watching the B&O's progress, and it took the company three years to build up passenger numbers to three hundred per day. It almost seems as though every country and its population required a collective taking of breath before embarking on constructing a major rail network, perhaps because people were aware of the irreversible changes that such a network would entail. The railways in Europe linked existing communities; those of the United States had a far more ambitious goal—spreading development and "civilization" across the vast continent—although many early railroads did start out with the far more modest ambition of merely trying to reach the next town.

Within a couple of decades, once the real potential of this invention had been realized, ambitious schemes for a transcontinental line, which had been the dream of some of the very early pioneers, sprang up. As the author of the preface to a history of the U.S. railroads eloquently put it, "while railroads in Europe were commonly the servants of established communities, in America railroads were often their creators."[11] Quite simply, as another historian, Stewart H. Holbrook, wrote, "the main achievement of the railroads was to help enormously to build the United States into a world power and do it well within the span of one man's lifetime."[12] There were, too, more practical obstacles: the mountains, and, particularly in the early stages, the vast rivers that traversed the country, which forced many early railroads to be built from one river crossing to another, obliging passengers to transfer to a boat to continue their journey. Freight also had to be taken across the rivers on boats in these early days. Several railroads that were built as competition to the canals nevertheless provided faster transport for passengers than had previously existed. For a time, a series of eight railroads interspersed with sections of canal speeded up the journey for passengers between Albany and Buffalo, but this must have made life difficult for anyone with substantial baggage to carry.

Once the success of the Baltimore & Ohio and other early lines seemed assured, railroad promoters popped up throughout the eastern seaboard, pushing plans for a dense network of railroads. The early 1830s were a period of experimentation and testing, but by the end of the decade the rapid growth of the railroad was in full swing, and it would continue until the end of the century. More often than not, these initial promoters struggled to raise the capital, even though the advantages of rail seemed all too obvious in an age of few alternatives. As in Britain, there had been a canal boom in the twenty years preceding the opening of the railroads in America, but waterborne transport was slow, expensive, and subject to paralysis during times of cold weather. The turnpikes were barely worthy of the name, mere dust roads prey to the elements, and the railroads therefore represented a major improvement in transportation infrastructure.

It would take time, though, for the railroads to achieve domination in America. The battles between canal owners and the early rail pioneers that had occurred in Britain[13] were replicated in the United States, where the railroads had to compete with the well-established steamboats. Moreover, just as in the United Kingdom, the process by which promoters could obtain a railroad charter and build a line in America was haphazard and fraught with difficulties—and, certainly in the early days, largely outside the purview of government. Even so, beware of the American myth of free enterprise untrammeled by government interference. Just because government did not set out a comprehensive plan for the railroad system does not mean it was not involved in a significant way. As the U.S. network grew and demands to extend it mushroomed, governments at both the state and federal levels invariably did become involved, eventually playing a vital role in subsidizing, regulating, promoting, and even vetoing railroad proposals. As one historian of the U.S. railroads has suggested, the rail companies and their promoters were at the mercy of the politicians: "Government subsidy, regulation and control have encouraged the rise of railroads, [have] shaped their maturity and have helped make their lives difficult, feasible, profitable or impossible."[14]

Like many canals, several early railroad schemes, such as the Portage Railroad and the Philadelphia & Columbia, were government enterprises, and three states—Michigan, Indiana, and Illinois—experimented with public construction, mostly with scant success. A more important aspect of government involvement was the financial assistance provided by several states to private schemes, which in the 1840s amounted, on average, to more than 10 percent of their capital investment, a proportion that rose in later years[15] without even taking into account help in kind, such as the free route surveys carried out by army engineers. Later, as the railroads spread westward into virgin territory, the provision of land grants became the principal means of government support for the railroads.

There were, though, several alternative ways of helping nascent railroads other than through direct grants. Railroads such as the Camden & Amboy were offered long-term monopolies, exemption from state taxation, or even, in Georgia, the right to operate a lottery to help finance construction. The most crucial measure, however, was the notion of "eminent domain" (or compulsory purchase), which gave the railroad freedom in the selection of its route after the charter was obtained. This way of doing things contrasted with the method used in the United Kingdom, where the route had to be approved by Parliament beforehand, putting landowners in a strong position to hold out for large compensation payments.

Nevertheless, few railroad projects obtained their charter without a fight. Invariably, there was opposition from all kinds of vested interests, from tavern keepers and turnpike owners to bridge companies, stagecoach operators, and canal officials. There was, too, the usual skepticism from people who could see no reason for the iron road. One Boston newspaper editor suggested that a project from his town to Albany in New York would be as useless "as a railroad to the moon."[16] And in a deeply religious country, there were always those who saw the railroads as an instrument of the devil. In Ohio, a school board called them "a device of Satan to lead immortal souls to hell."[17]

Although American railroads were, of course, based on the same scientific principles as those in Europe—and indeed, several of the early lines used British-built locomotives—the technology soon diverged to accommodate the scale of the task of running trains across such a vast country. The circumstances of the United States, where land was cheaper and the emphasis was more on access than speed, dictated different requirements: The track tended to be lighter than in Europe, and the surveyors generally chose routes that would require a minimum of engineering because of the lack of capital available to fund major work. The expensive tunnels and cuttings that were a feature of British and other European railways were avoided, as the lines skirted around obstacles rather than passing through or under them, and the engineers were content to design the railroads with gradients that were far steeper than those on the other side of the Atlantic. Bridges were basic affairs made of wood and used the standard design of a trestle bridge, an invention first drawn by Leonardo da Vinci, with the result that they resembled massive pieces of garden furniture standing precariously over rivers, but they were in most cases more robust than they looked. There were exceptions, though, as in India, and bridge disasters were far more frequent on the U.S. railroads than in Europe, continuing with awful regularity up to the end of the nineteenth century. The locomotives, too, were different, as they needed to overcome those steep gradients and long distances. The size and power of U.S. steam engines were never surpassed anywhere else in the world, but they were not as efficient as those built in Europe.

Consequently, the cost of building these early railroads was relatively modest, an average of around $34,000 per mile for the 9,000 miles completed by 1850, and as little as $17,000 per mile on railroads in the South, where the terrain was easier and where low volumes of traffic required lighter engineering standards.[18] During the same period, the cost of railways in the United Kingdom averaged £31,000 per mile.[19] Given that the exchange rate at the time hovered just below five U.S. dollars to the pound, the American railroads were markedly cheaper.

And yet, there were some aspects of the American railroads that were better than Europe's right from the start, notably locomotive design. For example, the engines were always fitted with cabs for the crew, a "luxury" that British railway companies felt would induce lax standards in their enginemen, but a measure made necessary by the rigors of the U.S. climate. The locomotives were also very soon built with four front bogie wheels[20] to add enough stability to cope with the curvier, cruder tracks in America.

Two other distinct features of the American locomotive were the bulbous chimney, designed to reduce the number of sparks, which was essential in locomotives that mainly relied on wood for fuel, and the cowcatcher—or, more crudely, a cow killer—an essential piece of equipment perfected by the evocatively named Isaac Dripps, chief mechanic of the Camden & Amboy, to limit the damage to trains that passed through open country where cattle ran free. In America, many regions had large areas of open range—in fact, this is still common in the American west—unlike in Europe, where fenced-in fields were the norm. At times, the cow proved to be a crucial weapon of farmers who were hostile to the railroad or who simply wanted a good price for their old stock. The Michigan Central, a line that crossed much cattle country, was originally state owned, and the politicians, concerned about public support for the railroad, ensured that there was generous provision to any farmer whose animals were struck by a train. When the line was taken over by private interests in 1837, the new owners made great efforts to reduce the cow casualties by putting up fences, but they remained surprisingly high. As Holbrook explained, "the simple, honest agrarians along the road were quick-witted to know Opportunity when it loomed on the steel rails in the form of the Iron Horse. They began to feed their oldest and poorest stock handy to the tracks and often, it is said, plumb between the rails."[21] The company, aware of the farmers' tactic, reduced the compensation to half the value of the cows. Incensed, the farmers launched a full-scale war on the Michigan Central, putting obstructions on the lines, burning stations, tampering with points, and even throwing stones at passing trains. It was only when the company infiltrated the farmers' organization and raided their homes just as an arson attack was being launched on a train in 1851 that normal traffic could be resumed. A dozen conspirators were convicted and jailed.

Hitting cows was a frequent source of delays, and rides on these early services were not always comfortable even if the train managed to avoid hitting them. Indeed, they were rarely comfortable. As one early account reports, "passengers discovered they not only had to pay to ride on the cars; they often had to help lift them back on to the rails and to push entire trains over slight grades."[22]

As the railroads spread, technical innovations allowed them to run faster services and to reduce delays, but the higher speeds they brought in their wake, combined with lax safety standards, resulted in more frequent and increasingly spectacular disasters, to which the official response was tardy and begrudging. The most important early change was the development of the "T rail," named for its shape, a rail made of iron that replaced the wooden rails, which had been topped with a sliver of iron that had been a constant source of problems. U.S. rail historians also claim the "sandbox," a device that improves grip by dropping a small amount of sand in front of the lead driving wheel, as an American invention. They report that it was made in response to a plague of locusts in Pennsylvania in 1836, when the mass of insects was so great that the several short railroads in the state could not get their locomotives to grip the rails because of the squashed insects. Brooms were fitted to the engines but soon wore out, and the solution only came about when a clever railwayman filled a box with dry sand, fitted it to the top of the locomotive, and ran pipes down to feed it onto the rails. However, British historians claim the invention was made in England. Perhaps it was genuinely one of those cases where the same solution to a common problem was discovered in two separate places simultaneously. Technical progress, too, proceeded in the United States, which saw great improvements in the efficiency and power of locomotives, and even brakes became more efficient, although, as in the United Kingdom, continuous braking throughout the train would not be made mandatory in America until the last years of the nineteenth century.

The telegraph was the most crucial development in allowing the more intensive use of tracks as it provided a method for train managers to communicate with each other across the system, greatly reducing the potential for cornfield meets. Until then there was no knowing at what time a train would arrive until it was spotted by a lookout at the top of a water tower, the railway equivalent of a crow's nest. The telegraph, using Morse code, which used a series of dots (short signals) and dashes (long signals) for each letter of the alphabet, was first developed in the

1830s, but it was not until the symbiosis between the railroads and the telegraph companies became obvious that it was widely used, enhancing the value of both inventions. Running the telegraph alongside the track might not have taken the wires along the shortest route, but the line provided a clear path through the countryside and ensured that train drivers would be able to report the location of fallen poles in a timely way. For the railroads, the big advantage was that they got the telegraph service free, and, crucially, that their messages had priority over those of the general public. It took a while, though, for enginemen to trust the telegraph, as electricity was a new discovery and there were doubts about using "lightning" to convey vital information.

An apocryphal story on winning over the drivers relates how Charles Minot, the superintendent of the Erie Railroad, was traveling on a westbound train at Harriman in New York state when the train was delayed waiting for a late eastbound service. Minot had persuaded his railroad to install a line of poles next to the track, but they were not being used by the railroad. An impatient fellow, Minot barged into the telegraph office and told the operator to transmit the message to hold the eastbound train until further notice. He then ordered the driver of his train, Isaac Lewis, to proceed, but the engineer, who was more conservative, did not want to take the risk. Minot dispatched Lewis to "ride on the cushions";[23] the terrified fellow perched on the back seat of the last wagon, and Minot took over the controls himself. By telegraphing ahead, Minot was able to proceed, and they finally met the eastbound service three stops down the line, saving an hour. The Erie officials were quick to see the advantages of this new method of communication, but it was not until the 1860s that its use became universal.

Despite all the discomforts and dangers, nothing stopped passengers from flocking onto the Iron Horse. Like the railways of the United Kingdom, the American railroads not only put the rival stagecoaches out of business, but also stimulated massive numbers of new journeys. One example can be found in the growth of traffic between two towns in South Carolina, Charleston and Hamburg, which had long been served by a two-seater stagecoach running a mere three trips per week. In the first six months of 1835, the Charleston & Hamburg Railroad carried 16,000 passengers,[24] an average of 2,500 per month, compared with just 50 per month that had been common on the stagecoach previously.

The railroads did more than merely attract passengers—they captured the imagination of the nation. Americans dreamed of far-reaching rail lines that would span rivers, plough through forests, and cross deserts to reach the other side of the continent. Between the opening of the Baltimore & Ohio in 1830 and the end of the century, the United States was buffeted by a series of railroad fevers that saw the mileage increase in fits and starts, reaching 250,000 miles by 1900. The railroad manias of the United Kingdom and other European countries were mirrored in the United States, but because the country is so vast and its economic regions so diverse, they occurred in different states at different times. Britain experienced a mini-mania in the 1830s and a major one in the following decade, while in the United States the rush in the early 1830s to promote railroads was brought to a temporary halt by a panic in some states in 1837; others continued without a pause, and, in any case, the whole country soon recovered its momentum by the end of the decade. The spread of the early railroads was not the result of any grand design or central plan but was instead stimulated by the desire of local citizens to be connected to the network, or simply to reach a neighboring town speedily, in much the same way that happened in Britain in the second half of the century when branch line construction was at its peak. This phenomenon is eloquently captured by Holbrook, who related the story of how the citizens of "Brownsville," an imaginary town, would go about getting their railroad:

> Some up and coming individual or simply a fanatical dreamer said forcibly that what his home town of Brownsville needed, if it were to share America's great destiny, was a steam railroad. He talked the idea to anyone in Brownsville who would listen or could not get away, and the more he talked . . . the better the idea seemed to him. It grew and blossomed and burgeoned and even soared, meanwhile taking on all the beautiful hues of the sky in the Land of Opportunity. It also dripped with gold, gold for all of Brownsville, soon to be a mighty metropolis, teeming with commerce, with industry, with the stir and bustle of countless travelers.[25]

The local notables would buckle under pressure, beginning to dream the dream too, and soon an application to the state would be made for the "Brownsville Railroad Corporation," probably with the words "& Western" added to demonstrate ambition. Indeed, if there was one factor that united many of these schemes, it was the idea that eventually the Atlantic seaboard cities of New York, Philadelphia, Boston, and Baltimore would link up with the growing Midwest towns for the trade in agricultural produce and minerals. There was, on the other hand, little incentive to build major north-south railroads along the coast, since there were well-entrenched shipping services that, unlike the canals, could undercut the railroads and almost match their timings. Therefore, it was the canal owners and their hired creatures who were in the forefront of objections to proposed railroads, but usually they would lose out, as their self-interest was just too naked, and the cannier ones would invest in their rival, as the steward of the canal-owning Duke of Bridgewater, Robert Bradshaw, had done when he bought shares in the proposed Liverpool & Manchester Railway before its construction.

Once the charter was obtained, then began the difficult bit of trying to raise money for the scheme: "Some local practitioner of letters was engaged to write a splendid pamphlet outlining the opportunity offered in the stock of the Brownsville & Western Railroad . . . widows and old men, and guardians of fools and minors were told how a thousand dollars would not only help to make Brownsville a leading city of the nation, but would also return a multitude of rich dividends, now and forever."[26] In some places where there was goodwill and excitement at the prospect of getting a railroad, the task was easy and the sum raised quickly. In others, it was harder, as the public was more skeptical, quite possibly because similar promises had been made some years previously by promoters for canals that had never been built. Then the local publicists would be called upon to write florid prospectuses, and, since support had to be galvanized at public meetings, the best local orators were on occasion called upon to rouse up interest. Or even, as Holbrook suggested, "needy pastors were hurriedly converted to steam, and they presently could see God's hand on the throttle." As ever, religion could be called upon to support either side.

The mania stimulated by the public clamoring for railroads saw a huge variety of lines promoted, many of which never saw the light of day. In America, there was no ubiquitous figure like George Stephenson, who seemed to have a finger in every railway in Britain and many overseas, but, instead, a whole host of promoters, big and small, corrupt and straight. By early 1837, at least two hundred railroads were being promoted, ranging from the overambitious Alabama, Florida & Georgia and the grand-sounding Western & Atlantic to more modest-sounding but evocative enterprises such as the Bangor & Oldtown, the Bath & Crooked Lake, the Corning & Blossburg, and the Beaver Meadow. On occasion their very names suggested that the promoters themselves were unsure of the viability of their schemes unless activities other than railroads were written into the title. Thus, there was the Atchafalaya Rail Road and Banking Company, the formidable Tuscarora and Cold Run Tunnel & Railroad Company, and the possibly tongue-in-cheek Towanda Rail Road & Coal Company. The early railroads were funded mainly by little people, such as businessmen and merchants in the terminal cities and farmers and tradesmen living along the route, all of whom would hope to benefit from enhanced land values and increased trade. Very little of this early capital came from the prosperous eastern banks or the European money markets, but both would play a crucial role later.

It was this kind of local interest that resulted in the creation of what, for a few years, was the world's longest railway. The merchants of Charleston, South Carolina, hoping to secure the trade of the rich local cotton-growing area, realized they needed a river outlet and promoted a 136-mile-long line from Charleston to Hamburg, just across the Savannah River from the rapidly growing Atlanta in Georgia. Horses and even sails were initially mooted as the power source, but the developers soon accepted that steam engines were the only realistic option and purchased the *Best Friend of Charleston* locomotive from the West Point Foundry, the first locomotive built in the United States speculatively, to use for the initial section, which opened on Christmas Day 1830. The journey that day along a short section of completed track was the first scheduled steam-hauled journey on the continent, just beating the Baltimore & Ohio. Although the line was completed within just three years, further expansion then stalled, stimulating one of those circular arguments about whether economic development was stymied by the lack of a railroad or vice versa. The inspiration behind the line, Robert Y. Hayne, a South Carolina senator, had envisaged a railroad running all the way from Charleston, the best port in the South, to Memphis, Tennessee, 700 miles away. His sudden death in 1839 killed not just his ambitious project for a Louisville, Cincinnati & Charleston Railroad, but much of the whole South's entrepreneurial spirit as well, slowing down growth in the southern states, a factor that would weigh heavily against them in the Civil War two decades later. The South, though, suffered deeper disadvantages than the paucity of entrepreneurs. As one historian put it, "nothing could make up for the inferiority of the South's ports, the lack of adequate markets for imports and diversity of exports, and a slave-labor system that immobilized her capital resources."[27]

In contrast, the major northern towns were all stimulating the development of railroads, both linking with each other and reaching westward. In New England, Bostonians had flocked to the display of English locomotives at the nation's first Railway Exhibition in 1827 and were aware of the need to connect with the Erie Canal and the railroads growing up parallel to it. Efforts to make the Massachusetts railroads a public works project failed, but three lines quickly obtained their charters: To the north, the Boston & Lowell, linking two important towns, served the booming textile industry; the Boston & Providence connected Massachusetts with Rhode Island; and the most notable, the Western Railroad, eventually ran 150 miles from Worcester to Albany. The city fathers of both Boston and Albany made the first trip on the line on Christmas Day in 1841, and the Boston delegation came back with a barrel of Rochester flour, which was used to bake bread for the customary banquet.

To the south, it was in Pennsylvania that railroads had the most significant early impact, but not along the "main line," where the misconceived mixed canal and railroad project was underway. It was the anthracite coal producers, eager to ship their fuel to both domestic and commercial users in New England and New York, who supported the creation of several successful railroads, including the Philadelphia & Reading, the Delaware & Hudson, and the Delaware & Lackawanna, which, according to one historian, "soon made antiques of the canals that had created the market for anthracite coal in the previous two decades."[28]

The spur for the construction of these early lines was, most often, the guaranteed freight flows of coal and other minerals. Typically, up to the time of the Civil War, the transportation of goods provided around two-thirds of the income of the average railroad. The most exciting development for passengers in the 1830s was the connection between New York and Philadelphia, which later connected to Washington, a line that would become the most used long-distance journey in America in ensuing years, although, since the bridges were built much later, initially there were river trips at both ends. The Camden & Amboy provided the crucial link between the New Jersey steamboat pier at South Amboy, where passengers disembarked from New York, and Camden, a small town across the Delaware River from Philadelphia; and by 1838 the through route to Washington, D.C., via the Philadelphia, Wilmington & Baltimore was completed. Like several other railroads mentioned above, in order to encourage investment the Camden & Amboy was granted monopoly rights in its charter. This concession proved controversial, as its early carriages, based, like those on many other contemporary lines, on stagecoaches, provided an uncomfortable, bumpy ride that irked the passengers, who had no alternative, although the journey was still far better than being shaken in a horse-drawn coach along the turnpike.

It was, of course, not just the passengers who benefited from the advent of the railroad. The Camden & Amboy became known as the "Pea Line" because it enabled New Yorkers to obtain fresh vegetables from New Jersey. As Nicholas Faith pointed out, "it could . . . just as well have been called the 'Peach Line' after another major speciality."[29] The American diet changed quickly as a result: By the early 1850s, fresh New Orleans peas were being consumed by Chicagoans a thousand miles away, and New Yorkers could enjoy that great delicacy, strawberries, for four months a year rather than for just a few weeks because of the far greater catchment area. Nicknames became commonplace for the railroads, with great efforts being made to ensure they fitted the acronyms, which were mostly gently affectionate. Thus the Baltimore & Ohio became the "Beans & Onions" and the Chicago, Burlington & Quincy the "Carbolic & Quinine," while the Chicago & North Western was known by its employees as either the "Can't & Never Will" or the "Cheap & Nothing Wasted." The New York, Ontario & Western, meanwhile, was called as the "Old Woman."

The rapid expansion of the railroads was taking place in a country that was sparsely populated compared with Europe. The population of Baltimore was just 63,000 in 1820, around half that of either Liverpool or Manchester when those towns set out to promote their first railway. Yet, the United States was already outstripping the rest of the world in rail miles, with 3,000 miles by the end of the 1830s, compared with just 1,800 on the other side of the Atlantic, a lead achieved despite numerous disadvantages, as European railways had better engineering, made use of more sophisticated metalworking techniques, and enjoyed easier financing.

The Americans, according to the historian John F. Stover, more than made up for these handicaps "with the assets of a greater need for improved transport, a relative freedom from long-entrenched customs, prejudices and monopolies and cheaper land for railroad right of way."[30] The latter, together with the use of the simpler techniques mentioned previously, meant that despite having built two-thirds more mileage, America had invested just $75 million,[31] far less than the European countries. There was still much to do. Although only four[32] of the then twenty-six states did not have some track, for the most part railroads were heavily concentrated in the New England and middle Atlantic states. Pennsylvania led the way by far with a quarter of the total mileage, while the southern states together could muster only 1,100 miles.

By 1840, the United States had as many miles of railroad as canals. Interestingly, just as in Britain, where canal building continued beyond the end of the 1830s, new canals were being built in America, but it took until the late 1830s before new mileage of railroads began to outstrip that of new canals. The railroads were, in fact, not universally embraced by a society that had very strong conservative elements—many people instinctively favored the gentler technology and pace of the waterway. When the very first sod was being turned on the Baltimore & Ohio, the president, John Quincy Adams, was doing the same for the Chesapeake & Ohio Canal, which would soon come into conflict with the new railroad. Waterways were seen by some as more natural, more in keeping with "the hand of the Great Architect," as a government report[33] as late as 1873 put it, than the railroads. There was, too, a hostility toward rail companies among many of the railroads' customers that was similar to a common attitude in Britain; they were seen as exploitative monopolists, and at times it was a criticism that was well merited. For example, early passenger fares were high, but they soon fell in the 1840s, when competition between lines forced them down to around 2.5 to 3.5 cents per mile in the east and no more than 5 cents elsewhere. Where there was no competition, however, the railroads were quick to exploit the situation by charging pretty much whatever they wanted.

As well as paying high fares, both passengers and freight shippers suffered from the lack of coordination between rival railroads, even where cooperation would have been mutually beneficial. First, there was the matter of gauge. Many of the early railroads chose the standard gauge because their locomotives came from the United Kingdom, but others, such as the Erie Railroad in New York State, whose trains ran on a 6-foot broad gauge, had deliberately selected a different one in order to prevent loss of traffic to rival lines. In most southern states, the 5-foot gauge was common, though not universal, and, perversely, two early railroads, the Camden & Amboy and the Mohawk & Hudson, chose slightly variant gauges, 4 feet, 9 inches, and 4 feet, 10 inches, respectively. Ultimately, the power of the northeastern railroads, which had nearly all chosen standard gauge, prevailed, and other lines followed suit. Not, though, before there had been a disaster caused by the pigheadedness of the early engineers. When the *New York Express* of the Lake Shore & Michigan Southern derailed and plunged into a creek in December 1867 near Angola, New York, with the loss of forty-nine lives, the investigators found that the problem was that the trains had been adjusted to ride on both the standard gauge (4 feet, 8.5 inches) used by the New York Central and the 4-foot, 10-inch, gauge of the Lake Shore in order to avoid the expense of changing the gauge on one of the railroads. On the sections with the wider gauge, not all of the wheel would ride on the rail, which was fine until a stretch of slightly misaligned track appeared, causing the disaster.

Besides the occasional accident, this lack of coordination between railroads was to cause untold difficulties and expense. The deliberate lack of involvement of the state, together with the competitive nature of the railroad developers, meant that even where two or more railroads terminated in the same town, there would be no rail connection between the two. In Richmond, Virginia, the map suggests that the town

could have been a junction for four separate railroads; there was no local link or branch between them, however, let alone a unified station
such as the *Hauptbahnhöfe*, which became the norm in Germany. This
was not just a result of lack of planning and coordination; rather, it re-
flects the fact that there were many vested interests that stood to gain by
ensuring that passengers and freight continued to suffer delays and inconvenience. Local haulage and carting companies were happy to keep
on transshipping the goods, and hotels and bars thrived by serving the
inconvenienced passengers. The war would change that: Philadelphia
had four unconnected railroads all booming with Civil War traffic, and
it was only then that two of the companies were embarrassed into making a connection, enabling passengers to travel between Jersey City (on
the other side of the Hudson River from New York) and Washington,
D.C., for the first time without changing trains—a journey that, incidentally, took nine and a half hours then but can be completed in just
two and a half today.

The opposition of the canal owners and steamboat operators came to
a head over the crossing of the Mississippi, the biggest barrier after the
Appalachians to westward travel until reaching the Rockies. Illinois, to
the east of the river, had been the most forward-looking state in terms
of railroad development. Apart from the Indians, who were being
quickly and cruelly displaced, the state's population numbered just a
few thousand, with people concentrated in a series of villages mostly
separated by vast distances. The paucity of the population did not stop
the state legislature from passing the most ambitious railroad bill yet,
however. It authorized the construction of 1,300 miles of rails, together
with improvements in river navigation and the construction of turnpikes across the state, all to be funded by the state of Illinois. It was, of
course, not promoted entirely for noble or visionary reasons, since the
main protagonist was Stephen A. Douglas, an Illinois senator who, like
many of the scheme's other supporters, was a landowner who stood to
make substantial gains personally from the improvements.

Progress on the Illinois scheme was delayed by the financial panic of
1837 and the revelation of massive fraud on the part of state employees
who had feathered their own nests by turning themselves into surveyors, land buyers, and estimators to profit directly from the rail bonanza.
Consequently, the state of Illinois found itself with just a few miles of

railroad and a massive debt, and for a while immigrants passed by the impoverished state on their way further west. Within a decade, however, Illinois had recovered, and it was laying down more track than any other state apart from Ohio. In the 1850s, 2,867 miles were completed; Chicago, a modest town of just 30,000 people in 1850, boasted no fewer than eleven railroads ten years later, marking its debut as the hub of the country's rail network.

The spine of the state's railroad network was the 700-mile-long Illinois Central, running north-south, which enjoyed brief fame as the largest system in the world. Although the state government was no longer directly involved, the federal government played a crucial role in stimulating that growth, as the Illinois Central was a pioneer in the system of land grants that was to be of great importance to the development of American railroads further west. In 1850, Douglas[34] maneuvered the first land grant act through Congress,[35] supporting the construction of a railroad that would allow the Illinois Central to sell thousands of acres of its land to settlers. This ensured the company's profitability and stimulated growth of the state's population, which reached 1.5 million by the end of the 1850s.

The railroads, however, had no intention of ending their westward expansion in Illinois. The Rock Island & Chicago Railroad, completed in 1854, deliberately ran its tracks right up to the Mississippi facing westward, with every intention of fording the river, using an island as support, and within two years, despite much local opposition from the Mississippi rivermen, the bridge was completed. The first train crossed over into Iowa in April 1856, but the triumph was short-lived. Two weeks later, a packet boat, the *Effie Aflon*, traveling from New Orleans smashed into the bridge at full speed. The ensuing fire from the ship's engines quickly spread, destroying the whole structure. Any thoughts that this may have been pure mischance were quickly dispelled when the next riverboat to pass the site displayed a banner announcing: "Mississippi Bridge Destroyed. Let All Rejoice." It soon emerged that, rather suspiciously, it was the first time that the *Effie Aflon* had ever ventured that far up the river.

This strong whiff of suspicion did not prevent the boat owners from suing the railroad for putting the bridge in their way, when it might have been expected that the railroad had a rather stronger case against them for destroying its structure. Here Abraham Lincoln enters the story. Lincoln in fact played a significant, but largely forgotten, role in U.S. railroad history, first as the lawyer representing the railroad in this case, and later as a major political supporter for the creation of a transcontinental railroad. The *Effie Aflon* case inevitably became a test of the much larger issue of whether the right of railroads to traverse rivers prevailed over the rights of those who used the waterways. The case pitted whole communities against each other, with the city of Chicago naturally supporting the railroad, while riverside towns such as St. Louis opposed it. Lincoln, who had previously been a river pilot, spent days surveying the site and talking to those who worked on the river. In court, he tactfully acknowledged that the accident had been down to "pilot error," but argued that the need for Americans to travel between east and west to populate the vast country should be paramount. The jury was tied, which meant the railroad had won; despite a later successful appeal in the Iowa courts, the railroad eventually prevailed in the U.S. Supreme Court, which declared that bridges were not a hazard to navigation.[36] The way to the West was opened up, and the dream of a transcontinental railroad soon became a reality, but not before the country was rent apart by the Civil War.

By the 1850s, the railroads had become more than just a way to get to the next town. They had established themselves as the basic means of transportation whatever the distance. The small size of the average railroad, and the lack of joint working arrangements between railroads, however, meant that longer journeys involved much inconvenience. Soon, at last, voices were being heard in favor of unifying the system. The creation of the first consolidated railroad, the New York Central, unified no fewer than fourteen lines, which then allowed a direct, continuous link of nearly 300 miles between Albany and Buffalo. This was pioneering on a grand scale, but other railroads were slow to follow the example, largely because of the absence of strong direction from the federal government in what was still very much a decentralized nation. The parochial politicians and promoters who had created the railroads were generally happy to retain control, and they lacked the imagination to see that they could profit from improved links to distant points of the nation. As a recent history of the railroads says, "the success with which local interests managed to hobble their lusty steed for over thirty years is a testament to the vigor of state and local rights in an age in which the powers of federalism were hardly realized."[37]

The outbreak of the Civil War in 1861 blocked further progress toward consolidation. By the start of the war, railroads had spread throughout the eastern part of the country and much of the South, but the lines heading west still stopped at the major rivers. The great project for a transcontinental railroad, first conceived by the early rail pioneers thirty years previously and actively promoted in the mid-1850s, would have to wait. The Civil War was to provide another sort of test for the railroads, one which they passed remarkably well, demonstrating, as one historian said, that "the railroads did more to change the art of war than anything since gunpowder."[38]

As we have seen, the military value of the railways was recognized in Europe right from the start; they were used early on to move troops quickly in order to quell revolts or tame rioters. In the early 1840s in Britain, both the Chartists and the troops that were called up to put down their protests traveled by train. The railways also enabled the rapid deployment of large numbers of troops in the revolutions that swept across Europe in the 1840s, with the earliest recorded use being the 12,000 Prussian troops rushed to Cracow in Poland to put down a nationalist revolt in 1846. It was the American Civil War, which lasted for four years, however, that saw the first sustained use of railways for military purposes.

Even before the outbreak of the war itself, the railroad played a vital role, moving troops during a preliminary skirmish in October 1859 when a small group of abolitionists, led by one John Brown, briefly captured the federal arsenal at Harpers Ferry in West Virginia as a protest against slavery. The raiders were subdued when a detachment of U.S. Marines, under the command of Colonel Robert E. Lee, rushed to the arsenal using the Baltimore & Ohio Railroad. Nearly three years later, the best-known bit of derring-do in the Civil War occurred, which has gone straight into the annals of mythology, and it involved a railroad. In April 1862, a group of twenty Union soldiers, operating in mufti as spies without uniforms behind enemy lines, boarded a northbound train of the Western & Atlanta Railroad in Marietta, Georgia. At Big Shanty, 8 miles up the line, they took over the engine while the crew ate breakfast; amazingly, their leader, James J. Andrews, managed to bluff his way through the next two stations, claiming he was carrying gunpowder for the front. Of course, the Yankees cut the telegraph wires as they went along, but the train's conductor, William A. Fuller, and an engineman, Anthony Murphy, had set out in pursuit, first on a handcar and then in an old locomotive that happened to be in steam and available. Fuller, despite having to remove obstacles placed on the track by the Yankees, gradually managed to catch up with Andrews, who therefore did not have time to burn the bridges and wreck the railroad, which had been the raid's mission. Eventually Andrews's engine, appropriately called *The General*, ran out of steam, and he was captured and hanged. It is not difficult to see why this adventure has repeatedly been filmed by Hollywood,[39] starting with Buster Keaton, who rewrote history for his 1927 film by giving his hero, surprisingly the Confederate engineman, the eponymous engine, *The General*.

Throughout the war, it was the Unionists who benefited most from the railroad, not only because their lines were better developed than those in the South, but because they learned to exploit the full potential of the railroad. They realized it was not just a method of transportation but a weapon as potent as artillery, whereas the southerners failed to see the central nature of the railroad in the war effort. Nicholas Faith summed up perfectly why the North was able to do this: "Railways were the supreme symbol of the gap between the industrial North and the largely agrarian South."[40]

Within eighteen months of the start of the war, the Union had organized the railroads on an integrated war footing, over the objections of the railroad companies, creating links between the different railroads to enable through travel. A separate department of government was created to run the railroads and build lines where necessary, and the man in charge was an experienced railroad man, Daniel C. McCallum, who was the general superintendent of the Erie Railroad. Crucially, whereas McCallum was given authority over the generals whose troops and materiel were being carried by his trains, in the South "the so-called director of railroad transportation was pretty much a nullity."[41] Military traffic in the North was given absolute priority, with routine freight and passenger trains literally "sidetracked" to allow the fighting men and their equipment through. Whereas the Unionists were well supplied by the lines from the coastal ports, the railroads in the South had not yet been joined up to create a viable system and were built to a lower standard. Crucially, too, the Confederate government did not take over responsibility for running the railroads in its territory.

The Baltimore & Ohio, along with several other lines, came to the rescue of the Union Army at a crucial point in the conflict. In the fall of 1863, General Ulysses S. Grant was laying siege to the Confederates in the strategically vital city of Chattanooga, Tennessee, but he needed urgent reinforcements to maintain the blockade. A complicated route starting on the Baltimore & Ohio, then transferring onto other lines through Ohio, Indiana, and Tennessee, was worked out, and in a remarkable operation lasting twelve days, two army corps, totaling 25,000 men, together with ten batteries of artillery with their horses and a hundred cars of baggage, were taken 1,200 miles over lines that were regularly being sabotaged at night by Confederate gangs, but repaired during the day by Union engineers. It was not, as routinely claimed by some U.S. historians, the first major troop movement by train in the world, since, as we have seen, there were several European precedents, but it was certainly the most ambitious to date. It was not only that the Unionists proved to make excellent use of the existing railroads; they also created and controlled a network of 2,165 miles of military railroad, much of it requisitioned from its southern owners. The skill of their engineers proved vital, too, as they learned how to repair war-damaged track and structures with amazing speed. The story of how a 400-foot bridge over the Potomac, destroyed by the Confederates, was rebuilt in just nine days by a team led by the remarkable Herman Haupt became another legendary war tale when an admiring President Lincoln described the bridge as "but beanpoles and cornstalks" because all the material used for its construction had been taken from local forests.

Even when it came to destroying each other's railroads, a tactic used routinely by both sides, the Union soldiers were rather better at causing irreparable damage than the rebels: "Confederate raiders never acquired the pure destructive skill of the more mechanically-minded northern soldiers,"[42] scoffed a military historian. The scorched earth policy of the Union Army, together with the destruction meted out by retreating rebels, resulted in over half of the southern railroads being unusable by the end of the war. In contrast, the northern railroads had not only suffered little damage, but had been greatly enhanced, both in terms of investment and mileage, owing to their vital role in the conflict.

Overall, the role of the railroads in the Civil War was not all positive. By enabling the rapid deployment of troops, the railroads spread the conflict across a far greater part of the country than would have otherwise been affected, turning it almost into a continent-wide conflict instead of a localized war. Moreover, the vital role of the railroads in the Civil War did not go unnoticed, ensuring that railways around the world would come to be viewed from a military point of view. Among the spectators of the operation that relieved Grant's troops was a group of Prussian officers who had crossed the Atlantic to observe the latest tactics and techniques of warfare, a surprisingly common habit of the leaders of neutral armies. Just three years later they used their experience to good effect. At the Battle of Sadowa, they beat the Austrians, thanks to the rapid deployment of troops on five available railway lines, and the chief of the Prussian general staff, Helmuth von Moltke, put the success down, unequivocally, to the railways: "We have the inestimable advantage of being able to carry our field army of 285,000 men over five railway lines, and of virtually concentrating them in five days on the frontiers of Saxony and Bohemia. Austria has only one line of rail, and it will take forty-five days to assemble 210,000 men."[43]

Railways, though, have their limitations in war, as would be discovered in later battles. They are easily sabotaged and therefore need considerable numbers of soldiers to protect them; they, obviously, only allow goods to be transported where the lines are, as it takes time to build new track, and are therefore dependent on the logistics available at the railhead; and they require excellent management, particularly of single-track lines, if bottlenecks are to be prevented. Even in the Civil War, the Union Army had to be flexible. General William T. Sherman from the North realized the limitations of using the railroad in hostile territory; having arrived in Atlanta by train, he used the river for his march through Georgian territory because "they can't stop the Tennessee River and each boat can make its own game."[44] Nevertheless, the railroads went on to play a vital role in numerous conflicts, and they became a backbone of the supply chain in several countries over the next hundred years and during both world wars.

With the Civil War over, and the importance of the railroad reemphasized, the dreamers who had sought to build a transcontinental line could now see their ambition realized (see Chapter 6). Meanwhile, however, Europe was getting the railway habit, and the iron road was becoming commonplace elsewhere as well.

FIVE

Joining Up Europe

B Y 1850, RAILWAYS WERE ESTABLISHED IN ALL THE MAIN EUROPEAN countries, but it was in the second half of the century that the iron road established itself almost universally as the principal means of transport for both passengers and goods. The spread of the railway was uneven, characterized by periods of boom and bust, with the various nations ready to intervene when things started to go wrong. Governments became aware that a healthy and extensive rail network was essential for development.

In Britain, the railway mania of the 1840s was short-lived, but its impact was long-lasting. It reached its peak in 1844, when Parliament was inundated with 240 petitions to build railways, and in three years the lines authorized approached 10,000 miles, of which about two-thirds were eventually built. As a result, by the early 1850s Britain had over 7,000 miles of railway, a stunning achievement given that only twenty years had elapsed since the opening of the Liverpool & Manchester.

The railways on the Continent developed in a different way from those in Britain and the United States, largely because of the contrasting political and legal frameworks. As we have seen, the various continental governments generally set out plans for the networks they wanted to see built; they also tended to intervene in the operation of the railway companies to a far greater extent than their Anglo-Saxon counterparts. Continental railway companies generally did not have the same freedom to set passenger fares and freight rates without government approval. This was particularly true for lines that crossed borders; the respective governments negotiated the rates for international traffic. There were differing approaches across Europe, with the unified and tightly controlled network in Belgium representing the strongest contrast to the United Kingdom and the United States, while other countries, such as France and Germany, had a mixed-economy railway, with involvement from both private and public sectors. All the continental systems, however, had one aspect in common: They were far more tightly regulated than the Anglo-Saxon railways. As one historian pointed out, "sometimes the state intervened in the smallest details of the day to day operation of railways."[1]

There were good reasons for the differences. Continental Europe in the nineteenth century was an unstable maelstrom, riven by regular wars and marked by a history of conflict and large movements of people. The likelihood of further wars influenced the establishment and planning of the railway networks, and some lines were built with military and strategic purposes rather than commercial imperatives in mind. States were bound to be more intimately involved in the planning there than in Britain and the United States, where, apart from the split that took place in the American Civil War, there was no question of territorial breakup. On several occasions, after a war, European railway companies on the Continent found that part of their network was now in another country, requiring substantial adjustments. The Alta Italia Railway Company, for example, operated railways in Austria as well as Italy, an arrangement that arose by accident when the Italian state claimed part of its territory from the Austrians. The company was pressured by the Austrian-Hungarian government to split itself in two; when it did so, the Italian part of the company soon fell into the hands of the Italian government.

The railways on the Continent were tied into long-term relationships with their respective states because of the way they were funded. First, the land on which the tracks were laid tended to be leased for a defined period before reverting to the government, in contrast to the situation in England and the United States, where railway companies bought the land. Second, construction of railways in continental Europe was frequently subsidized, either through direct support or through guarantees of rates of return, and states would seek to get some of their money back through taxes or levies on fares and freight charges.

The very different public attitudes toward railways on the Continent had roots in recent historical circumstances. The Anglo-Saxon railways were developed by pure, raw capitalism, remarkably untrammeled by government intervention. With little regulation, they were free to set rates for the carriage of goods largely on the free-market principle of what the market could bear. This resulted in rates varying considerably from route to route, which led, in turn, to accusations from the public that the railway companies were exploiting their monopoly positions, which was undoubtedly true (though another way of looking at it would be that the railways offered cheaper rates where they were eager to attract business, and that their strategy was a logical capitalist position). In continental Europe, rates were largely set by government, and they were perceived by the population as a kind of tax that should be kept as low as possible, since railways were considered to be a public service provided by state or quasi-state organizations. Even the type of operation was different on the Continent. Whereas in the United States the railways opened up huge swathes of virgin territory, in continental Europe railways served existing industrial centers. The journeys were inevitably shorter and the goods they carried far more varied.

On the Continent, the growth in railways lagged a decade or two behind Britain. In France, for example, after 1849, the network grew steadily at the rate of around 450 miles annually for the next four decades. Louis Legrand's national plan (mentioned in Chapter 2) took some time to materialize. In 1848, there was only a handful of disparate lines linking Paris with the Channel ports and a few nearby towns, such as Chartres, Le Havre, and Lille, and some major provincial services, linking, for example, Marseille with Nîmes, but there was nothing approaching a national system. That was to change over the next few years, with the creation of six major companies that would build the great majority of the new lines by obtaining concessions from the government.

Although the establishment of national networks was the primary aim during this initial period of railway construction, both politicians and railway promoters were aware of the potential for an international network crossing the continent of Europe. For example, even before the first Austrian railway was built, the idea of a line from Hamburg to

Trieste had been mooted. The adoption of the Stephenson gauge of 4 feet, 8.5 inches (or 1,435 mm),[2] throughout the European continent, with the exception of Russia and the Iberian peninsula, ensured that through traffic was able to run between countries. However, just having the same size of rolling stock was not sufficient to allow easy access between systems, which also developed different signaling technologies and working practices. The growth of a pan-European network was hindered by these technical variations (and also later by the use of incompatible electrification and safety systems). The differences had to be ironed out to ensure safety. Officials began to address these issues in a series of conventions held in Berne between 1878 and 1886, thrashing out agreements on both technical and legal matters, such as responsibility for damage or delay on international trains. There were, too, political issues that reduced free access to the railways; even today, despite more than a century of European cooperation between railways, drivers and engines often have to be changed at border points as a result of remaining technical differences and union agreements.

The rush to build railways in the second half of the nineteenth century was motivated by a combination of political expediency and economic necessity. Seeing the rapid economic benefits generated by the early railways, governments quickly understood their importance in stimulating further development. As absolute monarchies gradually gave way to states with wider political participation, particularly after 1848, politicians tended to respond more positively to the clamor from their populations to be connected to the railway network, which they saw as crucial to their region's economic future. The railways could serve a variety of purposes, by no means mutually exclusive: linking major towns with each other; transporting minerals, and particularly coal; taking people to and from work; or even connecting two relatively small towns that were traditional trading partners. As the networks expanded, many sections of railway were built simply to ensure that through routes became possible. And quite often, the main purpose was the one that had stimulated a myriad of horse-drawn wagonways at coal mines as far back as the seventeenth century—providing a route to the sea or a navigable waterway. As Allan Mitchell, a historian of the French and German railways put it, "by mid-century all of Europe's main ports had become railheads."[3]

This was certainly the stimulus for the first transalpine railway, the Semmering, a project given urgency by Austria's need to connect its capital, Vienna, with the sea and discussed in the very earliest days of Austrian railway development. In 1837, the Austrian government had commissioned preliminary surveys to investigate the possibility of a railway through the Semmering mountain range in the south of the country to provide a connection with the port of Trieste on the Adriatic, the only sea harbor of the Austrian Empire.

The Austrian government even consulted George Stephenson about the project, but he was deeply skeptical of its feasibility. The scheme was given impetus by the involvement of Karl Ritter von Ghega, a man with considerable experience in building Alpine roads, who was sent across the Atlantic to examine the U.S. experience of constructing mountain railways. Ghega came back convinced that crossing the Semmering mountains was feasible.

It was a daunting task. While the pass had been established as far back as the Middle Ages as a route for people and horses, and it was the lowest of the Alpine crossings, it still required the railway to climb to 2,880 feet. After Ghega surveyed the route, the scheme was given the go-ahead, in 1848. It was not, however, the technical arguments that convinced the emperor of the necessity of the project, but the political situation. The revolutionary uprisings of that year across Europe had frightened the Austrian government, especially as growing unemployment in the country risked stimulating unrest. Establishing a cheap transport route to the sea from landlocked Vienna became an economic and political necessity. By then the Südbahn had covered the Austrian plains with its tracks and was ready for the ascent.

The track on the Semmering had to be laid through thick forests on the side of precipitous mountains and across mountain ravines, perched perilously on boulder-strewn ledges that had to be protected by avalanche sheds and walls. The risks were made all too apparent when rocks sliding down the mountain killed fourteen men in one incident in October 1850. Ghega designed a route through the Alps between Gloggnitz (in Lower Austria) and Mürzzuschlag (in Styria), which was precisely twice as long as the 13 miles that separated the two towns as the crow flies. This was an enterprise that required new and untried techniques, such as building sharply curved viaducts over the valleys and blasting huge tunnels out of the mountains. There were fifteen tunnels and sixteen viaducts, including the elegant Kalte Rinne Viaduct, with its ten arches resting on a smaller second set, made even more attractive by Ghega's decision to use only bricks rather than iron and steel. All this was carried out largely by hand, with the help of much gunpowder, since no other explosives had yet been devised.

Ghega limited the gradient to 1 in 40, but even that was too steep for contemporary locomotives, and in 1851, a competition, similar to the Rainhill Trials for the Liverpool & Manchester Railway, was organized on a completed mountainous section of the railway. But in the event, when the line opened two years later, all competitors, including the winner, the *Bavaria*, proved useless, and a new type of locomotive was commissioned from Wilhelm Freiherr von Engerth, the professor of engineering at Graz University.[4]

An accident just before the opening of the line almost changed history. The German representative was Otto von Bismarck, who would later become the Iron Chancellor and bring about the unification of Germany. Walking up the mountain to inspect a tunnel, he came within a whisker of being killed when a temporary gangway over a ravine broke beneath him. He survived only by clinging onto the ledge as he fell. Many of the workers on the site were not so fortunate. The death toll from accident and disease among the 20,000-strong workforce of Germans, Czechs, and Italians was very high even by the lax standards of the day, and 700 perished, mostly from typhus and cholera, but also from accidents. To this day, each year on All Saints' Day, Austria's railway workers lay a wreath in their memory at the cemetery near the railway.

The first freight was carried on the line in October 1853, and passenger traffic began in the following July. The all-important connection between Vienna and Trieste[5] was completed in 1857, and though the cost of the line was four times the original estimate, the line proved its worth as a vital trade link for the empire in its declining years. Like many early railways, the Semmering was completed with a bit of a flourish through the provision of fifty-two houses in classic style for the trackmen, who were required to make continual checks to ensure that the line remained safe. The robustness of Ghega's design is demonstrated by the fact that most of the line survives virtually intact today as a main line, with very few alterations apart from being electrified in the 1950s, unlike several later Alpine routes that have either been closed or greatly modified.[6]

Since Trieste was at the time part of the Austrian Empire, the Semmering was a domestic rather than an international railway. So was the second line across the Alps, the Brenner, which opened in 1867 and was relatively easy to build as it required no major tunneling. The line, operated by the Südbahn, linked the Austrian Empire with Venetia, still in Austrian hands, and was militarily important in defending that territory. It would not be long, however, before all the major European networks spilled over their frontiers, first with small incursions to border towns and then as part of major routes crossing much of the European continent.

The French state, far more controlling than its Anglo-Saxon counterparts, was never going to allow the rail network to build up piecemeal and be solely reliant on the whims of promoters and local interests. An 1842 act had created a railways charter defining precisely the way future lines would be developed. The state took on responsibility for the groundwork and the major structures, while the companies would lay the rails and provide the rolling stock. The legislation encouraged the creation of large companies with the resources and the ability to create regional networks, a trend that would be seen in other countries, such as Britain and Italy. The first to emerge in France was the Chemin de Fer du Nord, established in 1845 with money from the Rothschild Bank. It entailed the consolidation of a variety of local lines, which, interestingly, stretched into neighboring Belgium, but only reached Paris the following year, when the line from the capital to Amiens and Lille was completed. The owners of the Nord were always likely to do well. Not only did it serve the rapidly industrializing region of Nord-Pas de Calais and the Channel ports of Boulogne, Dunkerque, and Calais, but it also catered to the burgeoning tourist attractions just north of Paris, such as Senlis and Chantilly, where Parisians flocked for a day out. Helped, too, by the relatively compact region it served, the Nord was always a profitable railway.

Some of the other six companies were not so fortunate. The Ouest, created in 1851, covered the much less populated area of Brittany and Normandy, which had no large industrial conglomerations but, instead, innumerable villages, which became linked by tiny lines creating a dense but inherently unprofitable network. The Midi, too, struggled. It was the only one of the big six that had no connection with Paris, as it was centered on Bordeaux on the west coast. Despite its financial difficulties, the Midi built some steeply inclined railways in both the Pyrenées and the Massif Central in the second half of the nineteenth century, making it France's "king of the mountains" railway. Another struggler was the Est, formed in 1853 by the amalgamation of the Paris-Strasbourg company, which ran trains to Basle across the Swiss border, and a smaller line that operated between Montereau and Troyes southeast of Paris. Again, the company was hamstrung by the low population density of the area it served, and the large distances between the capital and Strasbourg, the largest city in its territory. Later, too, the Est would suffer by losing Alsace, including Metz and Strasbourg, to the Germans after the 1870–1871 war, making its operation even less economic. The Paris-Orléans, whose name was a misnomer since its services extended well beyond the eponymous town, and was consequently on a far grander scale than might be assumed, was rather better placed, serving major provincial centers, such as Tours, Bordeaux, Limoges, and Clermont-Ferrand, all of which were reached before the outbreak of the 1870 war with Prussia.

The sixth company was also a more ambitious operation, the famous Paris-Lyon-Mediterranée (PLM), which became the biggest French railway company, operating the main line from the capital to Lyon. This main line had opened in 1854, and the railway quickly spread over the border to Switzerland and Italy. There had been a slight lack of urgency in completing the Paris-Lyon line because there was not only an excellent navigable alternative for freight, through the Canal de Bourgogne, which linked with the Yonne tributary of the Seine, but also, thanks to Napoleon's legacy of good roads and an army of 40,000 horses to be stabled along the way, an efficient stagecoach service enabling travelers to make the journey between Paris and Lyon in just three days. However, once built, it quickly established itself as France's premier line, as was apparent when Europe's first high-speed line, the Train à Grande Vitesse (TGV) Sud-Est, opened between Paris and Lyon in 1983.[7]

Created in 1858 through the amalgamation of various smaller railways, the PLM also took over France's oldest railway line, which now extended from St-Etienne to Lyon. The PLM was a highly successful passenger railway, benefiting not just from the high population density of its territory, which included the nation's three most important cities—Paris, Lyon, and Marseille—as well as several other major towns, including Grenoble and Dijon, but also from the tourist traffic to the fashionable Riviera resorts of Cannes, Nice, and Monte Carlo, which it was instrumental in developing (see Chapter 9). It was France's most important freight carrier, providing the main link with its biggest trading partners in the Middle East and Africa through the port of Marseille, which earned the route its grand name of *ligne impériale*. Moreover, as soon as the various tunnels through the Alps were opened, the PLM ran trains through to Rome and various other destinations in Switzerland and Italy.

The rapid growth of the French railways during the period of the Second Empire (1852–1870), stimulated by the 1842 legislation, was becoming a burden on state finances by the 1860s. The problem for the government was that once it had made a commitment to supporting the construction of lines, the clamor from communities not yet connected to the network became louder and louder. This was the dawning of the democratic age, and people expected the promises for new railways made by politicians at election time to come good. Everyone, even in the tiniest village, wanted to be connected to the railway, and the state was seen as responsible for providing it. As the standard work on French railway history says: "These minor lines were keenly sought after by local populations of areas that were not yet connected, and many imprudent parliamentarians had promised during their electoral campaigns a direct rail link to civilization and progress."[8]

Having stressed that the growth of the railways to the far-flung corners of the Hexagon (the French people's pet name for their country) was the key to the economic success it had promised to deliver, the government could not simply ignore these demands. An 1865 law, therefore, gave the *départments* and *communes* the power to construct these little branch lines, which would link up with the main lines over distances of 20 to 30 miles with the guarantee of state funding. This measure was passed to overcome the reluctance of the big six companies to build lines from which they were never likely to see a return. They had reached a position of almost total dominance, controlling all but about 1,000 miles, spread among twenty-one little railways, of the approximately

13,000 miles that had either been built or let as a concession. The new law was remarkably successful in achieving its aim of encouraging the development of small railways: Within ten years, there were nearly 2,800 miles of railway outside the control of the big six, run by thirty-five companies. Narrow-gauge railways were built to countless villages in areas where a standard-gauge line would have been too expensive. However, even with this attempt to reduce costs, their viability and usefulness were open to question. Just as in similar periods of railway mania elsewhere, the speculators soon piled in, promoting lines that had no economic rationale but nevertheless attracted state subsidy from a government that, through fear of alienating local opinion, was loath to turn down even the most outlandish projects. As in Britain, where lines that promised rather unlikely "direct" services between obscure towns were in fashion, schemes popped up in France for routes crossing the whole of the country, but avoiding Paris, such as Calais-Marseille or Dunkerque-Perpignan. These were, in effect, little more than a series of links along slow departmental railways.

This plethora of often hopelessly uneconomic railways led to the beginnings of the creation of the nationalized network that would later become the national rail company, the Société Nationale des Chemins de Fer Français (SNCF). A particularly impoverished group of railway companies in Charentes and the Vendée joined together with a series of companies covering regional routes between western towns, such as Poitiers, Orléans, and Rouen, to try to persuade the large Paris-Orléans company to bring them into its fold in order to guarantee their survival. When the company rejected the idea, on the basis that there was no hope of making a profit, the directors turned to the government of Georges Clemenceau, who agreed to take the company over in May 1878, arguing that the state could better manage such a network by imposing rationalization in a way that would be impossible for private concerns. It was the precursor to full nationalization, but that would be sixty years later, when the rest of the network was amalgamated to form SNCF just before the outbreak of World War II.

In the meantime, the role of the state in the railways gradually became more intrusive and overt. In early 1878, Charles de Freycinet, the public works minister, announced an ambitious plan to create a nationwide transport system with the railways at its heart. He wanted to classify some 12,000 miles of track, some of which had not yet been built, as being of national importance and to be controlled by the state. In the words of the Larousse history of the railways, "from now on, the government would ensure that the state of the railways would be its business and would never relinquish that prerogative."[9] It would, though, be a bumpy ride, as the big companies fiercely defended their independence while recognizing that they needed government support for many services. Freycinet was not able to impose his plan completely, since, as ever, there was insufficient money, and eventually a deal was thrashed out with the big companies. They agreed to provide a minimum of three services per day on every line, even those with very little traffic, in return for state subsidies to support their networks.

In Italy, Rome only obtained its first railway, a suburban line that ran 12 miles, from the edge of the capital to Frascati, in 1856. It was built by a British contractor, John Oliver York, but promoted by a French company with an Italian name, the Società Pio Latina. More significantly, another French company, the Società Pio Centrale, constructed the line out to Civitavecchia 50 miles from Rome. This line formed the start of the Mediterranean coastal route, which would eventually stretch into Tuscany and beyond. The pope was quick to give his blessing to these new lines. A galaxy of cardinals attended the official opening of the Frascati railway, and the pope himself (Pius IX) paid a surprise visit to the bridge being built across the Tiber River. Consisting of two lattice-girder spans supported on cast-iron columns designed and assembled by York, it survives to this day, accompanied by a more modern stone neighbor. The pope was greeted by an Englishman, describing himself as the British minister of public works,[10] who had been reluctant to be seen by the pontiff, as he was carrying an umbrella and was wearing work clothes, a grey coat, and a straw hat. The pope informed him that his attire was of no importance since he had been taken unawares and added: "Now you can say in London that the Roman pontiff is not always at prayer surrounded by incense, monks and priests. Tell the Queen that Her Majesty's Minister of Public Works found the old Pope surrounded by his engineers while helping to finish a new bridge across the Tiber."[11]

Italian railways lagged behind those of the other major European nations. There were still only 1,500 miles of railway in Italy by 1860, but

the pace of construction increased greatly over the next few years because the newly unified state started using them as a catalyst for nation-
building. In Italy, as in France, the logic of having big companies proved
overwhelming, but the railways did not grow organically; it was the
state that legislated for the merger of the railways into four large concerns and then, bizarrely, twenty years later, three.

The war that would eventually lead to the creation of the Kingdom
of Italy began in 1859, and its progress was speeded up by another
early example of using railways for military purposes. King Victor Emmanuel of Piedmont, Savoy, and Sardinia and his prime minister,
Camillo Benso, Count of Cavour, won the support of Napoleon III to
clear the Austrians out of northern Italy, and the French troops reached
the battle site rapidly by taking the Paris-Lyon-Mediterranée railway to
Marseille, where they embarked on ships to Genoa and then went on to
Alessandria by train. The Austrians, for their part, were hampered by
the lack of a railway and were defeated near Lake Garda, suing for
peace by ceding all territory east of Verona. With Giuseppe Garibaldi
launching his legendary march to take over the south of Italy, only the
Papal States were holding out from the new Kingdom of Italy that was
proclaimed in 1861, delaying the completion of unification until 1870.

All these wars and political changes had little effect on the rate of
railway building, since the various governments were anxious to see
railway projects brought to fruition, not least because they recognized
their value in pursuing political and military aims. Railway administration, however, was not made easier by the fact that a couple of Italy's
railways found themselves straddling two countries. The Victor Emmanuel Railway (Società Vittorio Emanuele), created by Cavour in
1853 to bring about the great project of crossing the Alps, was particularly badly affected when Savoy was transferred to France as
Napoleon's reward for helping in Italy's unification. By the late 1850s,
the railway had completed a substantial line between Aix-les-Bains and
St. Jean de Maurienne near the Mont Cenis Pass on what is now the
French side of the Alps; on the Italian side it connected Milan with
Turin, and further west it reached Susa, the starting point for the climb
toward the tunnel.

After the war against the Austrians, the railway was split, with the
French section being bought by the Paris-Lyon-Mediterranée and the Ital-

ian line becoming part of the Società delle Ferrovie d'Alta Italia, the Upper Italy Railway. The Lombardy-Venetian Railway (Société IR Privilégiée des Chemins de Fer Lombards-Vénitiens et de l'Italie Centrale), owned by the Rothschilds, also found itself split between two countries, as the Austrians had held onto a part of what is now Italy west of Verona. Controversially, the Austrians amalgamated their section of the railway with the Südbahn, the largest railway in the Austro-Hungarian Empire, a merger that was never recognized by the Italians, who were not going to have one of their strategic railways controlled from Vienna. This process of railways being broken up when the states in which they operated were rent apart was to be repeated numerous times in Europe over the next century, notably when the Austro-Hungarian Empire broke up after World War I.

Surprisingly, perhaps, the link across the Alps at Mont Cenis[12] was not much delayed by the separation, or, indeed, by the onset of the Franco-Prussian War of 1870, which broke out just before its opening. The tunnel was always going to be a mammoth task, the most ambitious undertaken by railway builders apart from the climb up the Western Ghats in India described in Chapter 3, and the Semmering railway, the first transalpine crossing. The Mont Cenis line presented many of the same problems as the Semmering, with the added challenge of an 8-mile tunnel built through the mountain at a height of over 4,000 feet. It was the first time that a railway tunnel was to be drilled through a mountain across an international border. The size of the tunnel, the longest in the world up to that point, inevitably presented an enormous challenge; initially, when work started at both ends in 1857, it was expected to take twenty-five years to complete. Indeed, at first progress was painfully slow. Excavation was by hand, and it took five years to dig out just a mile at each end. But then a crucial technological breakthrough was made. The engineer, Germaneo Sommeiller, who had gained experience by working on the Genoa line, Italy's first mountain railway, came up with the idea of a pneumatic drill to take over the time-consuming task of making holes in the rock in which to place explosive charges. Suddenly the pace accelerated, and the Italian and French teams joined up on Boxing Day 1870, barely half a meter out of alignment. The railway, which provided a route from Milan and Turin to Grenoble, Lyon, and the Rhône Valley, opened in October 1871, giving Italy its first railway link to France. The second followed almost immediately, with the extension of the Paris-Lyon-Mediterranée to Ventimiglia across the border from Menton and Nice on the Côte d'Azur, and from there to Genoa in 1872.

As mentioned before, the topography of Italy was not conducive to easy or cheap railway construction, with the Apennine Mountains down the middle presenting a natural barrier. As a result, two parallel mainline railways were built on either coast, joined at times by expensive routes through the hills. The poverty in the south ensured that large sections of the network would forever operate at a loss. Costs generally were high because Italy was dependent on imports for rails, coal, and rolling stock, and prices would rise as the weak lira's value fell. Susan Ashley, author of a book on Italian politics of the nineteenth century, summed it up brilliantly, explaining that "the new government of united Italy could not afford to build and operate railroads, but nor could it afford not to."[13] Railways were seen as essential to economic development and to cementing political unity. The prevailing liberal ethos was that the private sector should provide the investment, but in practice Italian railways were not an attractive proposition; consequently, while the private companies carried out the work, the bill was picked up by the state.

The difficult relationship between the state and the railways in France was therefore mirrored in Italy and compounded because of the foreign, largely French, shareholding in many early railways. In the forty years after unification, there would be a railway crisis every decade in Italy, and all of them had their roots in the fundamental lack of viability of the rail network. Invariably, each of these crises was followed by legislation to try to sort out the mess, and ultimately the result was the full nationalization of the system in 1905.

As one of its first measures, the new Kingdom of Italy supported a major program of railway building with the twin aims of not only uniting the country but also stimulating economic development in the impoverished south. As a result, 1,200 miles were completed in the first five years following the state's creation, ensuring that the gaps left by the pre-unification companies were now filled and making it possible to travel by rail between all of Italy's major cities by 1866, except in the sparsely populated south. To bring about this expansion, the government granted concessions to the railway companies, providing funds for the cost of building that were supposed to be paid back once the line was open. Although this government support succeeded in stimulating the rapid completion of projects, it also encouraged the railways to cut corners. By building lines on the cheap with many curves and few tunnels, they pushed up the costs of operation and made traveling on the system slower and longer. Consequently, most of these railways operated at a loss and were dependent on subsidies from the state right from the outset. In an effort to make them more efficient, in 1865 the government decided to rationalize the network into four large new companies, with a guaranteed, but low rate of return on their shares.

The merger took place in July 1865, with the largest company being the Alta Italia, which had more than 1,000 route miles in the north of Italy, including both the Lombardy-Venetian Railway and the Piemonte. The Società delle Strade Ferrate Romane, the Roman Railways, was the second largest, with 826 route miles, incorporating both Roman companies and a few lines in Tuscany. The company's main line ran from Naples through Rome to Foligno, where it split, with one section going to Ancona on the Adriatic coast, and the other to Florence and then Livorno on the Mediterranean. This meant it had to share Florence's main station, Santa Maria Novella, with the Alta Italia Railway, which led to the sort of territorial dispute that is commonplace in railway history but utterly incomprehensible to the passengers who have to suffer its consequences. The Roman Railways had the eastern side of the station, but most of its trains approached from the west and therefore had to cross the other company's tracks. Eventually an arbitrator was appointed, a young man named Riccardo Bianchi, who would later become the first head of the state-owned Ferrovie dello State upon nationalization in 1905. Bianchi made the obvious decision to order the companies to swap sides, and peace reigned. The Roman Railroads focused less on building branch lines than its northern rival, instead erecting palatial stations in both Rome and Naples.

Heading south, the third railway was the Strade Ferrate Meriodionali, comprising 578 route miles, with a main line running down the Adriatic coast from Bologna to Brindisi. Each railway was less viable than its northern neighbor; the Meriodionali struggled to make money, and railways further south were perpetual basket cases. Indeed, although there were many schemes to build railways in the south, few had actually seen the light of day by the time of the unification. The government had been reluctant to provide the financial support that the private companies needed to build lines in these sparsely populated areas with little industry. Therefore, the fourth, the Strade Ferrate Calabro-Sicule, covering Sicily and the south, was tiny. Progress on the old Victor Emmanuel Railway, with its plans to build lines in the deep south, had been painfully slow as a result of the ruggedness of the territory and the ubiquity of the malarial mosquito (see Chapter 8).

By the mid-1870s, the Italian government's hopes of keeping the railways in private hands through rationalization of the system had proved to be overly optimistic. The companies fell one by one into financial difficulties and had to be rescued by the state. Initially, the newly formed state had created these companies with the idea of seeking foreign capital, which had dominated early investment in the railways. This, however, became politically unacceptable, given the widespread opposition to the idea of paying dividends out of state funds to companies based abroad. Indeed, foreign investment in the railways had been a contentious issue even before unification, when the Rothschild Bank had been granted concessions by the pre-unification government for railway construction in southern Italy. The arrangement had been blocked because the government insisted on investment from Italian capital, thus changing the course of European railway history by preventing the Rothschilds, who already owned the Südbahn, the Chemin de Fer du Nord, and the Paris-Lyon-Mediterranée, from extending their empire and establishing their dominance over the Italian railway system.

Unable to call on foreign capital, the railways were forced into seeking state help, even if that meant losing their independence. The first to be nationalized was, strangely, the most successful, the Alta Italia, which became one of the first railways in the world to be taken over by the government, sparking a trend that would be repeated numerous times. In countries as far apart as Peru and Japan, or Russia and South Africa, lines were started by the private sector and nationalized, or, sometimes, they were built by the state and sold off. Ownership of the railways has always been a deeply politicized issue, as every nation has sought, but rarely found, the right compromise between entrepreneurship and state control, government funding and private investment. Politics has always played a part in defining this balance, but not necessarily in the way that may have been expected. A curious aspect of the nationalization of the Alta Italia is that it was promoted by the right-wing liberal government of Marco Minghetti, but strongly resisted by members of the left-wing opposition, who wanted to see the railways continue in the private sector, a bizarre reversal of the subsequent positions normally taken by Right and Left. The arguments for nationalization were best put by Silvio Spaventa, a leading member of the Right. Railways were a public service, like the post, he said, and therefore should be run by the state: "To abdicate this responsibility betrayed the public trust. Besides, big business could not run the railways either responsibly or honestly. Corporations served stockholders not citizens; sought profit, not public benefit."[14] Similar statements can be found in transcripts of parliamentary debates from the 1990s on British rail privatization, mostly, but not exclusively, from politicians of the Left.[15]

The counterarguments are familiar, too. The state was inherently inefficient because private managers had a vested self-interest in ensuring that their companies performed well. The railways would be run in a bureaucratic way, and the state would want to use the services to increase its power. There were even fears that state involvement would make people think that the government should do everything for them and expect it to do so. Again, these points differed little from those made by the supporters of privatization in Britain 120 years later.

Nationalization of the railways in Italy, however, eventually proved inevitable. The Minghetti government fell in 1876 when Minghetti's efforts to push through the wholesale nationalization of the network were rejected by Parliament, giving him the dubious distinction of being the first, but by no means the last, ruler to fall over a railway issue. Despite the rejection of an all-encompassing nationalization plan, the Alta Italia was eventually purchased by the Italian government (in 1878) under the scheme agreed to by the Minghetti government. The structure of the railways under the 1865 act had become unsustainable, not least because the government did not stick to its commitment to pay subsidies and, worse, because it introduced new taxes that made it even more difficult for the companies to remain profitable. The state, too, could not resist getting something back from the money it had invested in the railways. The companies moaned about the government interfering in their detailed operations, another complaint about the ever-fraught relationship between states and railways that would be heard around the world. Nationalization therefore came as something of a relief for the companies.

The government already owned the tiny Calabro-Sicule railway, because it had required massive subsidies. The Alta Italia, the ailing Romane, with mounting losses, was taken over in 1880. That left only the Meriodionali in private hands, which also was dependent on ever larger subsidies. An 1885 act regularized the situation, but it was passed only after the conclusion of a lengthy debate covering sixty-five sessions, the longest in the Parliament's short history, rehearsing all those now-familiar arguments at great length. The resulting act was, though, something of a compromise, because, while the state took over the ownership, the now left-wing government insisted that private companies manage the network on a concession basis. The mainland railways were divided into two companies, each covering one side of the Apennines. They were named after their respective seas: the Mediterranea and the Adriatica. Two much smaller railways covered Sicily and Sardinia. Again, the detailed arrangements set out in the contracts were phenomenally complex. The state agreed to subsidize the railways and share in any profits, rather than charging for access to the lines, and the rolling stock was sold off to the private operators, but with the agreement that it would be bought back at the end of the concession. The government also agreed to fund the construction of additional lines and to control the timetables, but not fares and freight rates, which were set by the legislation and could be altered under certain conditions.

But how do you value a railway? That was another area of contention. The government's opponents argued that it had overpaid for its purchase. There was no independent assessment of its value; the company had used the old trick of putting reparations and maintenance onto the capital account, increasing the book value of the company. Nevertheless, the state probably paid less for the railway than if it had built the whole rail system itself.

The purchase of the Alta Italia Railway Company from the Rothschilds in 1878 highlighted the complexities governing the financial relationship between states and their railways as well as the political issues that are still being thrashed out across the world in the debate over rail privatization. Railways are by their very nature different from conventional companies because they provide a public service that generally has to continue to operate regardless of whether it is profitable or not. This was particularly true until the advent of the motor car and the development of alternative forms of public transport, such as buses and aircraft, because the railway was effectively a monopoly providing an essential service. Moreover, railways have enormous fixed assets that in good times can be used intensively and therefore profitably, but in bad times cannot easily be mothballed or temporarily shut down. Therefore railways are particularly vulnerable to the vagaries of the economic cycle or to the emergence of a strong competitor, which is why so many railways across the world were eventually nationalized when they began to lose business to cars and trucks, and later, airplanes. It was precisely because railways in Italy were not profitable, as a result of the topography and poverty of the country, that the question of ownership and state control emerged there earlier than in other countries. The hybrid system in Italy survived a couple of decades, but in 1905, the railways were fully nationalized.

Italy was not, however, the first country to nationalize its railways. That was Prussia, where Bismarck, the Iron Chancellor, felt that they were too important militarily to be left in private hands. The railways in Germany, which was only unified after the 1870 Franco-Prussian War, had been built on a state-by-state basis, although the German Union of Railway Administrations had done much to ensure that the same technical standards, including, most importantly, the gauge, were maintained by all the railways. There were, too, a series of looser regional alliances that worked together to set rates and negotiate over rights of way on each other's tracks. With the states enjoying total autonomy, their railway networks had developed at different speeds and with varying levels of state involvement. Saxony, for example, where Friedrich List had enjoyed the most success in persuading the government to pursue the idea of developing a rail network, boasted the densest network of lines anywhere in Europe apart from Belgium. The Leipzig-Dresden route with its hugely profitable operations was the system's lynchpin, but that did not prevent the state from trying to take over the railway in 1854. That attempt failed, but the state imposed a tight set of regulations, turning all railway employees into civil servants. Bavaria started off by encouraging private enterprise but, eager to cash in on the profits of the early railways, took over the Munich-Augsburg company in 1844, and it became the cornerstone of the state network run by the Bayerische Staatsbahn. By the early 1840s, there was already talk of the great east-west railway that would link Paris with Vienna and become the western section of the Orient Express. The famous train would not start running until 1883 (see Chapter 8), although by 1860 it was possible to travel between the two cities by rail via a circuitous route passing through Frankfurt, Dresden, and Prague. Another crucial international link, the bridge across the Rhine between Kehl in Germany and Strasbourg in France, was completed in 1861, although a condition of obtaining German permission to build it was that Germany's section had to be mined for self-destruction in the event of a war.

Prussia, though, was the powerhouse of both Germany and its railways. Bismarck, having avoided death on the Semmering, became prime minister of Prussia in 1862. At first he continued with the laissez-faire policy toward the railways that was long established in Germany's largest state. At the time, the government only had an interest in five of the state's twenty-two railways, and over the next few years there was a railway boom in which twenty-seven concessions were granted to private companies for new lines. As was often the case with such booms, dubious characters were attracted by the potential for quick profits, and Prussia spawned its own dodgy "railway king," Bethel Henry Strousberg, who resembled the notorious and equally corrupt George Hudson, the principal villain to emerge from the British railway mania of the 1840s.[16] Strousberg built up a huge railway empire, including a cluster of tracks around Hanover and the main line between Berlin and Halle, but in the climate of rampant speculation he overextended himself, and he lost out when his contracts dried up as a result of the Franco-Prussian War. His crookedness, once exposed, brought about the downfall of the Prussian minister of commerce, Count Itzenplitz, over suspicions of bribery, and Strousberg fled to Romania, where he was imprisoned over unfulfilled railway contracts.

Until the war, Bismarck was content to stay out of railway affairs, apart from letting large numbers of concessions, but once the French had been beaten and he turned his attention to the unification of Germany, he decided that the railways needed a single efficient administration. In contrast to France and Italy, where the railways had consoli-dated into a handful of big railways by 1870, Germany still had sixty-six different companies split among eighteen state railway admin-istrations. Prussia, already the biggest state, incorporated northern Ger-many under the unification scheme, giving it a dominant role in the new Kaiserreich over the rest of the semi-independent states, including their railways. The vital role the railways had played in the wars with Austria and France convinced Bismarck that they could not be left in the hands of the private companies. However, bringing together eigh-teen different railway administrations was never going to be an easy task, especially as the remaining semi-independent states, including Bavaria, Saxony, and Württemberg, resented the dominance of Prussia. Bismarck disliked both the chaotic nature of the Prussian railways, which were partly state-owned, and the lack of coordination between the different state railway administrations. The Reich initially imposed itself on the railways by establishing uniform rules, with no issue too big or too small to apply standardization. The Imperial Railway Office, the Reichseisenbahnamt, stipulated that the departure of a train should be announced by the stationmaster with two long blasts of his whistle, and that compressed air brakes should be installed on all rolling stock.[17] Such *diktats* had their roots in military requirements; standard-ization was greatly favored by the generals in Bismarck's administra-tion, as they realized that railway systems using the same components for signaling, switches, and lights would be far easier to repair in the event of damage during a conflict.

Even the powerful Bismarck could not easily overcome the vested in-terests of the private railways and the state administrations, or the lib-eral zeitgeist, which fostered a climate of deep suspicion of any state involvement in business. Eventually, though, after a decade of debate and political maneuvering, he managed to push laws through the Prus-sian legislature in the winter of 1879–1880 that empowered the Reich to compulsorily purchase the private railways. Albert Maybach, the head of the Reichseisenbahnamt, was eager to take over as many of the railways as possible, as quickly as possible, and sought to do it "not one by one, with long intervals, but rapidly with broad strokes."[18] Many of the railways, particularly those that were financially stretched, meekly acquiesced to their purchase by the state, while others resisted and were in effect confiscated, although shareholders were paid a reasonable amount in compensation. Within six years, the Royal Prussian Railway had extended its holdings from 4,000 to 13,000 miles to become the largest industrial enterprise in the world, overtaking the British London & North Western Railway.

Bismarck's motives were partly political, as he aimed to reinforce the power of central government over private companies and state administrations, but his trump card in forcing through nationalization of most of the rail companies was military. He saw a strong Germany as the only way to maintain peace in Europe, and he was fearful that France might be able to gain an eastern ally to coordinate an attack against Germany on two fronts. A railway that could deliver troops quickly around the country, without interference from private interests, was essential in order to ward off the constant threat that France was felt to be: "Only [the railways] could connect the two fronts and bring the weight of the Reich's military machine victoriously to bear."[19]

The symbiotic relationship between railways and the military had deep roots. Railways, by necessity, are run in a disciplined way, with uniformed staff and rigid regulations, in order to ensure safety and prompt running. But that did not prevent conflicts from emerging between the German military and the remaining railways controlled by the local states and private companies. The military, for example, wanted to double the tracks on lines heading toward the French frontier, but these were little used and therefore not very profitable, which made the private companies and state governments reluctant to subsidize them. Eventually compromises were reached, with the Reich paying most of the cost of expansion.

It would be difficult to overemphasize the importance of the role of the railways in the political process and the frequency with which railway issues turned into full-blown constitutional crises. Debates over nationalization and expansion of the railway system, which were both expensive endeavors, occupied much of Bismarck's time and were fundamental to his efforts to create a strong Germany. As one of his war ministers, General Bronsart von Schellendorf, put it, "the security of the Reich rests in no small measure on the expansion of the railway network in the interest of national defence."[20] Bismarck was not averse to using the regular war scares, which later invariably proved to have arisen from false rumors, to ensure that he could raise the money to build up the railways in the manner required by the military.

On the other side of the frontier, the French were improving their railways with a similar military imperative, but their task was more difficult because they were preparing to resist an invasion, whose direction was uncertain, rather than preparing to launch one, whatever the German propaganda suggested. Therefore the French needed their railways to be flexible and adaptable, while the Germans were more interested in speed, or in the words of another historian: "The German priority must be to develop a first strike capability, whereas France would need to prepare a quick response from several fixed points."[21]

Most of the smaller European countries were expanding their rail networks at the same pace with similar state involvement. In the Netherlands, for example, as we saw in Chapter 2, the system had been developed by two major railways and several smaller ones, leaving large gaps. It was not a coherent network. By 1860, the Dutch Parliament, worried about foreign intervention in the railways, accepted a plan to construct 500 miles of line that would fill in the gaps in the system. The money to fund the network would come from the colonial revenues from East India, which, at the time, "flowed more richly into the government coffers every year."[22] Just as in Germany, the liberals, who opposed state intervention, were defeated; the need for the rapid development of the railway was recognized by the government and a majority of MPs. In a compromise, private organizations would build and run the lines, but the state would back the capital raised by the companies and pay for the great bridges that were needed to ford major rivers. The most notable would be the elegant Culemborg Bridge, designed by Gerrit van Diesen over the Lek, a branch of the Rhine, which became the longest single span in the world, nearly 500 feet, when it was completed in 1868. Indeed, as the engineers had learned how to cope with the marshy areas and soft ground for the long, straight stretches of railway, these were completed rapidly, but the bridges presented a formidable challenge. The rivers were full of drifting ice in the late winter and early spring, and they swelled enormously from the snow melt of Germany and Switzerland. Today, both thermal and chemical pollution have caused these patterns to change; then, the real danger was that if a thaw set in too quickly, a drifting mass of ice could build up, breaching the dikes.

After the first of the new state-built railways—the short line between Breda and Tilburg—was completed, a fierce debate ensued about who should run it. The liberals would not even allow consideration of the case for the government to run the trains, but various bids from existing railways were rejected, necessitating the creation of a new company, Staatsspoorwegen (State Railways), which, despite its name, was a private business rather than a government body. The intervention of the state in railway construction did, however, ensure that after the country's slow start, the Netherlands had a complete mainline network by 1890.

As elsewhere, military requirements had a major impact in the Netherlands, but in a rather surprising way. The country was littered with old-style fortress towns, which, in a flat country, were essential for defense in the event of a war. These fortress towns, such as Breda, Deventer, and Groningen, were not allowed to have a permanent brick-built station because this might provide cover for an enemy; instead they had to make do with low wooden structures that could be destroyed quickly to ensure that the artillery in the town would have a clear range of fire. As ever, the military was fighting the last war, and the Fortress Act of 1874, which repealed the archaic restriction on building, put an end to that bit of nonsense.

Holland, too, had its international railways. The Rhenish, whose directors included James Staats Forbes, a ubiquitous railway manager who ran the London, Chatham & Dover as well as the Metropolitan District railways in Britain, began operating trains between Amsterdam and Cologne in the early 1860s. For a while, some of its trains—without their locomotives—even crossed the Rhine by ferry. A dispute with the company's German partner, the Köln-Mindener Eisenbahn, over rates had led to its cooperation with another railway, the Rhenische Eisenbahn, and for a time the two were linked by a boat service between Spyck in Holland and Welle in Germany.

All over Europe, the pattern was similar. The main lines, by and large, were well established by the 1880s, and thereafter the rate of construction slowed down considerably, with only bits of infilling and the odd link to an inaccessible provincial town. And everywhere, there was a fraught debate about the extent of state involvement, given the recognition of the essential need for railways. Even in Britain, the state, at one stage, had almost become involved in running the railways. Following the 1866 banking collapse, several major companies that found themselves in financial difficulties had approached the prime minister, Benjamin Disraeli, to ask for government support, but they were rebuffed.

By the end of the century, the major European powers—France, Germany, and Great Britain—had networks that were similar in size in relation to their populations, all with just under one kilometer of track per 1,000 inhabitants.[23] For different reasons, railway growth was slower in Italy and the Austro-Hungarian Empire.[24] In Italy, the chaotic political situation, the poverty of the south, and the whole country's relatively undeveloped state all hindered growth. Austria was constrained by its mountains and by a lack of interest from its political masters, though the pace of development picked up in the last decades of the century, when its mileage almost doubled. Much of the increase in mileage was in Hungary, which, when separated off after World War I, boasted the second densest network in Europe[25] after Belgium. Other laggards included Spain and the Scandinavian countries, and a particularly late starter was Switzerland, but there was good reason for this— the almost impossibly difficult mountainous terrain. However, by the end of the century Switzerland would become the hub of a pan-European system, a crucial crossing point for Germany and other northern European countries, including parts of France, and provided a link with Italy via the St. Gotthard Tunnel.

But Switzerland had, at the beginning, made very slow progress. Besides the mile or so of railway that reached Basle in 1844, an extension of a French line, the first line built entirely within Switzerland was the one between Zürich and Baden, which was completed in 1847. Switzerland at the time was a poor and disunited country; it was ruled by its cantons and had little industry. The line, therefore, was not very successful, despite doing a roaring trade in the Spanish rolls that were Baden's claim to fame, and which could now be delivered fresh and crisp to Zürich breakfast tables, earning the railway the nickname of the Spanish *Brötli* (roll). No other line was opened for seven years, but the ubiquitous Robert Stephenson was asked in 1849 to draw up a railway development plan for Switzerland. He proposed a long dorsal line linking Geneva with Zürich and Lake Constance, crossed by another between Basle, Thun, and Chur. Wisely, he did not suggest a way through the Alps, and therefore his plan was implemented without too much difficulty, though it took twenty years to complete. The federal structure of the country meant that the cantons had the responsibility to grant concessions, and they were at times reluctant to do so.

Switzerland was, by then, beginning to develop industrially and realized that its place at the center of Europe would offer great economic prospects if the mountains could be breached by the railways. The Mont Cenis Tunnel showed that there were no insuperable technical barriers, but since its location meant it could not cater to German traffic to the south without a great detour, its successful completion prompted demands to build a tunnel linking the Reich with Italy. The most obvious route was through St. Gotthard, a pass that had linked south and north Switzerland for seven centuries.

Switzerland alone could not fund such a grand scheme, but both Italy and Germany, egged on by Bismarck, were eager to support it, because of the obvious advantages for trade links, and the project was approved at a meeting of the three governments. It would offer a new route for the transport of Ruhr coal to Italy, as well as providing far easier access for German goods heading for the Mediterranean ports. Approval for the scheme was granted by the Swiss Parliament in 1871. Although the whole railway was in Swiss territory, Italy paid just over half the cost, with the rest being split equally between Germany and Switzerland. For Switzerland itself, the railway would open up the country during the winter months, when previously much of its industry ground to a halt because it was impossible to cross the mountain passes.

At 9 miles, the St. Gotthard was to be the longest tunnel in the world, and several others were needed to bring the railway up to the height of the main one. As with the Semmering and Mont Cenis tunnels, an engineer of quite exceptional skill and courage was needed to attempt such a massive undertaking. That man was Louis Favre, the largely self-taught son of a carpenter from Chêne near Geneva, who had much experience as a tunneling engineer and had made a fortune working on a variety of civil projects. Work started in 1872 soon after he had been given the contract to build the tunnel, and he introduced a crucial innovation that was particularly daring. In order to bring the railway up to the height of the main tunnel, more than 3,600 feet above sea level (but

3,000 feet below the old pass used by stagecoaches and wagons, which was closed in winter), he decided to cut enormous circular tunnels with a gentle gradient into the mountain. Each tunnel would raise the railway so that it emerged at a point above where it had been before. At Wassen, for example, passengers heading south would emerge from a loop to find themselves well above a church spire they had seen from below just a few minutes before. The railway between Lucerne and Chiasso on the Italian border is 141 miles long, a fifth of it consisting of these loops, which, like hairpins on steep roads, add considerable distance to the journey.

In the dozen years since work had started on the Mont Cenis Tunnel, drilling technology had improved and, crucially, dynamite had been invented. Progress therefore could be made more quickly than on the Mont Cenis Tunnel, but construction still presented enormous dangers to the workers. The death toll was high, with at least two hundred workers killed in explosions and rock falls, although the precise number is not known. And poor Favre was one of the casualties. During an inspection visit in 1879, two years before the completion of the project, he suffered a heart attack and died on the spot, aged only fifty-four. His body had to be hauled out by one of the wagons used to carry out debris. A writer for the magazine *Scientific American* who had met him shortly before his death was in no doubt that the work had killed him, writing: "The man of broad shoulders and with head covered with thick hair in which here and there a few silver threads showed themselves, and who was as straight as at the age of twenty years, had begun to stoop, his hair had whitened and his face had assumed an expression of sadness that it was difficult for him to conceal. As powerful as it was his character had been subjugated."[26]

The St. Gotthard route, which finally opened in 1882, lived up to its expectations. After the opening, German exports to Italy tripled in less than five years. Consisting mostly of coal and iron, virtually all of it went by the railway. Interestingly, although Italy paid for half the railway, "the early commercial advantage was all to Germany, though northern Italy was being provided with the raw materials to develop its industry and, eventually, to take advantage of rich markets in northern Europe."[27] It was a classic example of a railway spreading wealth to both its extremities. It changed the face of European commerce and was the most significant engineering feat to affect world trade other than the Suez Canal, which opened in 1869. Its success stimulated the building of other Alpine tunnels, notably the Simplon, which, at over 12 miles, became the world's longest. Begun in the summer of 1898, it was designed to link the upper Rhone with Italy and took only eight years to complete, thanks to further technological progress on drilling equipment. The third giant tunnel, the 9-mile-long Lötschberg, which connects Berne with Italy, was beset with problems owing to unexpected geological conditions, including a rockfall that killed twenty-five workers, and was eventually opened just before the outbreak of World War I. By then, electric traction had become the norm because of the perpetual problems caused by smoke in these long tunnels with their inadequate ventilation shafts.

There was, of course, one European international crossing that would have to wait another century to be built. That was the Channel crossing, which had been proposed as early as 1802, long before the first railway was completed. A variety of projects had been put forward in the mid–nineteenth century by Thomé de Gamond, a mining engineer, and one was even presented to Queen Victoria and Napoleon III in 1856. John Hawkshaw, who later built the 4.5-mile-long Severn Tunnel near Bristol completed in 1886, at the time the longest in Britain, also took an interest. The warmer relationship between Britain and France after the French humiliation in the Franco-Prussian War stimulated interest in the project as well as the creation of a joint company to build the tunnel. Trial tunnels were started at both ends in 1881, but rather fanciful fears of a French invasion in the deeply conservative British military establishment killed the scheme off two years later. The tunneling experiments had proved successful, but the problem of ventilation for a tunnel over 20 miles long had not been resolved. It is doubtful whether the project would have overcome either this technical difficulty or the inevitable financial hurdles of what would have been at the time the world's most ambitious engineering project. The current 31-mile-long rail tunnel under the Channel opened in 1994 and accommodates only electric trains, avoiding the ventilation problem that steam locomotives presented.

As early as the 1860s, there had been proposals, too, for the idea of a tunnel linking Spain with Africa, which at their closest point are only

10 miles apart. Unfortunately, the sea is deep there, reaching 1,800 feet, and any tunnel built in the rock beneath it would have to be very long, since it would require gentle gradients at both ends to reach that depth. Therefore the tunnel would probably need to be at least 20, if not 30, miles long. Although this was at the limits of the technology of the time, the French encouraged the idea, eager to have a land link with their colonies in Morocco and Algeria, but nothing came of the project despite a stream of proposals by both French and Spanish promoters.

There was no shortage of similar grand ideas, several of which were actually built, such as the tunnels through the Alps, but there were countless others that never saw the light of day. Although this desire for expansion was partly driven by the sheer entrepreneurship and sense of adventure of the more dynamic railway managers, there was a baser reason, too. Railways suffered from diminishing rates of return as they faced competition from each other or, if they were monopolies, pressure from governments to keep rates down. During the period of rapid initial growth, big companies emerged that established their rights over lucrative routes that yielded easy profits. They were able to reward shareholders generously with that money as well as investing in the latest technology. However, once these main routes were established, they were under constant pressure from both local people and governments of all levels to build routes connecting smaller towns or branch lines, which were invariably less profitable or even had to operate at a loss. At other times, routes were built to prevent competitors from obtaining access to their routes; sometimes, particularly in England, rival railways tried to cream off business by building alternative lines, which is why there are three railways connecting England and Scotland. Moreover, as we shall see in Chapter 8, increased patronage on their tracks would lead to rises in costs as the railway companies had to invest in better signaling systems, additional rolling stock, station facilities, and so on. The list is almost endless.

The railways therefore could never stand still and always had to keep investing. As Allan Mitchell said of the Saxony railways, "like the other states of Europe, Saxony soon had to deal with diminishing returns. Once the main lines had been spoken for, investors became reluctant to part with their capital in order to promote ancillary projects that promised little profit."[28] Nationalization was not a result of the kind of ideological discussions cited above. Rather, it reflected the simple fact that it was difficult for the railway companies to make long-term profits out of such a capital-intensive industry. And yet they could not be allowed to fail because their services were an essential part of the growing economies of nineteenth- and early twentieth-century Europe. Moreover, governments were always ready to step in with extra regulation if there was a suggestion that a railway was exploiting its monopoly position. Add in the fact that the military was increasingly aware of the importance of the railways, and a strong involvement of the state, if not nationalization, was inevitable. The railways, by their very nature, could never be run in the way that the free marketeers favored.

Their continued growth was, however, inevitable. It was, of course, not only in Europe that the clamor to be connected to the rail network was loud. The European continent was relatively easy to cross, with only the Alps and the occasional large river as natural barriers. It is also small compared with other parts of the world, where there were far more ambitious projects involving the crossing of whole continents. All the other inhabited continents, with the notable exception of Africa, would eventually be breached, opening up virtually the whole world to the iron road.

SIX

Crossing America . . .

THE IDEA OF A TRANSCONTINENTAL RAILWAY FOR THE UNITED STATES was debated very soon after the pioneers of the iron road had laid the first lines. And so it was on other continents. Crossing continents and other vast tracts of land seemed a natural aspiration for railways, and plans were proposed to traverse not only the United States but also South America, Central America, Canada, Russia, Africa, and even Australia in the second half of the nineteenth century. And all would be achieved, apart from the Cape to Cairo, a project whose colonial intent was simply too ambitious in the face of natural and political barriers. These massive projects, the most ambitious in the history of humankind, faced seemingly insuperable difficulties and invariably ended up costing far more than the original estimates. Transcontinental railways were the ultimate *grand projets*, which inevitably made them the creatures of government, as they were not even remotely feasible without state involvement. Often inspired by little more than imperial swagger, they were grandiose undertakings that sought to connect distant corners of the globe with the main population centers in the quest for wealth, power, and national unity. They attracted adventurers and visionaries, and, though their purposes varied from country to country, they were more often rooted in the ambitions of politicians and railway promoters than in practical transport economics, though, ultimately, several proved profitable.

The story does, of course, start with America, which built the first transcontinental railway, but, surprisingly, not in the United States but in Central America through Panama. In a way the development of that railway had the same driving force as the one that would be completed in Utah fourteen years later—the opening up of the West—as it was the incorporation of Oregon and California into the United States that stimulated the promotion of the Panama Railway.

In the 1840s there were three ways to travel between the east and west coasts of the United States, none of them easy: the direct way overland by wagon, running the risk of attacks by Indians; a ship via the dangerous waters of Cape Horn; or trips up and down the coasts linked by a 50-mile hike through the jungle of the isthmus of Panama. Demand was growing. The California gold rush of 1849 attracted many prospectors to the West and there were also the prosaic, but more permanent, needs of the United States Post Office, keen to deliver mail to every state. The U.S. Post Office established shipping routes along both coasts down to Panama, which at the time was under the control of its southern neighbor, Colombia,[1] but the journey across the jungle created a terrible bottleneck for both goods and people. A railway was the obvious answer. U.S. shipping magnate William Aspinwall, who already had the contract to operate ships for the postal service between Panama and Oregon, won the concession to build the 47-mile railway across the isthmus.

Optimistically, Colonel George Hughes, the surveyor, and the two engineers, George Totten and John Trautwine, who won the contract to build the line, reckoned construction would take twelve months and cost just $1 million. Both figures ultimately would be completely wrong: The line took five years to build and cost $7.4 million to complete, making it the most expensive railway yet built.[2] Apart from the routine optimism from the promoters occasioned by the need to win over investor support, the main reasons for the cost and time overruns were that the conditions for construction were more hostile than for any other railway project yet attempted and, indeed, probably for almost any thereafter, apart from some equally murderous schemes in equatorial Africa (see next chapter).

It was the railway from hell. Not only is it hot all year round in that region, since it is on the equator, but in the long wet season from June to December there are frequent tropical downpours that can last up to three days. Supplies of both basic materials and labor were lacking. There was no durable timber to use on the railway, and the local natives were unwilling to work on its construction. All equipment, such as rails and railroad ties, had to be imported, mostly from the United States and Britain, and, to make matters worse, the best alignment could not be used because the landowner wanted to extract too much rent. Consequently, the line was started from Manzanillo Island, a swampy area on the north coast (which, in fact, is on the eastern side of Central America, as at that point the land runs on an east-west axis) that was fraught with danger, both from the local wildlife and from the risks involved in construction. It was almost as if the isthmus were consciously resisting this human incursion as the swamps swallowed up the seemingly infinite amounts of earth dumped into them to try to establish a stable surface. The poor laborers toiled in glutinous mud that at times reached up to their necks.

The shortage of local labor was to lead to one of the saddest episodes in the history of the construction of the railways, the terrible plight of the eight hundred Chinese men shipped in on virtually slave-labor terms to work on the railway. Indentured to a contractor, they received but a small portion of the $25 per month that he was paid for each man; worse still, they were deprived of the opium to which they were all addicted. Suffering from the ghastly conditions as well as withdrawal symptoms, and hated by their fellow Irish workers, they became severely depressed and started committing mass suicide in bizarre ways. While some hung themselves from trees and others paid the Malays to chop their heads off with their machetes, most simply walked out to sea or tied stones to their feet and jumped into the river. Finally, an overseer reckoned that there were 450 dead bodies of the Chinese lying in the jungle, and the survivors were quickly rounded up to be taken to the Chinese community in Jamaica.

The Chinese were not the only ones to die in droves. Even Irish navvies, with experience building the railways in Britain, proved unequal to the task; they were decimated, with the few survivors being transported to settle in New York. There was no shortage of ways to die: malaria, cholera, dysentery, smallpox, and unknown other infections for which there was no cure were the most common killers, but snakes, alligators, poisonous insects, and the inevitable accidents also took their toll. The death rate was staggering. At the nadir, in 1852, 20 percent of the workforce died each month, and only two out of fifty American technicians sent that summer to provide expertise on the railway lived to see Christmas.

Even those who did not succumb to these various diseases were debilitated by them since, on average, only a third of the labor force, which varied between 1,500 and 6,000, was able to work at any one time. The hardiest workers were, not surprisingly, the Afro-Caribbeans from the nearby Caribbean islands, whose physiology and temperament were best suited to the tropical conditions, and their survival rate was higher than for the other ethnic groups. The final death toll is unknown because the railway company only kept records of white fatalities, but a conservative estimate suggests that 6,000 men ultimately died, 120 per mile of railway, a rate worse than for any railway project before or since, and some historians suggest it was at least double that number.[3] Disposing of the bodies in the swampy conditions proved to be a problem until an ingenious, but ghoulish, solution was found. The cadavers were pickled in barrels and sent off to medical schools around the world, a trade that paid for the running of the railway hospital. A reporter who visited the hospital during construction found "melancholy rows of sick and dying men,"[4] and he was rather shocked to find that his host, the medical director, had a plan to assemble a complete "museum" of dead bodies representing all the racial types found among the workforce.

During his visit, work had stopped because of a shortage of money, but the railway was saved by a fortunate incident. There were still hundreds of gold diggers heading west, and two paddle boats with 1,000 men aboard happened to arrive just as the first section of railroad had been completed. They were surprised to see the railway and were eager to use it to save themselves part of their trek. In an effort to discourage them, the exorbitant rate of 50 cents per mile and 3 cents per pound of baggage was set, but the men were happy to pay, despite having to travel on open wagons, and the railway earned $7,000. This not only became an established form of income but, more importantly, encouraged potential investors in New York, where the price of the shares had plummeted, to come forward. Ultimately, a third of the cost of the railway was raised by carrying passengers before its completion.

The railway, which included three hundred bridges and culverts in its short length—the price of laying down track in a half-drowned country—was completed in 1855 and opened immediately to traffic. George Totten was the driving force behind the line and after surviving a bout of yellow fever remained in charge once the railway was completed. The railway proved extremely lucrative for its investors, generating, in the first decade, $11 million in profits, with dividends of up to 24 percent being paid annually, and until completion of the Panama Canal in 1914, it was the most intensively used freight line in the world.[5] Almost as a jest, the railway company had set extremely high fares for the carriage of both goods and passengers, with tickets at $25 for first-class passengers and $10 for those in second class for the 50-mile journey; railway officials were rather surprised to find that their customers accepted these prices, presumably because the railway had considerable advantages over any other form of transport to cross the continent. The railway frequently carried as many as 1,500 passengers in a day, and it transported the freight of three steamship companies as well as the U.S. mail. In order to maintain a sufficient workforce in such a hostile environment, for the most part the company treated its employees very well, though, as was frequently the case with railways, in a very patriarchal style. The company provided them with everything from food and housing to free hospitals, a well-stocked library, and billiard parlors and encouraged them to go to the churches it had built.

An oddity is that Panama is one of the few places in the world where the railway preceded the construction of a canal and was ultimately superseded by it. Even more strangely, the arrival of the French Compagnie Universale du Canal Interocéanique on the isthmus to start building the canal in the late 1870s proved initially to be the savior of the railway, which had begun to lose traffic following the completion of the transcontinental within the United States in 1869. Ferdinand de Lesseps, the canal's first but ultimately unsuccessful engineer, realized that he needed to get control of the railway to ensure he had a monopoly control of transportation in the area, and his company bought up the shares at a far higher price than merited by the profitability of the railway. The French invested considerably in improvements to the railway, but de Lesseps, who had built the Suez Canal, which was also completed in 1869, was unable to repeat the trick in Panama, defeated by cost, disease, and the impossibility of realizing his plan to build a sea-level canal without locks. Work on a canal with locks started again in 1904, when the Americans took over the project, and the waterway was completed in 1914. This event signaled the decline of the railway, which, to accommodate the canal, had to be moved to higher ground, all the while being used to supply the canal and dispose of its spoil. The railway survives today as an adjunct to the canal and is enjoying a·renaissance, as it was upgraded in 2000 to carry containers. Its gauge was finally changed from the original 5 feet to standard gauge to enable the introduction of cheaper mass-produced railway equipment.

THE TRANSCONTINENTAL RAILROAD across the United States was a long time in gestation. When it first began to be seriously considered in the 1850s, it was by far the most ambitious railway project anywhere in the world. To get to the other ocean 3,000 miles away seemed an obvious endeavor to unite an as yet unformed country, but its precise purpose was not immediately clear. What was clear was that it could not be a purely private-sector project because it was unlikely to be able to generate sufficient profits to pay for its construction. Therefore, it was the federal government that took the initiative by commissioning surveys of five possible routes from the Missouri River to California in the mid-1850s. Although the proposal was obviously political, it was more than that; it was a project that went to the very depth of the American psyche: "For all the fanfare that accompanied the building of the first transcontinental railroad in the 1860s . . . the importance of this accomplishment for many years was psychological," wrote railroad historian Albro Martin.[6] While that may have been the case, pork-barrel politics still had to be overcome. Lobbying by the vested interests of southern Congressmen pressing for the route to go through their regions held up the legislation; the Civil War then intervened, blocking any hope of further progress. Abraham Lincoln, the transcontinental's most fervent advocate, pushed the Pacific Railroad Act through Congress in 1862—a year that was a low point for the Union in the Civil War—helped by the absence of the southern politicians who had stymied previous attempts at legislation.

Like much of the story of the U.S. railroads, the construction of the transcontinental line encapsulated pioneering America at both its worst and its best, a "curious mixture of the noble and the ignoble."[7] On the one hand, the amazing feat of building 1,780 miles of track across difficult terrain and overcoming the elements, notably the terrible winter of 1866–1867, which was the worst on record with temperatures plunging to minus 40°F, as well as attacks from Indians, was an achievement to match any human enterprise since the building of the Pyramids. On the other hand, the purloining of tens of millions of dollars from both taxpayers and investors was on a scale equally unprecedented in history, and, indeed, presaged a period lasting up to the end of the century in which U.S. railroads routinely lapsed into criminality on a huge scale.

The original funding from the government consisted of loans for the construction of the railroad at the rate of $16,000 per mile on the easy sections, double that in the Great Basin, the huge mostly desert area of the western United States, and triple that through the Rockies. The system cleverly set up the companies as rivals, forcing them into a competition to build the most mileage in order to maximize the land grants they received. Lincoln's bill allowed land grants and loans to be made to companies that would build railroads between Council Bluffs in Iowa and San Francisco in California, but did not specify how much each railroad should build. The railroads were entitled to claim land extending 10 miles from the track, though only on one side, and just in case there were any Indians in these parcels of land, the act conveniently extinguished any rights or claims they had. The act provided for additional help to the companies in the form of favorable rates for transporting mail, government supplies, and, of course, troops, but first they had to complete the railroad.

Two rival companies had been created to build the railway: the Central Pacific, the brainchild of Theodore Judah, who had long lobbied for a transcontinental railroad and created the company in 1860, and the Union Pacific, set up by Dr. Thomas Durant, which would work its way west from Omaha in Nebraska, just over the Missouri River from Council Bluffs. It was a confusing nomenclature because the Central worked in the West and the Union worked in the center of the country. The scale of the project was demonstrated by the simple fact that the two companies were the biggest corporations in the country at the time.

The Central Pacific, starting in California, was first off the mark. Judah had done much of the groundwork, surveying a route and raising the funds. Judah, as one biographer suggested, "was never considered an entirely normal man,"[8] and, indeed, the adjective "fanatic" was often applied to him. But that was probably true of many of the early railway pioneers highlighted in these pages. To force through a railway against all the inevitable obstacles generally requires an element of obsession bordering on fanaticism. Judah had worked on various engineering projects in the East when he went west to be the engineer of the first railway in California, the 21-mile Sacramento-Folsom line. The railway was not a financial success, as the gold mines it was designed to serve soon became exhausted, but that did not stop Judah from exploring the possibility of a line further eastward through the mountains. He tried to get a bill for a Pacific railway through Congress in Washington, but failed and returned to California, where he successfully promoted a bill for a line from Sacramento to the state boundary with Nevada, 115 miles away. There was a huge amount of money to be made with the discovery of vast silver mines in Nevada, and, with promises of access to that wealth, Judah was able to sell the idea of the railway to a key group of local entrepreneurs.

Judah's breakthrough came out of a seminal meeting in Sacramento in November 1860, later much reported through rose-tinted hindsight, at which he sought to sell the idea of his railway to any local notables with a few hundred dollars to spare. Judah did not attempt to set out a grand vision of a transcontinental railway but instead presented the project in a way that would appeal to his small-town audience, an opportunity to make a quick buck transporting the ore from the silver mines of Nevada. He told the merchants what they wanted to know: "how to sell more of their goods, how to make their property more valuable, how to expand their businesses and stifle competition."[9] It was, in truth, the way that railways had been sold right from the beginning, but this time all the promises would be fulfilled.

Famously, a quartet in Judah's audience responded to his entreaties, a decision they would never regret. These were the Big Four of the Central Pacific—Leland Stanford, Charles Crocker, Mark Hopkins, and Collis P. Huntington—a group of lucky but determined men who happened to be in the right place at the right time and were savvy enough to see the sense in Judah's sales pitch. Along with a couple of others, who later fell by the wayside, the Big Four together signed up to pay for Judah to undertake a survey. These four unlikely small-town merchants would eventually each number among the richest men in the United States thanks to their investment in the Central Pacific. "Modest" is possibly an overly kind description of these future railway moguls. Stanford was a wholesale grocer who was soon to become governor of California and would later serve in the U.S. Senate. Crocker dealt in dry goods; Hopkins and Huntington were the proprietors of the hardware store underneath the hall where the famous meeting was held. Stanford was president of the Central Pacific, Huntington vice president, Hopkins treasurer, and Crocker the construction supervisor and president of a subsidiary bearing his name created specifically to build the railway. A historian of the Central Pacific later wrote, rather cruelly: "In later life four of [Judah's] listeners accepted easily the roles of men of vision, who had perceived a matchless opportunity and grasped it with courage. It was a role none of them deserved."[10]

Judah had already undertaken a partial survey of the line, but with money from his new backers he now set about finding a route through the Sierra Nevada, a rugged mountain range that reaches 7,000 feet above sea level fewer than 20 miles from the plains, hardly ideal territory for a railway. Judah, though, ascertained that building a line was feasible without too much expensive tunneling, and his route was largely the one eventually chosen by the railroad. His survey convinced the Big Four that the scheme was viable, and they set about creating the Central Pacific Railroad Company.

Although Judah had presented the railroad project as merely a local scheme tapping into the wealth of the silver mines, from the outset his vision was for a line that gave California access to the wealth of the East. He was the most vociferous supporter of Lincoln's bill, and when the act was signed into law in July 1862, Judah famously commented: "We have drawn the elephant. Now let us see if we can harness him up."[11] Unfortunately, Judah was never to find out, as Panama did for him as it had done for so many railway workers. He fell out with the

Big Four over their crooked attempts to try to cover up the potential difficulties of construction and headed back east after being paid off with $100,000, but he never reached New York. Forced to travel via Panama, as that was still the quickest way, he contracted yellow fever while crossing the isthmus and died at the age of just thirty-seven.

Despite its remote location, which meant that all supplies had to come from the east by ship, either through Panama, where everything had to be transhipped onto the railway, or the long way around via Cape Horn, the Central started construction before its rival in the east did. In January 1863, with much fanfare and lots of speeches, the first shovelful of earth on the Central Pacific was turned. While the Union Pacific had relatively easy terrain until it met the Rockies, the Central Pacific faced major natural barriers almost immediately as it needed to cross the Sierra Nevadas, the range that had been surveyed so thoroughly by Judah. Charles Crocker, a large man weighing 250 pounds who had remarkable bouts of feverish energy, broken only by occasional sleepy periods when he was overcome by the burden of his bulk, proved to be a remarkable project manager. His master stroke was to overcome the chronic shortage of local labor by employing thousands of Chinese. Overcoming widespread skepticism that the Asians, who weighed, on average, just 110 pounds, could undertake such heavy work, or withstand the excesses of heat and cold in the Sierra Nevada, Crocker first hired fifty as an experiment. They proved so adept that he quickly took on hundreds more until, at the peak of building in 1866, 10,000 Chinese were working on the railroad, representing 95 percent of the workforce. It was the genesis of the Chinese community in California, as many of the railway workers remained in America after the line was built, creating Chinatowns in numerous local cities.

Snow was the big obstacle. It began to fall in October and continued off and on for the next six months, creating drifts of up to 30 feet and sometimes requiring five locomotives fitted with snowplows to clear the tracks. At times all work except tunnel boring had to be stopped, and avalanches were a continual danger, at one time wiping out a whole encampment of buildings and men. The summit tunnel was finally completed at the end of 1867. Seven hundred guests were taken in a special train to a temporary terminus, where, after the formal celebrations, an enormous snowball fight broke out, since for many it was their first sight of the white stuff. It was not until the summer of 1868 that the tracks reached the eastern slopes of the mountains and the much easier terrain across the Nevada desert.

By that time, the Union Pacific was also in the western plains beyond the desert. A crazy race for mileage broke out, the epitome of the American obsession with competition. The Union Pacific's early progress had been held up by the lack of a bridge between Council Bluffs and Omaha; supplies had to be shipped in on the Missouri in every season except for winter, when, oddly, they could come direct from Chicago by rail on temporary tracks laid across the frozen river.

Both railroads set up a bizarre arrangement to undertake the work, creating subsidiary supply companies that obtained all the materials, built the lines, and then billed the railroad. This was based on the idea that the two Pacific companies would always survive any financial collapse, since effectively they were government backed, while the investors in the supply companies would obtain a good return on their capital. These arrangements conveniently disguised the way that funds were spent and allowed those building the railways to enrich themselves. Both suppliers—the Central Pacific's baldly named Contract and Finance Company and Union Pacific's Crédit Mobilier—were the subject of major scandals, and millions of dishonestly earned dollars were creamed off both the government and investors.

The first part of the Central Pacific railroad was built by Charles Crocker's eponymous company and provides an early example of the sort of financial shenanigans that became routine on both railways. On the first section of line, stretching 18 miles east from Sacramento, costs soon exceeded estimates and Crocker's cash ran out. No matter. The extra money, $150,000 on a contract originally estimated at $275,000, was made up by Central Pacific's coffers, a clear breach of normal accounting rules that amounted to "a polite form of embezzlement."[12] But since the other three were silent partners in Crocker's outfit, they benefited personally, too.

The Crocker contract only lasted until the railway reached the state line, at which stage the Contract and Finance Company took over. It made little difference, since that, too, was owned largely by the Four and Crocker was its president. How exactly the Big Four, "financial illusionists *par excellence*,"[13] kept the construction project going, by fair means or foul, is unclear even to historians of the railroad, but some-how, despite plunging close to insolvency several times, work continued uninterrupted. After the completion of the railroad, any chance of a successful investigation into the Central Pacific's subsidiary was made impossible by the "accidental" destruction of the company's books in the early 1870s. There was, in fact, plenty of honest money to be made, but as the historian John F. Stover put it, "none of the participants wished to be satisfied with a modest profit."[14]

The extent of the Crédit Mobilier scandal came into the open in De-cember 1867 when the company declared its first dividend, which amounted to close to 100 percent. As one historian said, "even in that era of rubbery business ethics, that was considered more than decent."[15] The four Sacramento businessmen became fabulously rich. By the time of the completion of the line, the Big Four were reckoned by a govern-ment commission to have pocketed $63 million, and to have obtained 9 million acres of land between them, and the owners of the Union Pacific were not far behind in garnering purloined wealth. Stanford's name is the one that lives on the most, as he later founded Stanford University in Palo Alto, one of the top universities in the United States (and, in-deed, the world), in memory of his son who died young.[16]

An early indication of how the scams worked on the Union Pacific was the initial contract let by Crédit Mobilier to build the first hundred miles of line. Dr. Thomas Durant, the Union Pacific's vice president, asked for an estimate from his chief engineer, Peter Dey, who assessed it at $50,000 per mile. Durant proceeded to let the contract for $60,000 per mile to a contractor who effectively represented Crédit Mobilier, but recommended a longer route for the sole purposes of taking in more land that would eventually accrue to the company. Dey, an honest engi-neer, promptly resigned. This proved to be a great loss; his replacement did not last long because of his patent incompetence. He made unwise choices, such as using softwood for the railroad ties, which promptly rotted, and was unable to organize the work, with the result that the railway meandered ahead at barely one mile a week, rather than the mile per day that had been the target.

The company was saved by the arrival of the former Union general Grenville Dodge, who brought in two brothers as principal contractors, John and Daniel Casement. Dodge was an experienced railway engineer who, before the war, had persuaded Lincoln that Council Bluffs would be the right starting point for a railroad heading to the Pacific. More amenable to the corrupt ways of Durant and his backers, but nevertheless an excellent engineer, he was the man who built the Union Pacific, and he is widely reckoned to be the United States' greatest railroad engineer.

Dodge organized the workforce along militaristic, even Fordian lines. First the surveyors were sent ahead. While the broad route of the line had been determined, its precise path was by no means set, unlike in the United Kingdom, where parliamentary legislation determined exactly where a railway should go. Cheapness was the watchword. The track today is largely a straight line, but when it was first laid out, there were curves around the merest obstacle to ensure that the minimum of work needed to be done. A few hundred yards of extra track was far cheaper than blasting the way through a hill.

The work was assigned to different teams. The surveyors left wooden markers to show where the railroad should go. Next came the graders, who laid out the path for the track, moving mountains of dirt and blowing up any rocks that the surveyors, despite their best efforts, could not find a way around. They built the embankments and the bridges, leaving everything ready for the tracklayers who followed them. The rails, which were very short, and the railroad ties were delivered by train as far as the line went and then loaded onto horse-drawn cars to be taken to the railhead. Two gangs would then work in unison, one on each side of the track, lifting the rails and setting them in place. Another team would follow, spiking the rails in: three hammer blows per spike, ten spikes to the rail, four hundred rails to the mile.

As the Union Pacific's tracks were pushed west, the whole show moved along with them. Temporary company towns, "hell[s] on wheels," as they were called by a visitor, were created, complete with saloons, brothels, expensive railroad-owned shops, and, of course, the accommodation for the workers, mostly tents or flimsy shacks. These towns were probably more dangerous to the workers than the sometimes perilous conditions on the tracks, as gun battles and saloon fights were frequent and no less violent than those portrayed in Westerns; the bloodiest incident, in Laramie, Wyoming, in the eastern foothills of the Rockies, claimed five lives and resulted in fifteen injuries. There was, too, the ever-present risk of Indian attacks, since, not surprisingly, the

Cheyenne and Sioux who lived off of buffalo in the plains were deeply hostile to the iron road, which threatened their lands and their very way of life.

The military did not have sufficient resources to protect all the work-sites, so Dodge ordered all the men to be armed and ready to fight off raiders. On numerous occasions, the Indians attacked work camps or caused train crashes by removing rails, and killed and scalped railway workers and the soldiers protecting them. But, as a British reporter, Henry Morton Stanley, put it, the Indians were never going to be al-lowed to keep "half the country as a buffalo pasture and hunting ground."[17] The revenge of the white man was fierce with calls to exter-minate the native Indians, which fortunately were not carried out, though many Indians were killed in reprisals. Of course, in a wider sense, the iron road allowed the settlement of millions of acres of land where previously the Indians had roamed, and therefore the railroad promoters and advocates can be held partly responsible, albeit inadver-tently, for their awful fate. The Central Pacific had less trouble with In-dians than the Union Pacific did, partly because the natives in the west were far less warlike than the tribes in the center of the country; but it was also because Huntington cleverly bought off the chiefs with free passes on the railway and appeased the rest by allowing them to ride on freight cars whenever they wanted.

At the peak, there were 10,000 men working on the Union Pacific, with up to 100 surveyors and 3,500 graders operating as far as 200 miles in advance, several thousand carpenters and tracklayers at the main site, and up to 1,000 men working for the main civil engineering contractor in Chicago to produce prefabricated sections for the bridges. The workforce was a mixture of immigrants from the East, dominated by the Irish, and veterans of the Civil War, all lured to the West by promises of high pay, which could be as much as $3 per day. Dodge later commented that, had it not been for the war experience of the men who built and managed the project, it would not have been possible to complete the railroad. Interestingly, the workforce on the railway also included freed slaves. Although the workers mostly lived in the open air, as tents were reserved for the foremen and bosses, the conditions were as good as could be expected, varying according to the quality of the water supply and the state of the company's finances, and the occasional strikes were mostly prompted by delays in the delivery of wages.

The wasteful competition between the Union Pacific and the Central Pacific came to a head in the plains when the railways effectively bypassed each other, unwilling to recognize their rivals' existence. For long distances the two lines ran parallel, and while the Central Pacific was working out its route through Utah, a section already long completed by the Union Pacific, the latter, in turn, sent out surveyors almost to the California border. The ultimate absurdity was in the Promontory Range, where the Central Pacific crossed a deep valley with a large embankment only for the Union Pacific to ford it with a precarious trestle bridge 50 yards away. What caused this wasteful duplication was the generous government funding of $32,000 for every completed mile, which was far more than the actual construction costs on the flat desert section of the Great Basin, especially as huge curves were created in order to avoid expensive inclines.

In fact, the ease with which the railway could be built was demonstrated by Crocker, who, by adding a few beefy Irish crews to his Chinese gangs, won a $10,000 wager from Union Pacific's Durant by laying 10 miles of track in one day. It was more of a public relations feat than a realistic engineering exercise, but the names of the eight men[18] who were handpicked to carry and spike the rails were nevertheless recorded in the history books. Actually, relations between the mainly Chinese men of the Central Pacific and the predominantly Irish workers of the Union Pacific became extremely hostile when the railways started running parallel to each other. At one point, the Irishmen were blasting through without warning their Chinese rivals nearby, and the latter retaliated by deliberately cascading debris onto them. The two groups of navvies eventually called a truce, realizing their game was going to cost lives, but it took a government agreement to sort out the meeting point following a parallel process of lobbying in Washington.

Huntington of the Central Pacific worked on the president, Andrew Johnson, while General Dodge of the Union Pacific, who was an Iowa congressman, sought the support of president elect Ulysses S. Grant, who, soon after he took office early in 1869, agreed that the railways should be joined at Promontory in Utah.[19] This was a temporary shack town where, on May 10, 1869, trains from each company met, and the last spike—gold, of course—was driven in by Stanford and Durant, the Union and the Central working together at last. No excuse was needed for the stupendous celebrations, where, as one writer put it, "oratory and whisky flowed in almost equal measure."[20] It was far more than a railway event: The rail wedding at Promontory is celebrated today in the annals of American history as the day on which the country was unified and the diverse states truly became the United States of America. The American travel historian Seymour Dunbar was not exaggerating when he wrote: "If those of the future times should seek for a day on which the country at last became a nation . . . it may be that they will not select the verdict of some political campaign or battle field but choose, instead, the hour when two engines—one from the East and the other from the West—met at Promontory."[21]

Though the railway was celebrated as linking the two coasts, few people—or even goods—undertook the whole journey; in reality, the purpose of the railway was to create a myriad of shorter connections, which had far greater commercial benefit. Moreover, while the line may have been complete, it was a ramshackle affair, with enormous curves and poor track that slowed down trains, making long journeys tortuous affairs. Although the transcontinental railroad was vitally important for the development of the United States, opening up new parts of the western United States, other railroad companies, such as the Chicago & North Western, the Rock Island, and the Burlington, received comparable land-grant aid.

The first transcontinental service was an excursion in May 1870 organized by the Boston Board of Trade in a train formed of carriages from George Pullman, the inventor of the luxury sleeping car[22] (see Chapter 9). The journey from Boston to San Francisco, which was reached by a ferry from Oakland on the other side of the bay, took eight days, and the passengers were entertained with a daily newspaper published on the train. The early journeys involved four changes of train and some poor connections; a continuous route and train service across America was not possible until 1872, when the bridge over the Missouri between Council Bluffs and Omaha was completed. Since through journeys involved travel on the lines of different railway companies, it took some time before branded trains, such as the "Overland Flyer" or the "Pacific Express," emerged. Amazingly, the first regular through sleeping-car service between the two coasts was not inaugurated until 1946; before that passengers had to change at St. Louis or Chicago.

By the end of the nineteenth century, no fewer than five transcontinental routes had been built in the United States linking the Midwest or the Mississippi to the Pacific. Three routes would also be completed in Canada by 1915. The next two lines were both opened in 1883: the Southern Pacific to Los Angeles and the Northern Pacific to Seattle, adding to the complicated nomenclature, although this time at least both were broadly where expected on the map. These lines had been creeping across the country (westward, in the case of the Northern Pacific, but eastward in the case of the Southern Pacific) for many years. It was a rush to exploit the West, stimulated by land grants and promoted by eager railroad companies whose officials saw nothing but gold in the hills and made dubious and mostly unfulfilled promises of high returns for their investors. The Southern Pacific was the brainchild of the Big Four and stretched eastward from Los Angeles, hugging the Mexican border. It reached El Paso on the Rio Grande and from there linked with a number of local lines to complete a route to New Orleans.

The Northern Pacific, which started in Minnesota, was an achievement arguably even greater than that of its predecessor, a massive enterprise through territory that General Sherman described "as bad as God ever made or any one could scare up this side of Africa."[23] It made steady progress in the early 1870s, but when it reached the North Dakota town of Bismarck, whose name obviously reflected the origin of most of its inhabitants in 1873, the company was forced into financial receivership, and the government, burned by the Crédit Mobilier scandal, was reluctant to intervene. Work was eventually restarted from the other direction by a former Civil War newspaper correspondent, Henry Villard, who had gained control of the Oregon Central Railroad. Raising $8 million from investors and using no fewer than 25,000 men, half of them Chinese, at the peak of construction, the line was completed in September 1883, with the now traditional gold spike being driven in at the suitably named Gold Creek in Montana.

By that time, the sole transcontinental line built without government support was threatening the Northern Pacific's viability. The Great Northern was largely the work of a remarkable one-eyed frontiersman,

James Hill, probably the greatest American railroader,[24] who for thirty years pursued his vision of a line that would open up the vast prairies of Montana[25] to settlers and allow the export of grain to the Far East. Lithe and short, his rather unappealing looks—he had a big nose, deep-set eyes, and a small mouth—had been made worse by the loss of an eye following an archery accident in his youth, but he was tenacious, and despite being prone to emotional outbursts his calculating mind always enabled him to outwit opponents.

Hill's railroad was built to a far higher standard than the others, with no tight curves and smaller gradients, and reached Seattle in 1893. Hill had the satisfaction of seeing all the other transcontinental lines face receivership in the depression of the 1890s while he continued to pay dividends to his investors, because he controlled a through-route all the way to the Pacific and thus was not at the mercy of other railroads. He cemented his position by assuming effective control of the Northern Pacific at the same time that he completed his own line.

The fifth line, built by the Atchison, Topeka & Santa Fe Railway, which eventually became known as the Santa Fe, initially kept quiet about its transcontinental ambitions, starting out slowly, heading westward from Topeka, Kansas, in 1868. The Santa Fe funded itself through Colorado by setting up real-estate offices to sell the land granted to it by the government. This strategy brought in immediate cash and, crucially, stimulated demand for transportation on the railroad. The company fought battles—literally—with its rivals from the Denver & Rio Grande, who were also trying to build a line through the narrow Royal Gorge, and later through Raton Pass, the only easy route into New Mexico, resulting in two years of low-level guerrilla warfare until the Santa Fe triumphed. The railroad, which reached Santa Fe in 1884 and then connected through to Los Angeles, became, after a period of bankruptcy in the 1890s, the most successful of the five transcontinentals, pioneering an excellent meal service provided by waitresses who were the precursors of flight attendants. Chosen for their youth, character, and attractiveness, they were known as "Harvey girls" after Frederick H. Harvey, who had developed the concept. It was, however, the heavy freight that would turn the railway, which linked Santa Fe with Chicago entirely on its own metals, into the most successful of the five.

The Liverpool & Manchester Railway, which opened in 1830, was the world's first railway to link two major towns with double track using only steam power.

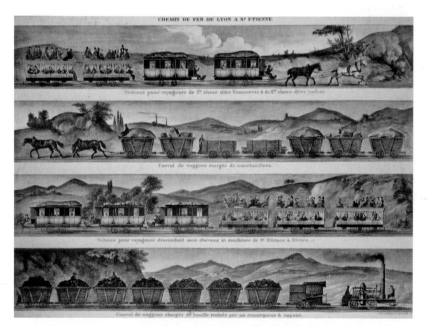

Early carriages on France's first railway between St. Etienne and Andrézieux, which used both steam and horses to haul passengers and freight. Note the double-decker carriages with the more affluent sitting on the lower deck.

One-horse race: In 1830, Peter Cooper staged a contest between his locomotive, *Tom Thumb*, and a horse to prove the machine's superiority for the Baltimore & Ohio Railway. Despite the fact that the horse won when the locomotive broke down, the railroad opted for steam rather than equine power.

Ireland's first railway: The Dublin and Kingstown Railway, built in 1834, gave passengers a wonderful view of Dublin Bay.

A drilling machine with a diamond bit powered by compressed air – invented by the French civil engineer Rudolph Leschot – was used to bore the Mont Cenis tunnel connecting France and Italy. Technical improvements such as this enabled tunnels to be dug out far faster than before.

The Golzschthal Viaduct built in Saxony, Germany, in the 1840s was typical of the many elegant structures built for the railways, enhancing the countryside through which they passed.

Cuba, which opened its first line in 1837, was the first country in Latin America to join the railway age and developed a very widespread network to serve its sugar plantations.

The first transcontinental line, the forty-seven-mile-long Panama Railway, was completed in 1855 after eight years. It was built at great cost to human life because of the high incidence of disease among workers but proved tremendously profitable for its promoters.

Hard labour: Australia's first railway opened in 1836 in Tasmania and used convicts to push the carriages along. Happily, as this picture shows, they were able to rest on the downhill sections by sitting on the wagons.

Commodore Perry sailed to Japan in the 1850s with the intention of opening up the country to trade with America, bringing with him a model of a train, which greatly impressed his hosts.

The meeting of the Union Pacific and Central Pacific railroads at Promontory, Utah in 1869 not only completed the transcontinental line but, by linking the two oceans, was a seminal moment in the history of the United States.

Workers brought in from China proved invaluable in building the Central Pacific, even though they frequently faced hostility from fellow laborers.

American locomotives – with their bulbous chimneys, to reduce the emission of sparks, and their cowcatchers, essential in a country where cattle roam unfenced – were easily distinguishable from their European counterparts. They also tended to be larger and more powerful, although they rarely travelled faster than European locomotives because of the poor condition of the track.

Mountain Creek Trestle Bridge, part of the Canadian Pacific Railway: Canada's first transcontinental railway was completed in 1885. Trestle bridges, which are more stable than they appear, were used extensively on railways throughout North America.

Sunday worship aboard the Union Pacific Railroad during the 1870s.

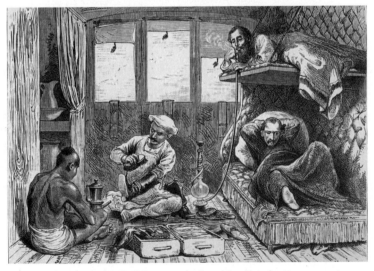

Traveling by rail in India was a comfortable affair for Europeans in first class, even for their dogs, but the locals fared far worse, prompting widespread complaints throughout the British period of rule.

The Gare de l'Est in Paris with its *beaux-arts* frontage was one of numerous large stations built across Europe in grand style by railway companies eager to demonstrate their pre-eminence.

In sharp contrast to today's high-speed services, the early Japanese railways were sedate affairs with very low speeds and infrequent trains. This woodblock shows the country's first railway – part of the Tokyo-Yokohama line – which opened in 1872.

Elephants were used at times as shunters in Indian railway yards.

Crossing the Ghats was the most difficult engineering challenge facing the early Indian railways in the 1850s. It took eight years to build the fifteen-mile pass through the mountains using the innovation of switchback sections which reduced the need for expensive tunnelling.

Above: The world's most magnificent railway station? The Victoria (now Chhatrapati Shivaji) terminus in Bombay took ten years to build and was completed in 1887 in a style that incorporates both western and eastern influences.

Right: Rail disasters, like this one in India in 1908 that killed twenty people, became increasingly frequent as speeds increased and more trains crowded on to the tracks. However, safety gradually improved as the lessons from such disasters were learned by railway authorities around the world.

THE RHODES COLOSSUS
STRIDING FROM CAPE TOWN TO CAIRO.

Left: The Cape to Cairo railway project was conceived by Cecil Rhodes as a way of consolidating British power over the length of the African continent from Egypt to South Africa. However, his overarching ambition was not always welcomed back home by the people or the government who refused to fund the project.

Above: The spectacular bridge over the Victoria Falls opened in 1905 as part of the never-completed Cape to Cairo railway.

Left: Launched in 1883, the *Orient Express* eventually had several routes linking Europe with the Turkish capital Constantinople. This poster outlines the schedule in 1895.

The Trans-Siberian Railway, the most ambitious railway project ever under-taken, took thirteen years to complete. It opened in 1904, just in time to transport soldiers to the front for the Russo-Japanese War.

The Trans-Siberian was built by a mixed workforce of local peasants, convicts, and even soldiers due to the shortage of labor over much of its route.

Engineers inspect the track of the toughest railway of them all: the Madeira–Mamoré railway in Brazil was laid through Amazon forest in the early years of the twentieth century after four previous failed attempts. Known as the 'Mad Mary' due to the high death rate among construction workers, it never prospered because the local rubber market collapsed soon after completion.

British capital was used to build railways across the world and, while many investors made considerable profits, others lost all their money, ending up with only lavishly produced certificates to decorate their walls.

The Peru Central Railway, one of the Andean railways whose construction was a testimony to the inventiveness of its engineers and workers, took more than fifty years to build. When it was completed in 1908 it was the highest railway in the world, reaching over 15,600 feet above sea level.

La Paz, Bolivia. The Andean railways reach places inaccessible by any other means.

Car or train? Many railways in South America used a bizarre assortment of vehicles adapted to run on rails because the roads for conventional cars and buses were so bad.

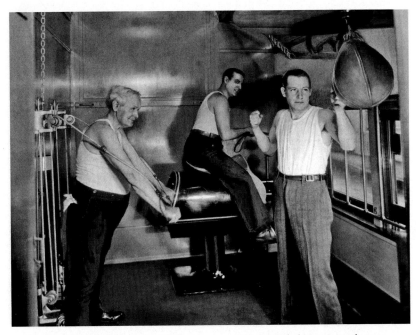

Between the wars American railway companies provided a variety of new facilities for their passengers, such as this gym on a Pullman train in 1927.

These lines were all built with degrees of corruption that ranged from deep to extreme. As before, the principal method of extracting illicit profits was to create a construction company that was supposedly independent but actually owned by the railroad developers; the mechanism provided a sure-fire way of extracting investors' and government money while superficially appearing to do everything legitimately. The business methods of these railroad promoters could have been straight out of the Enron handbook of ethics, but that is not to detract from their real achievement in stimulating the economic development of the western United States and ensuring the long-term cohesion of the nation. As a railroad official put it, "through its energetic railroad development, the country was producing real wealth as no country ever produced before."[26]

In Canada, too, widespread corruption and dubious politics inspired by national pride led to the construction of three transcontinental lines, filling the country with far more railway mileage than it could support economically. Their promoters and builders may have been honored with knighthoods, and in one case even elevation to the House of Lords, in keeping with British colonial tradition, but that did not mean their business practices were any more ethical than those in the United States or that the fortunes they made were any smaller. Canada had been a slow starter in the railway game, with just 66 miles of largely short lines connecting waterways completed by 1850, but railways were to be the way that the country distinguished itself from its southern neighbor. As Nicholas Faith explained, "the Canadians used railways not only to define their national boundaries, but also to repel the opportunities they offered the Americans to integrate the two countries' economies and thus, in Canadian eyes, subjugate them to the American yoke."[27]

In 1853, an international railway stretched from Portland, Maine, to Montreal in Quebec, and it later became part of the Grand Trunk that formed a mainline route through eastern Canada, which, by 1860, thanks to a railway boom in the 1850s, had 6,000 miles of line. The Grand Trunk had unified the eastern provinces and helped transform the disparate parts of the colony of British North America into the new nation of Canada, and the vision of extending its borders out to the Pacific would be the stimulus for the construction of a transcontinental line.

This idea had first been discussed by the British government, the colonial power, as far back as 1844, and the Grand Trunk was even offered the opportunity of building it, but, penurious and overstretched, the company, which already extended well into the United States, refused.

However, the construction of a transcontinental line went from being merely a visionary idea to being a necessity when, in 1871, it was made a precondition by British Columbia for joining the Dominion of Canada, rather than the United States, which had recently acquired Alaska from the Russians and was a keen suitor for the province, as a merger would extend its borders along the whole Pacific coast north of Mexico. Under pressure from Canadian politicians, the line was finally authorized by the British government in an act passed in 1872 that not only allowed its construction but specified that subsidies in the form of both money and land would be available. Such support was essential. The difficulties of building a line through the Canadian Rockies appeared even more insurmountable than those faced by the transcontinentals in the United States. It was not just the sheer length of the line, which had to stretch 2,700 miles from the developed regions of Ontario to the Pacific coast at Vancouver, that made building it an awesome undertaking, but also the size of the natural obstacles and the harshness of the weather.

At least the mistake of having two companies competing from each end of the line was not repeated, as the whole contract was given to the Canadian Pacific Railway, which was incorporated as a private enterprise despite the large subsidy it would receive. As with all the other grandiose projects, there were individuals whose drive and determination would see it through to completion. George Stephen, a Scot of modest origins who arrived in Canada at the age of twenty-one in 1850, had saved the St. Paul & Pacific Railroad, based in Minnesota, from bankruptcy in the 1870s. It was turned into a profitable line by extending it across the Canadian border to Winnipeg in Manitoba, which previously had only been served by steamboats. This success won him and his associates the contract to build the Canadian Pacific.

While the canny and devious Stephen (later Lord Mount Stephen) was the wheeler-dealer who ensured that the funding was in place for the line through a variety of negotiations, not all of which were on the right side of the law, William Cornelius Van Horne (later Sir) was the manager who saw through the construction of the line. He was one of those tireless characters, like George Stephenson and several other early railway promoters, who had the ability to work all hours of the day to finish a project. Van Horne was a "hands on" manager always "pushing, prodding and inspiring his crews to ever more Herculean efforts."[28] Corpulent himself, he ensured they were well fed but expected them to work as tirelessly as he did. Once he was advised not to walk across a rickety, unfinished trestle bridge, little more than a couple of planks strung across a 160-foot ravine where a few days previously several workers had plunged to their deaths, but he simply strolled across while his engineers followed crawling gingerly on their hands and knees. Van Horne played as hard as he worked, arm-wrestling with his men or taking part in endless hard-drinking poker games well into the early hours before heading back to the site without going to bed.

The Canadian Pacific went to the north of the Great Lakes, and though the initial sections westward were relatively flat, the granite shelf of the Canadian Shield required considerable blasting through hard rock. After the prairies, the line had to cross both the Rockies and the high Selkirk Mountains. The sheer difficulty of finding a route through these two ranges was illustrated by the fate of thirty-eight surveyors who died seeking a route through Rogers Pass, just one of three high passes that had to be crossed. Not surprisingly, given the difficulties of construction, corners were cut. Most dramatically, the 5,300-foot Kicking Horse Pass, with an awesome 1 in 22 gradient, required the service of four locomotives to be climbed and had a terrifying descent. Safety on this stretch depended on the rather limited braking power of the engines, and trap sidings had to be provided to capture runaway trains. Within a few years a series of tunnels was built to provide a more manageable incline of 1 in 45.

More seriously for the 15,000-strong workforce, half of whom were Chinese, the "build quick, build cheap" philosophy created deadly risks. At least eight hundred men were killed in the construction, many perishing for want of basic safety measures, such as the failure of the railway company to provide fused dynamite, which had recently become available for blasting but was more expensive than the unstable and dangerous nitroglycerine that the company insisted on using. Even with such skimping, money was a constant issue. In the difficult parts of the mountains, costs rose to a staggering $500,000 per mile; perhaps inevitably, the project ran out of money. Part of the problem, too, was that the land that was supposed to fund much of the construction proved more difficult to sell than expected. At the last gasp, when construction was almost complete, Van Horne sent a desperate telegram to Stephen warning that there was no means of paying wages. The company was bust, but the railway had already proved its worth to the federal government in a way that, as we have seen, several other countries, such as India and Prussia, had already discovered: the rapid deployment of troops to quell a rebellion.

In 1885, there was a revolt in Manitoba instigated by Louis Riel, the leader of the Métis people, a mixed-race aboriginal group. The rebellion was put down within days thanks to the arrival of troops by the new railway, a contrast from Riel's previous insurrection fifteen years before when it had taken three months for reinforcements to arrive, allowing the rebels to extract concessions from the government. The Canadian government, therefore, immediately passed a bill authorizing extra payments to the line, which was finally completed in November 1885. Within eighteen months, through-trains, named rather unimaginatively the *Atlantic* or *Pacific Express*, depending on the direction of travel, were covering the nearly 3,000-mile journey between Montreal and Vancouver. Van Horne refused to allow a golden or silver spike to be used to make the final connection at the ceremony on the remote Eagle Pass in British Columbia and confined his speech to just fifteen words: "All I can say is that the work has been well done in every way."[29]

Another thirty years would pass before the opening of the next Canadian transcontinental, the Canadian Northern, which, as its name suggests, took a northerly route through the mountains. If the Canadian Pacific brought the country together, the Canadian Northern populated it in the same way that James Hill had sought to use his Northern Pacific to help attract settlers to Montana. The Canadian Northern, like the later transcontinentals south of the border, crept westward incrementally, both by acquisition and construction. Again, it took the foresight and ambition of a couple of visionary men to realize such an ambitious project, William Mackenzie and Donald Mann, a pair of Canadians who in the 1880s formed a partnership in the railway contracting business. Mirroring the skills of Stephen and Van

Horne, Mackenzie was the money man, while Mann was the construction expert. Starting with the construction of a 125-mile line in the prairies between Gladstone and Lake Winnepegosis, they gradually built up a network of similar branch lines with the help of grants from the provincial governments. Grain was the economic raison d'être for the company, and they pushed the railway west into central Saskatchewan, which by the turn of the century had begun to attract settlers. Faster-growing strains of wheat that were resistant to frost and able to flourish so far north had been developed, making the area more viable for agriculture. As a history of the Canadian railways put it, "trains of the Canadian Northern brought in new colonists and supplies, and carried away the bountiful harvests."[30] By the early years of the twentieth century, the Canadian Northern had become the country's third-largest railway, after the Canadian Pacific and the Grand Trunk.

The ambitions of Mackenzie and Mann, however, were far greater: They wanted to see their iron road stretch out to the west coast, hoping to exploit the potentially massive profits from grain in the prairies to turn the transcontinental line into a paying proposition. Moreover, although land grants had been phased out, massive government subsidies, amounting eventually to 25 million acres of land (worth $50 million), were paid to support the construction of the line. The troubled Canadian Northern, however, never made money. Although it used the easiest route westward, through the Yellowhead Pass, which ensured far easier gradients than the Canadian Pacific's route, there was no shortage of engineering difficulties and obstacles, and the company had to be bailed out by the Canadian government in 1913 just before the line was completed. The railway opened in January 1915, but with the outbreak of the war, the flow of immigrants had dried up and the western Canada boom collapsed. And so did the railway, which within three years of its completion was taken over by the Canadian government; but that did not prevent both Mann and Mackenzie, the last great Canadian rail pioneers, from being awarded knighthoods.

Rivalry between the Canadian Northern and the Grand Trunk led to the construction of the third transcontinental Canadian railway, an extension of the Grand Trunk, owned by British interests and controlled from London. Its eastern network, concentrated in Quebec and Ontario, had grown both through expansion and acquisition to become Canada's second-biggest railway and by the 1890s was regretting its decision not to expand westward. In 1902, its ambitious manager, Charles Melville Hays, announced plans to extend its network to the Pacific coast. The line was built in two sections, east and west from Winnipeg in Manitoba, stretching from Moncton, New Brunswick, on the East Coast to a new town on the Pacific, Prince Rupert, 500 miles north of Vancouver. The financial arrangements were complex, with the government taking direct responsibility for the eastern section, called the National Transcontinental, and underwriting the western section with a bond issue. Neither section was particularly easy to build. The route went far to the north of existing settlements through a wilderness of forests, lakes, rivers, and swamps in order to open up the area to agriculture. Interestingly, little thought was given to the real potential wealth of the area, the vast resources of timber and minerals that later would be the mainstay of the line.

The surveyors on the eastern section faced a daunting task when they started their work in 1904. At the height, there were fifty teams of surveyors seeking out the best routes for the 1,800-mile section between Winnipeg and Moncton. The surveyors spread out far and wide in the wilderness, using ships' rockets to guide lost members back to their camps; not surprisingly, given these difficulties and the impossibility of surveying in winter, it took four years to establish the route. Work started on the easy sections in the Quebec prairies in 1906 and took seven years to complete, as progress was greatly hampered by the huge areas of muskeg—bogland—which at times seemed to swallow infinite amounts of spoil before being stabilized. As with all these projects, spending escalated out of control. By the time the eastern section opened in 1913, the original cost had soared from $60 million to $160 million. There was, however, a gap across the St. Lawrence River at Quebec following two disastrous attempts to cross it. In August 1907, the bridge—which was to be the largest cantilever structure in the world—suddenly collapsed, killing seventy-five workers. Then, when its replacement was being hoisted into place nine years later, the center span fell into the river, killing another ten people. It was not until 1917 that the route between Winnipeg and Moncton was finally complete.

The western section, the Grand Trunk Pacific, cost a similar amount, $140 million, by the time of its completion in April 1914. The folly of building a third transcontinental railway was highlighted by the fact that the Grand Trunk ran parallel to the first one, the Canadian Pacific, for 135 miles between Winnipeg and Brandon in Manitoba; for twice that distance, between Edmonton, Alberta, and the Yellowhead Pass, the tracks invaded the territory of another rival, the Canadian Northern. Through the pass itself the railways ran side by side.

The driving force behind the scheme, Hays, believed that only a first-class railway built to a far higher standard than its rivals would be viable. Instead of accepting the sharp gradients and long windy curves that characterized the other lines, he specified gentle gradients—the maximum allowed was a very demanding 1 in 166—and fewer curves, which, of course, meant more work in creating embankments and tunnels. He argued that the lower costs of operation would more than make up for the extra expense of construction. But, in fact, the line never made money. Although the Grand Trunk Pacific did not benefit from land grants from the government, it exploited development opportunities by creating new towns every 10 or 15 miles along the route, all laid out in an identical way to a standard street pattern and dominated, unavoidably, by a huge grain elevator. The western terminus was Prince Rupert, a new town 500 miles up the coast from Vancouver, created at the instigation of Hays, for which he had grand plans. Its name, which honors an obscure seventeenth-century Bavarian soldier and inventor, had been chosen by a schoolteacher, the winner of a competition with $250 as the prize, even though the rules said there should have been no more than ten letters. Hays wanted to create a port for passenger trips and develop a tourist industry, but his death as a passenger on the *Titanic* in 1912 put an end to these ambitious plans and the town of Prince Rupert boasts a population of barely 13,000 even today. Hays's death, too, meant the railway lost its main defender, and the company never recovered from the huge debts incurred in its construction. Within three years of its opening, 90 miles of steel rails had been ripped up and shipped to Europe for military use, forcing its trains to run on the parallel tracks of the Canadian Northern. And a couple of years later, in 1919, when the railway was declared bankrupt, it passed into the hands of the minister of railways.

The fact that three transcontinental lines were built in Canada, given the sparsely scattered population, was testimony to the overarching ambition of Canadian politicians and entrepreneurs. It has been said that "the illusions of grandeur, which made the Dominion, with a population of only ten million, build railways sufficient for fifty million people, seemed to have become an integral part of the Canadian soul."[31] The railway promoters and their investors had hoped for a Canada of 100 million people to ensure the profitability of the railways. Even today, the country has a mere third of that number, but at least the two surviving transcontinental freight railways of Canada—the Canadian Northern and Grand Trunk Pacific railways having both been incorporated into the Canadian National Railways, and privatized in 1995—are highly profitable, thanks to the huge boom in imports from the Far East.

THE FINAL AMERICAN crossing was the transandine railway in South America through Argentina and Chile. South America was, with a few exceptions, a late starter in the development of railways. Not only was the topography unfavorable, but most of the countries were poorly developed economically, had small populations, and were ill-governed. As we see later (Chapter 8), that delay did not prevent the most spectacular railways in the world from being built. While the Himalayas are higher than the Andes, the railways in the latter reach much further up the mountains and have steeper gradients and sharper curves than the railways of the Rockies. As the author of a history of the transandine railways argues, "from an operational point of view, no railroads could pose such problems as those of the Andes. None climb so high. . . . There is good reason to claim that the railways of the Andes are the most interesting in the world."[32] Although the transandine was by no means the most impressive of the Andean railways, its construction was nevertheless a great achievement.

The idea of a transandine railway had been suggested as early as the 1850s by a South American railway pioneer, William Wheelwright, an American who had built Chile's first railway, a 50-mile line between the copper and silver mines of Copiapó and the harbor at Caldera that opened in the last few days of 1851. The line, built by American contractors, was immediately profitable and soon spawned branches to various other mines. Wheelwright next envisaged a line between Santiago, the capital, and Valparaiso on the Chilean coast, barely 50 miles as the crow flies, but nearly twice that length along the difficult roads winding up the cliffs from the sea. He managed to obtain funding to start the work, but after just 4 miles had been completed the money ran out, partly because of the lack of suitable boring equipment to break through the rock.

The scheme was rescued by Henry Meiggs, another American, who was the most remarkable of the early Latin American railway pioneers because he had the knack of ensuring that schemes were completed on time and without massive extra costs. Like many other talented and adventurous men who built the first railways, he had a rather checkered past, having left San Francisco under a cloud of suspicion of fraud and shady dealing. However, unlike many of the others, he had the respect of the men who built the line, as he believed in feeding and paying them well. He took over the ailing Valparaiso-Santiago scheme soon after his arrival in the country in 1855. He contracted to build the line in three years for a fixed sum but managed to complete it in two, earning himself a handsome profit. However, the idea of extending the line over the Andes to the east of Chile and into Argentina proved too ambitious, and the scheme was abandoned until the 1880s, when it was revived by two other visionary railway promoters, a pair of Chilean brothers of British descent, Mateo and Juan Clark, who had built the first telegraph system across the Andes.

The story of the construction of the transandine is similar, in many ways, to those of other continental projects across the Americas. There were the usual financial problems, cost overruns, technical difficulties, and shortages of manpower, but with the extra complexity of crossing a frontier running between two countries governed by unstable regimes. The 155-mile route mostly followed an old mule crossing—the same one used by Charles Darwin in 1835—and took thirty-six years to complete after the concession had been granted to the Clark brothers in 1874. Work only started thirteen years later, at Mendoza, the terminus of the Argentine section, and in 1889 on the Chilean side, because the Clarks faced the almost inevitable financial difficulties, and the London bankers, unsurprisingly, took much convincing that their investment would be sound. The initial plan was for a standard-gauge railway with

gentle inclines that would be able to carry heavy freight, but the specifi-
cation set out by the Chilean and Argentine governments was for a far
more modest line, using meter gauge (3 feet, 3.375 inches)—which
meant breaks of gauge and consequently changes of train on both sides
of the Andes. Soon after construction started, however, revolutions got
underway in both Chile and Argentina, halting work in 1893 and resulting in a ten-year hiatus, which all but bankrupted the Clarks. Eventually
the project was funded from London through a new organization, the
Transandine Construction Company, and the scheme was completed in
stages by 1910.

The biggest obstacle was the summit (Cumbre) tunnel, almost 2 miles
long and at a height of 10,450 feet. It was largely cut by hand and took
twenty-one years to finish. Progress was painfully slow on the Argentine
side, with barely 1,000 feet completed after three years. The Chileans,
more experienced miners, were better, managing twice that rate, but it
was not until the introduction of machines that progress improved.
Work was impossible for three or four months a year when the site was
cut off by snow, but it was clearly carried out to a very high standard;
the eventual meeting between the two teams was virtually perfect, with
the two bores a mere 3 inches out of alignment.

The main difficulty on the Argentine side, where the gradients were
less fierce, was fording the dangerous Mendoza River, a raging torrent
prone to changing course and rising suddenly after a downpour. The
railway crossed the river in a narrow valley ten times in the space of 20
miles, and in one case an ingenious method was used to lay a long
bridge girder: An embankment was built up during the dry season on
which the girder was laid, and then was allowed to be eroded away by
the river when the rains came, leaving the bridge standing on its own.
Another clever idea was to ensure that the track in the mountains was
left exposed to the elements, with few cuttings, to allow snow to be
blown quickly away by the winds that invariably followed the bliz-
zards, though in a few other places snow sheds were built to guard
against the frequent heavy avalanches. It was a slow, winding railway
with no fewer than 22 miles of rack sections (where the locomotives, fitted with a special cogwheel, are helped up the hill through the use of a
toothed rack between the rails) as a result of the sharp gradients. Most

of the locomotives were imported from Britain, though one Peruvian and several German engines were also used on the line.

The line to Mendoza from Buenos Aires had been completed during the prewar railway boom in Argentina in 1907, and consequently a through journey, taking thirty-six hours from Valparaiso to the Argentine capital, was possible when the transandine opened in April 1910. Previously the 3,500-mile journey by sea around Cape Horn had taken eleven days. But even this huge time advantage did not guarantee the line's economic success. The railway always struggled, owing to high operating costs and the small capacity on the narrow-gauge line, which only allowed for relatively light locomotives. Hopes of large freight-carrying never materialized. The economics of the railway were further damaged by the hugely complex operating arrangements, with the two sections run separately by the individual countries, except for during a brief period between the wars.

The Chilean section, electrified in 1927 using Swiss technology, was nationalized in 1934, the year in which a massive surge of water in the Mendoza River washed away a section of the Argentine line, resulting in the closure of the line for a decade. It reopened and flourished for a while but eventually closed in 1984, though much of the track is still intact. There are plans by the two governments to reopen the line in the early 2010s. It was by no means the most spectacular of the Andean railways, rising to a height of 10,000 feet, compared with 15,000 feet at various points in Peru and Bolivia, where the tunnels and bridges represent much greater feats of engineering, as we will see in Chapter 8.

SEVEN

... and Other Continents

THE OTHER GREAT TRANSCONTINENTAL SCHEMES WERE STARTED
later than those in America but were equally epic projects that, in
their different ways, broke new ground, figuratively and literally.
As with the lines across America, the motives were mostly political, with
little real economic rationale despite the optimism of their promoters.
Indeed, of all the cross-continental railway schemes, the Trans-Siberian
Railway was the most political, a creation of a weak government intent
on holding together a vast but poor nation with little regard for the economics of the project. It was, too, the most remarkable engineering feat,
a railway that was twice as long as its American counterpart, stretching
nearly 6,200 miles between St. Petersburg on the Baltic and Vladivostok
on the Pacific, and running through some of the most inhospitable land
in the world—although, thankfully for its engineers, much of it, in western and central Siberia, was relatively flat.

Just as with the American transcontinental railway, there had been
dreams of improved transport to Russia's furthest lands for several
decades. However, Russia in the early age of the railways could not
have been more different from the country that would become its rival
in the Cold War for much of the second half of the twentieth century.
Russia's huge but largely inaccessible and infertile territory ensured
that the requirements of the military determined government policy
through the need to impose control over distant lands. Consequently,
the Trans-Siberian was a military and political project, rather than a

commercial venture. It had none of the romance and adventure associated with the American transcontinentals; rather, it was "a shabby bureaucratic affair, and its cost, for a poor country, was staggering," in the
words of a historian of the line.[1]

Railway development in Russia had not followed the European pattern because of its weak economy and the stringencies of the police state
imposed by the tsar's regime, which feared that the railways would be a
democratizing force. The line between St. Petersburg and Moscow,
completed in 1851, was one of the world's great early railways. It was
built by an engineer, Pavel Petrovich Melnikov, who was easily the
equal of a Brunel or Robert Stephenson. He produced a railway that
was almost dead straight, with gentle gradients, but despite its success it
did not lead to a network of secondary lines and branches as the early
railways of western Europe had. Partly, the reason was cost, allied to
the lack of capital in pre-industrial Russia, but crucially the railway
could not be profitable because Tsar Nicholas demanded that all passengers be subjected to a police check before traveling and required to
carry a passport. This stymied the development of a strong passenger
base by limiting travel to the rich and the politically sound.

The humiliating defeat in 1856 of Russia by Great Britain and France
in the Crimean War showed that Russia had fallen well behind its European rivals, and this acted as a spur for modernizing the state. Consequently, under Alexander II, a forward-looking tsar, there was a modest
railway boom in the 1860s, but the constraints under which the Russian railways had to operate meant that even by the 1880s there was
only a sketchy network of main lines. For example, there were main
lines connecting St. Petersburg with Warsaw, then part of the Russian
empire, and Kiev and Odessa, but nothing like the complex networks
being built elsewhere. Even in European Russia, west of the Urals, the
density of railways[2] before the start of the construction of the Trans-
Siberian in 1891 was less than one-twentieth of the United Kingdom's
system. The lack of sustained investment and the poor maintenance of
the track affected performance, as even express trains averaged, at best,
30 miles per hour. The journey time for the nearly 400-mile trip between Moscow and St. Petersburg, which was originally thirty hours,
was still a ponderous fifteen hours in 1880, at an average of just 26
miles per hour, far slower than contemporary trains in Europe (see

Chapter 9). As in Europe, the state's role in the railways increased as the private companies that had been encouraged to build them fell into financial difficulties.

The Trans-Siberian was much more than a mere extension of the existing railway network. Since the sixteenth century Russia had been growing in all directions from its heartland, the Duchy of Muscovy. After its victory over Napoleon in 1812, Russia had continued its expansion into the Caucasus Mountains and across the Urals into Siberia, but it had never consolidated its hold over these distant lands. Alexander III, although more conservative and cautious than his father (who was assassinated in 1881), saw the railways as a powerful unifying force that would allow him to impose the stamp of Russian authority in these regions, and he pushed strongly for the construction of the Trans-Siberian. Siberia was a grim land, with a small population largely concentrated on a few river arteries and roads, and those who lived there were mostly in the region because they were employed by the state either to maintain the roads or guard the territory. There was, too, an assortment of criminal exiles—being sent to Siberia was a long-established tradition—many of whom had escaped to scratch an impoverished living off the land, which frequently made life difficult for the few honest citizens.

Diplomats and bureaucrats dispatched to serve in the Pacific coast town of Vladivostok, the largest settlement in the region, though still little more than a collection of huts and only wrested from the Chinese in 1860, complained of the expensive rotting meat and vegetables and the lack of staples such as fresh milk and butter. There was, though, "an abundance of drinking houses, taverns and houses of pleasure,"[3] which added to the feeling that civilization's toehold in Russia's far east was precarious as a result of the harsh climate, poor land, and absence of transport links. This was not the burgeoning Wild West of the United States full of entrepreneurs and adventurers out to make a fast buck. There was wealth in the form of minerals, which gave the railway at least a small measure of economic purpose, but the true value of this vast hidden treasure only became apparent when a thorough geological survey was undertaken after the Trans-Siberian had been completed.

The impetus to build the railway was rooted in the history of the Russian occupation of Siberia. Russia had established a base on the Pacific as early as the seventeenth century, but it was only in the mid–nineteenth century that its territories there began to be threatened by Western interests. The development of efficient steamships in the 1840s, and the completion of the Suez Canal in 1869, made it easier for the major Western nations to reach the Pacific side of Russian territory than for Russia itself. The building of the American transcontinental railway and the beginnings of the Panama Canal project raised fears among the Russian elite that the country's territorial integrity would be threatened by the growing commercial opportunities afforded by improved transport routes.

There had been dreams of a Siberian railway as early as the 1850s, but the first serious discussion emerged soon after Alexander III became tsar. The growing threat of China, with which Russia shared thousands of miles of border, was the catalyst that focused attention on the need to strengthen the nation's eastern provinces. Russia and China had pretty much ignored each other for a couple of centuries, but a dispute over Turkestan in the 1860s led to a series of diplomatic disputes that threatened to break out into full-scale war. The Chinese began to colonize Manchuria, a largely empty buffer area between the two great states, and were even contemplating a Manchurian railway with European backing, giving further ammunition to the growing lobby for the Siberian railway.

It was not only China that was seen as a potential threat. There was also the growing military presence of Britain, which was in competition for territory in the Far East with Russia for much of the second half of the nineteenth century, and the relations between the two countries had been fractious even before the Crimean War. In 1885, Britain occupied Port Hamilton off the coast of Korea, from which Vladivostok could be attacked, and the Russians feared that the completion of the Canadian Pacific Railroad, which cut the journey between England and Japan by fifteen days, to five and a half weeks, would enable the British to transport troops there faster than they could. Building railways with the aim of securing distant territory had a precedent. In June 1885, the tsar ordered the extension of the Transcaspian railway along the border with Afghanistan, both to make it easier to subjugate the rebellious local tribes ('twas ever thus!) and to put pressure on British interests in India[4] through Afghanistan.

By the mid-1880s, therefore, pressure was growing in government circles to build a Siberian railway. There were domestic concerns, too, arising from conflicting views on the extent to which Siberia should be allowed to develop independently of Russia. The Siberian romantics who saw the region as separate and independent from European Russia, with its own strong identity and traditions, thought that a Siberian railway would be used by absentee mining firms, who would be shipping out the region's natural wealth in exchange for expensive imported manufactured goods that would flood the fragile local economy. In Moscow, these views were fiercely opposed by the dominant conservative and imperialist elite, who argued that any such talk of regionalism and separateness was a threat to the integrity of the Russian state. Siberia, they argued, was not a colony of Russia but part of Russia itself, and the railway was a way of ensuring that it would remain so forever. Alexander III, concerned about the growth of separatist thinking, responded by further restricting Siberia's autonomy and strongly supporting the construction of the railway.

As consideration of the Trans-Siberian intensified, the Russian railway network proceeded steadily eastward, reaching Orenburg at the southwestern border of Siberia in 1877, and in 1880 an immense bridge, modestly called *Imperator Alexander II*, spanned the Volga near Syrzan, bringing central Russia even closer to Orenburg and the Siberian steppe. In 1884, work began on the line from Yekaterinburg on the east of the Urals to Tyumin, 1,330 miles east of Moscow.

Starting such a massive project as the Trans-Siberian in a relatively backward country such as Russia in the 1880s was no easy task. It was built eastward from Chelyabinsk and westward from Vladivostok, and was by far the biggest railway project ever, at its peak employing more than 89,000 men. The eventual cost of the initial alignment would be at least 855 million rubles (say, around £85 million, the equivalent of $425 million using contemporary exchange rates[5]), a staggering sum for an impoverished government of a largely agricultural nation with a small tax base. First there were major administrative hurdles, with the finance and transport ministries both trying to assert control over the project, which led to major internal battles. Then there was the need to obtain the funding, which led to the sensitive issue of whether this vital part of Russian infrastructure should be financed by foreign interests, and thirdly, there were concerns about the choice of route through the steppe. It took almost a decade of wrangling over these issues before the ceremonial first stone for the railway could be laid.

Just as with the American coast-to-coast railroad, it would take an inspirational and visionary figure to push the project through. In the case of the Trans-Siberian, it was Sergei Witte, who, despite having a noble mother, had to work his way up from the lower echelons of government administration to become, first, transportation minister, and then minister of finance. Witte's father and grandfather had been colonial administrators in the Caucasuses, a land with the same type of frontier spirit as the American West. Given this upbringing, it is no wonder that Witte had an interest in colonizing Siberia. Witte, whose cleverness was matched by his ambition, started his career as an administrator with the Odessa State Railway but then enjoyed a meteoric rise, once jumping seven grades in a single promotion in the railway hierarchy, thanks to his adept use of contacts, including the press, and his canny networking, as well as a ruthlessness that several times saw him turn against former allies in order to further his own interests. A Russian historian described him as a man of "great ideas" who was "unscrupulous in taking advantage of any means available to make things happen his way. He was no stranger to playing a game concocted of brinkmanship, bribes, rumours and allegations published in the press."[6] A staunch monarchist with an almost filial devotion to the tsar, and a fierce nationalist protective of mother Russia, Witte saw the Siberian railway as a way of showing the rest of the world that the country was the equal of the major European powers, able to undertake the globe's biggest engineering project. In Witte's mind there were endless hopes for the railway: It would bring about an end to the isolation of the country's east, it would make Russia a key player in trade between Asia and Europe, and it would open up new horizons for world trade, with Moscow at its center. As a biographer of the railway sardonically suggested, this description of the Trans-Siberian Railway by Witte was "a modern equivalent of the medieval religious doctrine that proclaimed Moscow as 'the Third Rome.'"[7]

Witte first pushed the project through the extremely bureaucratic machine of Russian government and then ensured it had virtually limitless supplies of money from the exchequer. Witte's cleverest manipulative act was to have the future Tsar Nicholas II, Alexander III's heir, appointed in 1891 as chairman of the Committee of the Siberian Railroad, the government body overseeing the scheme; it was a brilliant move that "all but guaranteed the completion of the Siberian Railroad."[8] Nicholas had traveled to the far east earlier that year and had laid the foundation stone of the railway at Vladivostok, and the journey had put him strongly in the camp of the Orientalists, those Russians who saw a kinship between Russia and Asia that was distinct from Europe. Indeed, Nicholas's commitment to the scheme was such that he chose to retain the chairmanship of the committee following his accession as tsar on the death of his father three years later.

The railway was more than just an iron road through the Russian steppe. In order to provide materials for the railway and to create through transportation routes, the rivers in the region were improved to make them navigable, and the cost of the railway therefore included a vast array of associated works. The railway in effect connected 25,000 miles of navigable rivers to form a transport network for the summer months. The centralized nature of the Russian state meant that the route was determined with little regard to natural barriers or existing settlements, many of which were bypassed in an effort to keep the length of the railway to a minimum. Surveying was cursory, with the teams sent out by the Ministry of Transportation exploring a narrow band stretching just two kilometers on either side of a line that had been drawn on the map arbitrarily by the civil servants in St. Petersburg. Not surprisingly, as a result, the precise line of the route was frequently changed quite dramatically by the construction gangs that followed.

The awesome scale of the task of building several thousand miles of railway in one of the remotest and coldest environments on earth cannot be overestimated. When work finally started in 1891, it progressed relatively smoothly in western Siberia, but conditions worsened on the stretch between the Ob River and Lake Baikal in central Siberia. Beyond Krasnoyarsk, the foothills of the Saian Mountains are interspersed with bogs and the hills are high, with precipitous forested slopes. The surface remains frozen until mid-July, two months longer than in western Siberia, and when it melts the land becomes a swamp, with the result that the navvies frequently had to labor in 2 feet of water. Countless rudimentary wooden crossings were built to ford torrents and streams, and many substantial structures were needed, such as a half-mile-long bridge across the Enisei. Lake Baikal itself, a deep basin surrounded by steep, rugged mountains, provided the biggest engineering challenge.

After the lake, there was the equally unforgiving area of Transbaikalia where the Trans-Siberian became a mountain railway, its tracks cut perilously into ledges on the sides of the cliffs. Earthworks in Transbaikalia were heavier than elsewhere on the railway because of the permafrost, which could only be broken up with dynamite. Flooding was an ever-present danger here; in a particularly serious incident, a 200-mile stretch of completed railway, including fifteen bridges, was swept away.

Progress from the eastern side proved equally difficult. The easternmost section, named after the Ussuri River, was initially built too near the torrent and had to be rebuilt because engineers had underestimated the height of the water level when the snow melted. Moreover, heavy rains reduced the already short working season by two months annually, and there was the added threat from bandits. The railway initially went south over the Chinese border through Harbin in Manchuria; this was an easier alignment than the Russian route via the Amur River that was eventually completed in 1916. Most trains, however, continued on the original route until the Japanese, who invaded Manchuria in 1931, changed the gauge to standard.

The Trans-Siberian line opened in stages, with trains initially using a temporary line across the frozen Lake Baikal in winter and British-built ferries in summer. Building the railway around the lake involved the construction of two hundred bridges and thirty-three tunnels. When it was finally completed in 1904, there was a through route all the way from the Baltic to the Pacific.

Apart from the harsh conditions, the other difficulty was finding sufficient labor to build the railway. The Committee of the Siberian Railroad had calculated that 30,000 navvies would be needed to prepare the earthworks and another 50,000 men to construct the line during the first three years. In the west the demand was mostly satisfied by the local peasantry, who supplied 80 percent of the workforce, but once construction reached the sparsely populated forests of central Siberia, these laborers withdrew, reluctant to move too far from their homes. The solution was a very Russian one and proved highly effective: the use of prisoners. They had already been used on railway construction in the 1860s, but not on such a large scale.

In the spring of 1891, 600 convicts who were being sent by ship from Odessa to Vladivostok were rerouted to work on the Ussuri railway. A handful of the most vicious criminals escaped, causing a crime wave in Vladivostok, but the majority labored assiduously. Overall, the experiment was deemed a success, and the tactic was rapidly expanded to include political exiles and others who had been sent to Siberia. At the height of the construction, there were 13,500 prisoners and exiles working on the railway; by extracting the most likely recidivists, escape levels dropped to 1 percent. They had the added inducement that for every year they worked on the railway, two were knocked off their sentences. Initially, these forced laborers were paid just 30 percent of the wages received by their free colleagues, but later, in order to encourage them to work harder, they were granted parity. Their productivity rose accordingly. Battalions of soldiers, too, were enlisted to provide added manpower.

Work conditions were undoubtedly harsh, with thirteen-hour days (5 A.M. to 7:30 P.M., with a ninety-minute break for lunch), six days per week, with Sundays off. Living arrangements were spartan, with many sleeping in tents or thatched huts, which offered little protection against the torrential rains or the cold. The food was generally plentiful[9] but consisted of only the basics, with a hot dish of fish or meat at lunchtime, together with a vegetable portion, and thin gruel with butter or lard at dinner. Unlike the heavy drinking of British navvies, there was little alcohol, apart from a small amount of wine that was allowed on holidays.

As with all of these great railways, innumerable men paid for the project with their lives. In Transbaikalia, there were regular outbreaks of the potentially fatal Siberian boil plague (usually a cattle disease), and in 1895 cholera struck the workers on the Ussuri. However, overall, given the harshness of the conditions and the weather, the death rate—around 2 percent for both prisoners and free laborers—compared favorably with the rates on other major railway or canal-building projects, and was far better than those in India, Africa, and Panama.

While the sheer achievement of building the line over a period of thirteen years cannot be overestimated, the high cost, amounting to two and a half times the original estimate, and the poor quality of the railway were, as one critic put it, a "monument to Russian official bungling and laxity of administration."[10] The construction and operation of the line were bureaucratic and centralized to an absurd degree. The records of the Committee of the Siberian Railway, for example, chaired by the

tsar and attended by several of his ministers, meticulously noted its decision to grant a peasant, Fedor Koniakin, 500 rubles (£50, or $250) as
compensation for a disability that had resulted from an accident that
occurred during a surveying expedition. Despite such detailed central
control—or indeed, perhaps because of it—there was little clear accounting of the money spent on the project, and inevitably there was
extensive corruption. Basic standards of bookkeeping were never established, and the true cost will never be known. Even ten years after completion of the railway, investigations into where the money had gone
were still being carried out; the research was never completed because
of the outbreak of World War I and the ensuing revolution. The government turned a blind eye to these lax standards because it wanted the
railway built, and built quickly, regardless of the cost. This was not a
private railway accountable to shareholders, but one that was the crea-
ture of an authoritarian and undemocratic state.

And as such, it was a failure. The railway's impact on Siberia never
fulfilled the high hopes of Tsar Nicholas II and Witte. The colonization
of Siberia and the far east afforded by the railway had little in common
with events on the American side of the Pacific. Russian emigration
eastward was sponsored and supported by the state, with very little of
the spontaneous Wild West spirit that characterized the massive movement of peoples in the United States.

While, as we have seen, in the United States the federal government
subsidized the railways through land grants, soft loans, and surveys, in
Russia virtually every aspect of the process was controlled by the government. Cut-price rail fares were offered to family "scouts" who were encouraged to move to Siberia and later bring their wives and children. The
peasants induced to move east were given loans to help them settle, as
well as land and building materials, and could obtain agricultural imple-
ments at low set prices. It was hoped that the vast Siberia could become
the grain basket of the country with the help of subsidies, but the infertil-
ity of the land and the harshness of the conditions ultimately made this
impossible. Siberia was not the American Midwest, or even Ukraine. Virtually all aspects of the emigrants' lives, even their spiritual needs, were
catered to by the state in the effort to "russify" Siberia through the construction of the railway. A fund established by the tsar provided special
mobile railway-coach chapels staffed by priests, who would follow the

workers as construction progressed. The government ensured that churches would be built once stations and settlements had been established. It even tried to bring back Slavs who had emigrated to the United States, provided they were not "infected with socialist teachings."[11]

The settlement of Siberia was very much in the Russian tradition of trying to dictate where people should live, which was hardly surprising in a country where the serfs had only been emancipated a generation ago. Indeed, although the process was carried out by a monarchist government, it was more like the Soviet-type colonizations that would follow in the twentieth century than the free-market experience of the United States. Even decades after the construction of the line, Siberia's industry remained backward; for the most part, the small amount of economic development stimulated by the railway was confined to 30 or so miles on either side of it.

Therefore, while in transport terms the Trans-Siberian undoubtedly opened up Asian Russia, it had nothing like the same effect as the transcontinental U.S. railway. Partly, the harsh realities of the geography and weather were against it. The Siberian land and its climate were largely untamable, and the distances involved, mostly through incredibly hostile territory, were far greater than in the United States. As a historian of the line suggested: "The success of colonization and development depended on the success of the railroad,"[12] and the railway itself was not very successful. It was built, as we have seen, not in response to any great economic need, but by government fiat in order to fulfill the tsar's political aims. The tsar, of course, was doomed and would end up facing a Bolshevik firing squad. The painful irony about his greatest legacy, the Trans-Siberian Railway, is that it was more a Soviet project than a tsarist one: an expression of centralized power built at the whim of an imperialist regime. The same historian of the line summed it up thus: "The Trans-Siberian 'taught' Russia not managerial capitalism, as railroads did in the United States, but rather the possibility of perfecting a centralized economy."[13] In other words, the tsar's project had created the methods of his own undoing.

The motivations behind the railway's creation and the inadequacies in its construction would ensure that it was not, in modern parlance, "fit for purpose." The railway, which was single track throughout, with the occasional passing loop, had, unsurprisingly, been built to a deficient standard in virtually every way. The permanent way was flimsy, with lightweight rails that broke easily, insufficient ballast, and railroad ties often carved from green wood that rotted in the first year of use. The small bridges were made of soft pine and rotted easily (though fortunately the large ones were well constructed, and several survive in their original form today), the embankments were too shallow and narrow, often just 10 feet wide instead of the 16 feet prescribed in the design, and easily washed away. There were vicious gradients and narrow curves that wore out the flanges on the wheels of the rolling stock after as little as six weeks' use.

From an operational point of view, there were insufficient passing places and sidings, and initially the railway was expected to carry just three trains per day in each direction. Its inadequacies were highlighted in the Russo-Japanese War of 1904–1905. The Japanese, who wanted to establish power over Manchuria and Korea, had correctly calculated that the railway would not have sufficient capacity to bring large numbers of troops eastward; they launched their war confident that they would not find themselves outnumbered by the rapid deployment of fresh troops. Despite recent improvements to the line, including the completion of the route along the cliffs of Lake Baikal, the Russians were beaten both on land and at sea. When the defeated troops used the railway to return home, delays caused by lack of capacity led to a revolt that left the railway in the hands of mutineers and rebels for several days.

The railway did, however, have clear attractions for certain travelers. Soon after it opened, twice weekly international trains were launched, a very handy alternative to the long sea route for diplomats heading from Europe to Beijing or Tokyo. It cut their journey time from nearly two months to a couple of weeks, according to the timetable, although in practice the journey generally took longer. In theory, too, travel on these international trains was luxurious, with staff being instructed in meticulous detail about the emptying of ashtrays and spittoons and the need to keep carriage temperatures at 14°C (just under 60°F). In practice, though, customer service was not generally the railway's strong point. One early American traveler[14] reported a seventeen-hour halt with no explanation and described how at each station where the train stopped the peasants would scramble out of the carriages to make their tea and cook their soup, and "still the train would wait," even though there was no service waiting to pass in the other direction (the locomotive was probably taking water from the often very small and inadequate pipes characteristic of the line). Eventually, "for no particular reason," the stationmaster would ring a large bell, and after he had repeated the exercise five minutes later and the guard had blown his whistle a couple of times, the train would start lumbering off "only to repeat the process two hours later at the next station." For long stretches, the trains had to slow down to about 12 miles per hour, half the projected speed, because of the poor condition of the track, which inevitably caused extra delays.

The railroad's restaurant cars served meals by St. Petersburg time, oblivious of the seven time zones separating the Baltic from Vladivostok on the Pacific, so toward the end of the eastward journey breakfast would be dished up at 2 P.M. and dinner at 3 A.M. The tight security also added to the discomfort. The foreign diplomats and the few early tourists who ventured on the international service were banned from taking photographs of even the most harmless sights; nor could they communicate with the railway's staff, who invariably spoke only Russian.

After the war with the Japanese, millions of rubles were spent on improvements, increasing capacity from three daily trains in each direction to a dozen, and theoretically shortening the journey time. However, "people set out from Moscow in high hopes of reaching Vladivostok in two weeks but they always allowed an extra week for mishaps," according to a railway historian.[15] In the early days, there was an average of two derailments on every journey of the international trains. Most of them caused little damage or delay because they occurred on sections of the track where speeds were very low, but there were some serious accidents as well. In the worst year, 1901, before the line had fully opened, there were nearly 1,000 accidents and derailments, with 93 deaths and over 500 serious injuries, more than on the whole of the rest of the Russian network. The poor condition of the line when it was first completed was highlighted when the locomotive hauling the inaugural train between Maiinsk and Actinsk fell into a river below the tracks. The railway did have other uses, however. The infrequency of trains meant that it was safe for motor vehicles to drive along the track, which was better than any other route, as witnessed by the fact that the 1907 Beijing to Paris auto race was won by a car that had bumped along the Trans-Siberian Railway for much of its length.

AFRICA PRESENTED a rather different set of obstacles, which ultimately proved insuperable. Moreover, the rather insane ambition to create transcontinental railways on different axes by the two big imperial powers of the day—north-south for the British, east-west by the French—nearly precipitated a war. There was also a tense face-off between Britain and the Portuguese, who together with the Transvaal Republic wanted to build an east-west route linking the two Portuguese colonies of Angola on the Atlantic Ocean and Mozambique on the Indian Ocean.

The neatly alliterative but overambitious Cape to Cairo railway, stretching 6,000 miles, was an empire-building project promulgated largely by that great imperialist Cecil Rhodes who had established Britain's dominance in southern Africa. His dream was for a continuous line of pink[16] from one end of Africa to the other, and a railway was perceived as the means of establishing that dream and maintaining control over the continent. As a historian of the Cape to Cairo railway has argued, "the history of the railways is the history of the British in Africa. Everywhere that the Union Jack flew, railways appeared as the primary means of communication and imperial expansion."[17] The railways were the British Empire's equivalent of the roads built in lands conquered by the Romans, not least in Britain itself.

A railway between the Cape and Cairo was first suggested in a *Daily Telegraph* editorial in 1876, following Henry Stanley's exploration, but the British government was never in a position even to contemplate funding it. Instead Rhodes, who had made a fortune by creating the De Beers mining company and who in 1890 became the prime minister of the Cape Colony—South Africa was yet to be united—instigated the scheme as a way of extending British interests northward. As with the other transcontinentals, the concept for the project was not that many people would travel from one end of Africa to the other; rather, as he wrote, "the object is to cut Africa through the centre, and the railway will pick up trade all the way along the route."[18]

It is a complex story of a project that was diverted by discoveries of minerals, which led to other railways being constructed; delays caused by lack of government support; setbacks surrounding the loss of its main protagonist, Rhodes, in 1902; and ultimately, problems arising from the arcane politics of Africa and the sheer scale of the scheme. Despite its failure, the Cape to Cairo idea left a legacy of a string of railways throughout the continent, many of which would never have been built without the grandiose plan to cross Africa. The story of the project is therefore in many ways the history of the railways in general in what was then known as the "Dark Continent."

The first railway in the Cape Colony was completed in 1863, and the line was gradually extended northward over the next couple of decades, reaching Kimberley, the center of the diamond-mining industry, in 1885, thanks to a bridge over the Orange River, which Rhodes had persuaded the British government to fund. There was something of a railway boom in southern Africa in the final fifteen years of the nineteenth century; it was stimulated by the discovery of various minerals and by the outbreak of rinderpest, a disease of cattle, which wiped out 90 percent of the herd in South Africa. The consequent shortage of oxen to pull carts made the need for railways even more pressing.

Several of the lines spreading across what is now the Republic of South Africa were built by George Pauling, who, together with his brother Harry, and later his cousin Harold Pauling, formed the most successful railway contracting company in Africa. George Pauling was one of the great characters of African railway development, a fat man who professed that the only way to resist the local disease was through vast consumption of food and, especially, alcohol. Famously, on one two-day trip along the Beira Railway with its manager, Alfred Lawley, and chief engineer, A. M. Moore, the three consumed 300 bottles of German beer. Breakfast for three, a few days later, consisted of 1,000 oysters washed down with a modest eight bottles of champagne. The Pauling brothers were largely responsible, too, for building up the network of railways inside what was later Rhodesia. The drive northward was interrupted by the discovery of gold in the Transvaal; in a distraction from the main goal, a line was constructed to the east, diverting resources from the Cape to Cairo project.

Proposals were drawn up for another railway, running east-west, to provide the fast-growing Fort Salisbury (later Salisbury and now Harare, the capital of Zimbabwe), where there had been a gold rush, with an outlet to the sea and a connection to the partly constructed main Cape to Cairo line. Rhodes had his eye on the nascent colony, where the British flag had only been recently raised. The white presence there in 1890 amounted to little more than a few hunters, prospectors, and opportunists, supported by the odd administrator. Fort Salisbury was 1,000 miles from Kimberley, the nearest railhead, and the road up from South Africa was made almost impassable by the rebellious local tribespeople. A railway to the nearest navigable point, 50 miles up the Pungwe River from Beira in Mozambique, was clearly the answer, but the Portuguese, the colonial rulers, were reluctant to cooperate. It was only after the British deliberately provoked an incident with a Portuguese gunboat that a treaty led to the start of the construction of the railway. George Pauling, fresh from building a line between Pretoria in South Africa and Lourenço Marques (now Maputo) in southern Mozambique, was soon appointed as contractor.

The line, which was the longest narrow-gauge railway in the world at the time, used the tiny 2-foot gauge. Like the Panama and the Indian railway up the Ghats, it was also one of those projects with a claim to being the deadliest in terms of the toll on its workforce. During the first two years of construction, which started in 1892, George Tabor, a historian of the Cape to Cairo, has reckoned that "60 per cent of the white men—about 400 [out of a total of about 650]—died of fever [and] the 500 Indian employees almost all succumbed."[19] The Africans fared only marginally better, with a death rate of around 30 percent, because they were slightly more immune to the malarial mosquitoes. Malaria was, indeed, the main killer, as the benefits of quinine were only just being understood, but dysentery, cholera, and sleeping sickness also contributed to the high mortality rate, as well as the shortage of fresh food, which resulted from the lack of transport capability that the putative railway was intended to solve. There was, too, the ever-present danger from wildlife, particularly crocodiles and hippopotamuses[20] in the river and lions on the land. The lions quickly realized there were easy pickings at night, since the workers mostly slept in the open where they were easy prey: "In one month lions scoffed two of our white employees,"[21] Pauling is reported as saying. He omitted any mention of the numerous native victims of the feline predators.

The first 50 miles of construction, in particular, were a nightmare. Conditions were little different from those on the murderous Panama Railway built half a century before. There was even the same problem of dispensing with bodies, as coffins were in short supply, so the dead were unceremoniously dumped into the river, weighed down with stones: "Sometimes at night strange gurgling sounds were heard and bubbles appeared, as gas escaped from the fast decomposing corpses," wrote a historian, who added that the bodies "also had a nasty habit of being washed up after heavy rains on to people's verandahs."[22] It was "a monumental few months of misery in the worst fever country in the world."[23]

The railway had to be built on embankments because of the regular flooding of the Pungwe and Zambezi rivers, which turns the whole area into a lake, and thus progress was slow. All materials had to be brought in from Great Britain by tug up the Pungwe from Beira, with the result that the railway could only be built from that end, since there was no construction material or labor available in Salisbury. But despite this difficulty and the terrible conditions, the line reached its halfway point to the Mozambique frontier, 75 miles from the river, in October 1893, eighteen months after work had started. This hastily assembled railway, built to very low standards on its tiny gauge, was immediately put to use and was deemed a great success. Its little engines carried passengers in open-topped wagons rather like those on a fairground ride, with, if they were lucky, a tarpaulin to protect them from sparks, at speeds rarely exceeding 10 miles per hour. The male passengers had to be willing both to help lift carriages back on the track after the frequent derailments and to push the train up the switchbacks and spurs when a second engine was not available. By 1894, when the line reached Chimoio, nearly 120 miles from the riverhead at Fontesvilla, the rail journey took between fourteen and thirty-six hours depending on the number of delays, in addition to the two-day cart ride to or from Salisbury. At Chimoio, the usual financial difficulties, and the obstacle of a mountain range, which had to be crossed at Umtali[24] on the border between Mozambique and Rhodesia nearly 4,000 feet above sea level, brought construction to a halt. Eventually, with a workforce numbering up to 12,000, the line was completed to Salisbury, 200 miles from Umtali, in 1899; as befitted such an important railway, its gauge was widened to a far more manageable 3 feet, 6 inches. Salisbury was soon reached from the south, too, which meant there was a through line to South Africa from the Mozambique coast.

As with all such railways that establish a basic transportation link where none existed before, the line utterly transformed the economic situation of the country it served. It created a supply route for goods, especially vital mining equipment, and opened up new territory for passengers, connecting Salisbury, hitherto an isolated, almost wild backwater, with the modern world. The power of the railway was such that Umtali, a small town before the arrival of the iron road, had to be moved in its entirety when it was found that its location was inaccessible to the line because of a steep hill where the old wagon route crossed the range at Christmas Pass. If the railway could not get to the town, then the town had to get to the railway. Identical plots of land on the new site were allocated to householders, and the ramshackle buildings were transported down the hill by the railway company. With the arrival of the railway, it became "a trim little town with flamboyant trees and parks [and] run by that old school of landladies that took the strain while their husbands took the back seat, often behind the bar."[25] The Beira Railway, although going east-west, was very much seen as part of the Cape to Cairo project. When the first train from Beira reached Umtali in February 1898, the locomotive was decorated with flowers and sported a colloquial but ultimately misguided message: "Now we shan't be long to Cairo."

In 1898, Rhodes brought the main line of the Cape to Cairo up through Bechuanaland to Bulawayo in southern Rhodesia,[26] having persuaded Pauling to build the railway at great speed across the vast Kalahari Desert. Although the weather conditions there were very different from those in the Mozambique forests, they were equally harsh, with high temperatures and a shortage of water. The target of a mile per day was nevertheless achieved. It was an efficient operation over relatively easy territory, similar to the last stages of the first transcontinental across the United States. At one stage the surveying parties were just a day ahead of the platelayers; the division of labor was Fordian, with every man carrying out a specific task efficiently, from laying out the railroad ties and rails to spreading the ballast and banging in the spikes. Bridges were often crude temporary affairs that would have to be improved later, but speed was of the essence, as Rhodes was rightly concerned about the unstable political situation, which eventually, in October 1899, would lead to the outbreak of the Boer War.[27] As he foresaw, it would put a halt to further progress.

Rhodes's achievement in bringing the line 1,400 miles up from the Cape was recognized by Queen Victoria, who sent her congratulations during ten days of riotous celebrations that greeted the railway (the great man himself missed them, as he was convalescing from a bout of malaria). The question on reaching Bulawayo was where the railway would go next. Rhodes, confident that he could now achieve his goal, wanted to take the shortest route up to Lake Tanganyika, about 800 miles through Rhodesia, where the coaches and trucks would be carried 400 miles north on ferries. The route would have veered east between the borders of the Congo Free State and Portuguese-controlled Mozambique, north up to Lake Tanganyika, and then on ferries through three other lakes, Kivu, Edward, and Albert, with rail connections to cover the short land gaps between them. It would have continued through Uganda along to the White Nile, and then up to Cairo, again with long sections being covered by ships along the river.

This direct route, though, was blocked because of a mistake by the Foreign Office a decade previously that had allowed Germany to gain control of Tanganyika, thereby creating a barrier across eastern Africa where the map was no longer pink. Rhodes, realizing he would have to negotiate with the Kaiser to allow the railway to pass through East Africa, went to Berlin, but after several meetings it became clear that the Germans were adamant that a British railway would not be allowed to cross their territory. The Kaiser did offer Rhodes the concession of allowing a telegraph line to be built across the territory, and ultimately a rickety wire, held up by trees and thin poles, carried signals all the way through to the Mediterranean. Rhodes's hopes that this thin metal cable would open the way for a railway were misplaced, especially given the added problem of funding. The British government, which blew hot and cold on Rhodes's great project, showed no interest in providing the £2 million ($10 million) that Rhodes estimated was needed to build it.

Nevertheless, Rhodes did manage to continue northward from Bulawayo. George Pauling was a willing contributor to the enterprise, promising, as ever, to build a mile a day, and Rhodes cobbled together the financing from various sources, including De Beers, the diamond company he had created. Delayed by the Boer War and then by the death of Rhodes in March 1902, at the age of just forty-nine, the line reached Victoria Falls, after a relatively easy 300 miles from Bulawayo in 1904, amid the usual celebrations. The locomotive arrived bedecked in greenery: palms for Cairo, and proteas, the beautiful local flowering plants, for the Cape.

Like the other transcontinentals, the line soon attracted those adventurous—and affluent—tourists willing to brave the discomforts of the weather and the sheer length of the journey, as well as the indignity of having to down a daily dose of bitter quinine served with breakfast— in order to visit a part of the world not hitherto seen by westerners until it was "discovered" by David Livingstone barely a generation before. The *Zambezi Express* was soon a regular arrival, with the passengers using the train as a place to sleep until the opening of the Victoria Falls Hotel in 1906. There was, too, the *Diamond Express*, the *Imperial Mail*, and the *African Express*, all of which went to various places on the coast, provided by the Cape Government Railways. No expense was spared to attract these luxury tourists, who might otherwise have chosen to sojourn on the Riviera or take waters in the German spas. The coaches were "finished with oak panels, teak framing and mahogany mouldings; windows were hung with crimson curtains and tables were spread with startling white linen."[28] "Express" was not exactly an accurate description for these trains, as the average speed was only about 35 miles per hour, much slower than the 50-55 miles per hour common in Western Europe. This "African speed" meant that Cape Town to Bulawayo was scheduled at just under three days, but who cared when six-course meals of the highest quality, including excellent wines, were served three times a day, along with refreshments in mid-morning and afternoon? After dinner, the men retired to a rather cramped smoking room or could have a bath using the constant flow of hot water available from the engine's boiler.

The height of luxury, though, was simply to rent a private carriage that could be tacked on to a scheduled service. These became available in 1909. Each one comprised three bedrooms, a bathroom, a lounge (in which meals were taken), a writing or smoking room, and a reading room for the ladies, as well as a couple of spare rooms in which to dump the servants and luggage. For the price of £360 per week (about £20,000, or just under $35,000 in today's money), the food and chef were included.

It was not, however, these luxury travelers who were going to make the investment worthwhile and deliver a return to the shareholders. It was the riches of the minerals, hundreds of miles to the north, that would provide the real bonanza. But first there was the obstacle of the Zambezi River to be crossed. George Pauling had deliberately brought the line to Victoria Falls because Rhodes had always stressed that he wanted one of the greatest sights in the world to be seen from the railway, despite the fact that there appeared to be a rather easier crossing at a place called the Old Drift a few miles upriver. In fact, though he never lived to see the bridge, Rhodes proved unerringly prescient. A survey discovered that within sight of the falls, there is a gorge that allowed the crossing to be made with a single span of just 600 feet, and the banks were found to be of solid basalt rock that was sufficiently strong to provide foundations at both ends.

Breaching the chasm was hard enough, however, and there was much trouble merely connecting the two sides: When a kite failed to carry a line across, a small rocket attached to a thin cable was used, and then thicker ones were dragged across, until eventually an electric winch carrying men in a device resembling a ski-lift chair could be deployed. Unsure of the quality and ability of the local labor, the bridge was built in England in the yard of a successful contractor, Cleveland's of Darlington, which had bid lower than Pauling and used an innovative method. The bridge near the falls was transported in kit form to the shore, where it was assembled like a giant Erector set. The two halves of the single-arch cantilevered bridge were built out simultaneously from both banks until they were ready to be joined. The date was anticipated with trepidation. On April Fools' Day 1905, an odd choice, just fourteen months after work on the bridge had started, the two sections were joined, with an audience of thousands of tribespeople from across the country attracted by their chief's prediction of a spectacular collapse into the abyss. The Cassandra was proved wrong. After some difficulty caused by heat expansion, the two halves of the bridge were successfully bolted together; the ensuing celebrations lasted for several days, with the added feature of a regatta upstream from the falls. The bridge spanning the gorge is easily the most impressive and memorable structure of the Cape to Cairo project, offering one of the most striking railway images anywhere in the world and comparable with the Forth Bridge or those rickety trestles from the pioneering days of the U.S. railroads.

The completion of the bridge gave access to central Africa, and with Rhodes gone, it was George Pauling who was eager to push the railway northward, though for interests that were rather baser than those behind the notion of a Cape to Cairo railway. He was supported and funded by Robert Williams, who had obtained the concession to exploit minerals in a large area north of the Zambezi. Williams, rightly, had predicted that there would be extensive mineral wealth on the divide between the Congo and the Zambezi rivers, but as he needed a railway to exploit the deposits that his preliminary prospecting had uncovered, he took up the mantle of the Cape to Cairo project.

Even before the bridge had been completed, therefore, his contractor, the tireless Pauling, was busy at work north of the falls surveying the area to find a route up to the Kafue River in Northern Rhodesia (now Zambia) and a way across it. A locomotive to help with the work had been stripped down into small enough parts to carry across the Zambezi on the winch, and Pauling, with a surveying team, headed into the plains. This surveying party was a bizarre mix of the luxurious and the make-do. The surveyors were accompanied by a party of three hundred porters carrying both the necessities and the luxuries of camp life, including tables, beds, chairs, medicine chests, barometers, and compasses. And, since Pauling was in charge, large quantities of alcohol, especially champagne, were in the baggage train. The chef, according to Pauling,[29] was a "treasure" who "provided magnificent dinners" every night, comparable with those found in major European hotels, from a mixture of game, fruit, and vegetables gathered from the plain and tins carried by the porters. Working from dawn to dusk, with a break for the midday heat, the party trekked for up to 20 miles daily working out the route. At night, there was the occasional game of bridge, though mostly Pauling and his fellow surveyors were too tired, instead indulging in "a somnolent smoke" before turning in.

Along with the usual dangers facing railway builders in Africa, there was the added peril of hostile tribesmen, including cannibals with a particular liking for young women's thighs and babies. One night, Pauling was appalled to be told that a couple of his workers had stolen a two-month-old baby from a nearby village and were preparing to cook it.

He managed to prevent the feast just in time. The local tribesmen were a particularly fierce bunch, deliberately eschewing any clothes even on cold evenings in order to demonstrate their manhood, both physically and psychologically. In addition, the hungry lions in the area were extremely voracious and indiscriminate, gobbling up everything from the oxen used to haul carts to the various pets that Pauling and the other whites had taken with them—as well as any men unwise enough not to ensure that their tents were well secured. The large herds of roaming elephants proved to be a hazard to the supply trains running up the completed sections of the line, causing collisions that were fatal to both the beasts and the train crews.

The big obstacle was the Kafue River. Because of the flat terrain, the meandering river is up to a mile wide in parts. But a section where it was just 1,500 feet was found, and a bridge with thirteen steel spans was laid, with remarkable speed, in just six months. The spans were supplied from England by the prosaically named Patent Shaft and Axle Tree company and sent by rail from Cape Town. The aim was to reach the Congo Free State border 400 miles away, but the initial contract was to build the railway from just north of the Zambezi to Broken Hill, where Williams had already established a rudimentary town to serve the mine. Pauling achieved his usual target of a mile per day, taking just nine months to complete 281 miles of railway, admittedly over the easy topography of the great central plateau. On the way, almost casually, the town that would become the capital of Northern Rhodesia—Lusaka, named after the local headman—was created.

By the end of 1906, therefore, the line stretched more than 2,000 miles from the Cape. Pauling wanted to continue to the border, but there was no way to finance the project and no precise idea of where the railway would go. After a delay of three years, Pauling, largely funding the project himself, began building again toward the frontier with the Congo Free State, and later crossed it, taking the Cape to Cairo for the first time off British soil. The railway stretched far into the Congo, the personal fiefdom of the appalling King Leopold II of Belgium, whose legacy of oppression and exploitation for personal aggrandizement remains today. In 1918, at Bukama,[30] on the banks of yet another part of the Congo river system, which comprises 12,000 miles of navigable waterway, 450 miles from the border with Northern Rhodesia, the Cape to Cairo project finally came to an end. Trains with exotic names such as *Cape* or *Congo Express* ran twice a week from the Cape to Bukama, taking six days to cover 2,600 miles. From Bukama, steamers along the Upper Congo reached Lake Tanganyika, which was now in British hands, having been taken from the Germans as one of the spoils of victory in World War I. During the war, the political ambition to complete the project had briefly flourished in the Foreign Office, but after the Armistice the political momentum stalled, and it ultimately dissipated.

The project might not have fulfilled Rhodes's dream, but the southern section of the Cape to Cairo provided Africa with a spinal railway from which various branches could be constructed to reach the sea, either directly or via navigable waterways. King Leopold, for example, eager to exploit the resources of ivory and rubber, and later all kinds of minerals, had developed a railway in the early 1890s between Matadi, upriver from the coast, and Leopoldville (now Kinshasa) to bypass a section of river that was not navigable. The construction of this narrow-gauge line, just 750 millimeters (2 feet, 5.5 inches), was another of those desperate affairs with a mortality rate that was inconceivably high; it took three years to complete the first 14 miles and eight to finish the whole 240 miles. The navigable section of the river ended at Port Francqui,[31] and to tap the mineral wealth of the Katanga[32] region a line was started that eventually joined Pauling's at Bukama.

Further south there was the Benguela Railway across Angola, which eventually met the Cape to Cairo at Tshilongo, 100 miles south of Bukama. The Benguela was another epic railway, and it was largely the work, yet again, of Pauling's company. Built at the instigation of Robert Williams with British money, it ran solely through Portuguese and Belgian territory, as, four years before his death in 1909, Leopold had handed over his personal colony to the Belgian state. Williams's intention was to provide the shortest possible route for the export of copper from Katanga, cutting the journey via the Cape by 3,000 miles. Labor was scarce, as the coastal Angolans had been decimated by successive slave raids, and imported workers succumbed quickly to all the usual tropical diseases. The topography was difficult, too, with a mountain range having to be crossed within the first 50 miles. A steep 1.5-mile rack railway was built to reach the plateau 3,000 feet above the ocean, which was still visible from the railway. Williams had hoped there would be plenty of mineral discoveries along the way, but he was to be disappointed. The surveying teams, spreading out about 60 miles on either side of the railway, found nothing other than a small copper deposit. Stopping for the war and the inevitable shortages of funds, the project was not completed until 1932 when the connection with the Cape to Cairo was made. Apart from the modest line running entirely in South Africa, linking Cape Town and Durban, which was completed in October 1895, the Benguela thus became part of the only true trans-African railway, providing a through route linking the Atlantic port of Benguela in Angola with Beira in Mozambique on the Indian ocean.[33]

These railways in central Africa were built to exploit resources rather than to bring in development. They were not, in contrast to the other transcontinental lines, seen as a way of opening up the area for immigrants who would create self-sustaining communities. Nor did they stimulate local economic development for the existing inhabitants. The Africans were pushed aside in the same way as the Indians in the United States, but the climate and conditions were too harsh for permanent settlement by whites. Instead, the minerals, which were signed away by local chiefs either by subterfuge or by force, were removed at a minimum cost with the help of the railways. Although these lines inevitably created a few local jobs in the mines and on the railways themselves, their primary purpose was the extraction of minerals and other resources through a process that provided little benefit to local people and their communities. Their purpose, indeed, was the same as that of the original wagonways built in Britain and elsewhere in Europe from the seventeenth century onward: the linking of mines with the nearest navigable waterways, except, of course, that they were considerably longer.

If the southern section of the Cape to Cairo never got as far as Rhodes, or indeed Williams, had hoped, nor did the northern section. Starting in Egypt and running through Sudan, it was an equally ambitious exercise simply because of the sheer scale of the enterprise, even though it passed through relatively easy territory. The line had to traverse vast swathes of desert, and its construction in those harsh conditions was only made possible through military discipline. Indeed, it was intended as a military railway built to give the British access deep into Sudan, and was constructed at the instigation of Herbert (later Lord)

Kitchener, who at the time was the *sirdar* (commander-in-chief) of the Egyptian army.

Before the British occupied Egypt in 1882, the Egyptians already had an embryonic railway system, Africa's first. Completed in 1856, the 120-mile line connected the port of Alexandria on the Mediterranean with Cairo and was designed by none other than the ubiquitous Robert Stephenson. The railway was no mean achievement, having to cross the Nile twice, and the line was highly profitable, as it attracted travelers using the overland route between Europe and India, which avoids the long sea voyage via the Cape. They had previously had to use camels or rough horse-drawn carriages to cross Egypt. Despite the opening of the Suez Canal in 1869, which took away much of its revenue, the railway was extended to reach Assiut on the banks of the Nile by 1874 and Luxor, 340 miles south of Cairo, in 1898.

Sudan, south of Egypt, had been abandoned in 1885 by the British after the siege of Khartoum, which ended with the massacre of Major General Charles George Gordon and his army by rebels. The rebels had been led by Mahdi Muhammad Ahmad, a religious leader opposed to Western control of Egypt. A decade later, Kitchener obtained permission from the British government to build the Sudan Military Railway through to Khartoum in order to reconquer the country and defeat the Mahdi rebels. To get to Khartoum, hundreds of miles across the desert, Kitchener realized that a railway was needed from the Sudanese frontier at Wadi Halfa on the Nile, which could be reached by ships from Luxor.

Helped by an eccentric but capable young Canadian military railway engineer, Edouard Girouard, and supplies from the United Kingdom, work started in 1896 with the reinstatement of an existing railway that had been destroyed by the Mahdi's followers. At first progress was slow. Although the territory was flat desert, conditions were difficult because of the water shortage, which was simultaneously also making life difficult for Pauling across the Kalahari 3,000 miles to the south. Temperatures were regularly above 100°F during the day. Fortunately, water, which was essential for both men and machines, was found at two points along the line, which saved the day. The labor force, too, was somewhat unreliable, since a shortage of men had led to convicts and other misfits being taken on, but soon the military discipline and precision paid off. Construction reached a furious pace, with, at times, 3 miles being completed in a day. Girouard was greatly helped by the fact that he had cadged five engines originally from Rhodes, whom he had met on a trip back to London to sort out supplies.

Every day, two trains were sent down the line from Wadi Halfa, where all the engineering work was carried out, to the teams laying the track. The first would arrive at dawn, carrying 2,000 yards of rail along with railroad ties and the accessories needed to lay them, and, of course, water, while the second, arriving at noon, brought in more rail and equipment and also luxuries for the white engineers to keep their morale up: Along with food and water there would be whisky, cigarettes, and even newspapers. Only halted by occasional battles—which were won easily, thanks to the supply line—and intermittent problems with the availability of equipment, Kitchener and the railway reached Omdurman, on the outskirts of Khartoum, in September 1898. It was there that the decisive battle[34] for control of the country was fought, in which, incidentally, Winston Churchill took part.[35] A remarkable 576 miles of railway had been completed, in less than two years, amid conditions of extreme hostility both from the weather and the local population. No wonder the railway was feted by Victorian writers as "the greatest weapon against the Mahdi."

The Egyptian railways were standard gauge, but Kitchener built his line to the cheaper Cape gauge, 3 feet, 6 inches, demonstrating his ambition that one day it would meet up with Rhodes's railways down south—although the more practical reason may have been that the five locomotives Girouard had borrowed from Rhodes were Cape gauge. With rapid progress at both ends, by 1900 there seemed to be a strong possibility that the Cape to Cairo railway would be completed. That year, Rhodes had cheekily sent a telegram to Kitchener saying, "If you don't look sharp, I will reach Uganda before you." Kitchener replied, with equal chutzpah, "Hurry up." But there was a long way to go, and the Boer War, the lack of will on the part of the British government, the blockage by the Germans in Tanganyika, and the sheer ambition of the project meant it would never be completed.

As was the British way, with the exception of the Sudan line driven through by Kitchener, and small sections in the Cape Colony and East Africa, the vast majority of the completed sections of the Cape to Cairo had been built by the private sector. The Cape to Cairo may never have been finished, but its partial construction left behind a notable legacy, helping to establish a permanent British presence in much of central and southern Africa and effectively creating the two new colonies of Northern and Southern Rhodesia, which, understandably, were named after the railway's principal protagonist.

Had the line been finished, it would not, in any case, have been a complete railway like the other transcontinentals. It was never expected that a traveler would have been able to undertake the whole journey in a single train because of the difference in gauges and the various sections covered by boat. The river journey on the Nile extended more than 850 miles, and there were no plans to build a parallel railway.

Nevertheless, by 1928, with the construction of a section of line in Uganda, the whole journey became possible by public transport—buses, trains, and steamers—and entirely on British territory. Mombasa, on the Kenyan coast, could be reached by a combination of boat and train from Khartoum. The building of the Sudan Military Railway had stimulated further development of the iron road in Sudan to serve local interests, rather than as part of the grand design of a transcontinental railway. The railway reached Kosti, 240 miles south of Khartoum, in 1911, and from there a ship could be taken up the Nile to Juba. There followed a 100-mile journey in a bus over the frontier into Uganda, where, at Nimule, a steamer and another bus reached another railhead at Namagasali. Finally, after two separate lines had been linked in 1928, a direct train went from that point to Mombasa. It was not a trip for the casual traveler.

Mombasa, an old slaving port on the Indian Ocean, had been the starting point for the Uganda Railway, whose name referred only to its projected destination, since its entire 600-mile length was in Kenya, terminating at Kisumu on the lake. A boat could then be taken to Kampala in Uganda, although later the line was extended there. This was a truly imperial project, built by the British government with little purpose other than to cement its colonial power, and was dubbed by the press the "Lunatic Railway" because of the cost and difficulty of construction.

After much discussion in the British Parliament, the sum of £5 million ($25 million) was allocated, though Pauling, who was snubbed for the contract because of his association with the out-of-favor Rhodes, claimed he could have done it for a third of the price. As there were few local inhabitants, the labor was largely imported from India: Some 32,000 men were brought over from the Indian subcontinent soon after work started in 1896. Many died, such as the twenty-eight killed by a pair of ferocious, maneless lions on Christmas in 1898, prompting their fellows to strike until the beasts were shot. Overall, the meticulous records kept by the railway showed that 2,498 workers died during the construction. Although most of the surviving Indians returned home to the subcontinent, precisely 6,724 chose to stay after the completion of the line in 1901. They created a community of Indian East Africans who were to flourish as traders and merchants but later suffer the disgraceful expulsion from Uganda during the dictatorship of Idi Amin in the 1970s. The line swept through the gap between Mount Kilimanjaro and Mount Kenya and halfway along reached Nairobi, which was then little more than a water hole in a swampy marshland but, thanks to the availability of flat land, became the railway's headquarters. As Tabor put it, "no other capital city developed so readily and rapidly because of the railway, as did Nairobi."[36] After passing Nairobi and crossing the Great Rift Valley, the line climbed up the escarpment on twenty-seven viaducts to reach the astonishing height of 8,740 feet, to the coldest railway station in the British Empire—on the equator.

The Uganda Railway, which at £5.5 million ($27.5 million) went only 10 percent over budget because the British government had been pessimistic in its estimates, proved to be anything but "lunatic," as it was handed to a private company and quickly became profitable. Branches were soon built to help in the export of coffee, soda, and sisal, and tea became a viable crop thanks to easy access to Mombasa. Moreover, the railway effectively realized the original aim of establishing Britain's undisputed rule over the territory. Unlike the lines in the Congo, the Uganda Railway attracted considerable numbers of settlers who grabbed the land without any regard for the native population, creating the colonies of Kenya and Uganda.

Further south, the Germans did much the same for Tanganyika. They built a parallel line inland from Dar es Salaam that reached Kigoma, on the banks of Lake Tanganyika 750 miles away, in February 1914, just before the outbreak of the war that would result in Germany losing its colonies. The line took much longer to build than the parallel British one further north, partly because of understandable German hesitancy about the value of the railway to the mother country's economy, but also because of the bureaucratic nature of German administrative processes, born of the Teutonic emphasis on *Gründlichkeit* (thoroughness). Like the Russians on the Trans-Siberian, the Germans insisted on accounting for everything, however small, and reporting in great detail back to the center, which caused endless headaches for railway managers and created a weighty bureaucracy. The historian of the line, M. F. Hill, recounted a lengthy dispute between Berlin and the line's supervisor, who complained bitterly "that the typewriter ribbon sent to him did not fit his Remington machine" and that previous requests for a new machine "had been refused on the grounds that [his] office work was likely to decrease."[37] His letter was accompanied by a four-page memorandum listing the typewriter's faults, which Hill was able to uncover when he wrote the railway's history fifty years later, though he does not tell us whether the supervisor's request was granted.

Despite such obstacles, another line was built from Tanga, on the Indian Ocean almost equidistant between Dar es Salaam and Mombasa, northward toward the southern flank of Kilimanjaro, and then just east of the mountain at Arusha, its terminus. These lines, too, attracted settlers, particularly from South Africa, where many Boer farmers were eager to flee British control. The Germans briefly considered the idea of an extension to meet the Cape to Cairo and run a line through to the Congo Free State, but realized there was no economic justification, since the rich minerals from Katanga would never be transported in an easterly direction. Or at least not until sixty years later, when the Chinese took an interest, building the Tazara Railway to enable Zambia to export its copper without going through South Africa or Rhodesia. Sections of the line are actually on the route that the Cape to Cairo might have taken had the Kaiser not blocked its path, and as a result the dream of a Cape to Cairo railway is not entirely dead. As recently as 2004, the Sudanese government commissioned a study by German consultants for a rail link to Kenya connecting eventually with the Tazara line, which would almost complete Rhodes's vision. It was priced at $25 billion for a 1,500-mile railway, but war and political difficulties have intervened to put the idea on hold.

The third colonial power with trans-African rail ambitions was France, which, not to be outdone, had a strategy in the late 1890s to link its colonies from west to east across the continent, from Senegal to Djibouti. Southern Sudan and Ethiopia were in the way, but France sent expeditions in 1897 to establish a protectorate in southern Sudan and to find a route across Ethiopia. The scheme foundered when a British flotilla on the Nile confronted the French expedition at the point of intersection between the French and British routes, leading to the "Fashoda Incident." Kitchener and the leader of the French expedition, Major Jean-Baptiste Marchand, who had reached Fashoda on the Nile south of Khartoum with a force of 150 after an epic fourteen-month march from Brazzaville,[38] discussed the situation over copious amounts of champagne brought by the French across Africa. The French not only paid for the drink but had to accept defeat, as Kitchener had arrived with a far bigger force, and their respective governments eventually communicated a treaty that involved French withdrawal. As a result, the French were forced to abandon any hopes of establishing a presence in the Nile; instead, the British allowed them unfettered access to thousands of square miles of the Sahara Desert. Despite this setback, the French *Transaharien* continued to be part of the lexicon of its politicians for several decades; the Vichy government, the puppet regime of the German invaders in World War II, even began to build it in 1943, but that attempt was soon abandoned, finally putting an end to the whole madcap concept.

AUSTRALIA, TOO, would eventually boast a transcontinental railway. Amazingly, its north-south line, the Ghan, named after the Afghan camel trains that used to trek along the same route, was only completed in 2004 (see Chapter 13). The transcontinental east-west connection across Australia was, as in Canada, a bribe for a far-off province to join the club. The railway linking Perth across hundreds of miles of empty desert with Port Augusta in South Australia, and thence the rest of the country (though, of course, with different gauges), was one of the inducements offered to Western Australia to join the Commonwealth of Australia, which the five other states had established in 1901. It took ten years to sort out the survey, but the legislation was eventually

passed in December 1911. It authorized the construction of a line between the two existing railheads, Kalgoorlie, which had been reached
by a 400-mile narrow-gauge government railway in 1896 from Perth,
and Port Augusta, the South Australian railhead.

Although the enterprise was not on the same scale as the other
transcontinental railways as it was only 1,050 miles long, it was nevertheless remarkable, not least because it was undertaken quickly at a
time when the world was at war, which created severe supply difficulties. Admittedly, the terrain was flat, with a huge plateau covering over
half the distance, which allowed the line to include the longest stretch of
straight railway anywhere in the world, a distance of 297 miles.[39] However, the climate was harsh, with temperatures at times topping 120°F,
and, crucially, not a single watercourse was crossed. The arid conditions presented enormous difficulties during construction, and once the
line was open, the steam-hauled trains had to carry water equivalent to
half their load, making operations inefficient and expensive. Work
started in September 1912 from both ends and was completed five years
later. Although the adjoining railways at either end were narrow gauge,
the line was built to standard gauge, forcing through travelers to change
trains. Worse, passengers had to change trains again at Terowie, 136
miles north of Adelaide, because the line connecting with the state capital used broad gauge (5 feet, 3 inches) as a result of South Australia
having built railways in three gauges.[40] It was only in 1970 that Australia's intercity route was finally converted to standard gauge.

These transcontinental railways were the biggest projects of their
day, and the period in which they were being built, particularly the last
quarter of the nineteenth century, was the heyday of railway construction in most countries. Virtually every country in the world, with the
exception of a few African colonies—and even including some small
islands—had a railway by the beginning of the twentieth century, built
for a variety of reasons and purposes.

EIGHT

The Invasion of the Railway

B Y THE FINAL QUARTER OF THE NINETEENTH CENTURY, RAILWAYS were well established throughout the world. The sight of a steam locomotive puffing across the countryside had become commonplace, not only in towns and on the major routes between them, but also in remote regions penetrated by the burgeoning number of branch lines. In 1880 there were 280,000 miles[1] of railway, and that figure would rise to nearly 500,000 by the end of the century. Across the world, railways were growing at the rate of 10,000 miles per year, and they would continue to do so until the outbreak of World War I. This was the period when the railways were in their heyday, spreading everywhere with the misguided confidence that the railway age would define the twentieth century as it had the nineteenth.

The first underground railway was completed in London in 1863, and there was no shortage of other astonishing engineering feats, but the last quarter of the century would see railways go beyond what had been achieved before to conquer the most inhospitable parts of the world. There were no boundaries, physical, social, or topographical, that could prevent their progress or delay their dominance. The transcontinental railways may have been the headline stealers in terms of their ambition, but they were by no means the only ones to overcome remarkable obstacles. Mountains, rivers, deserts, and jungles, as well as sheer distance, were all comfortably ignored as the railway reached places that seemed inaccessible for such a cumbersome and complex invention. Since these regions were often little inhabited, the new railways were frequently politically rather than economically driven, just as the transcontinental lines had been. Even so, many were used for a precise economic purpose only, such as serving a particular mine or facilitating the export of a single agricultural product like sugar or wheat.

The global expansion of railways was at its peak in the final twenty-five years of the nineteenth century, when industrial development in many countries was stimulating unprecedented rates of economic growth. The railway was unchallenged as the principal means of transport for both passengers and freight, since the development of autos and trucks was still embryonic. But it worked the other way, too. For some countries, the opening of their first railway was the stimulus for a period of sustained economic growth and industrialization based on the development and spread of the railway network. Every European country,[2] except for Albania,[3] had a railway by 1869, when both Romania and Greece opened their first lines. Most Asian countries had joined the age of the iron road, at the behest of their colonial powers, by the 1880s, and in Central and South America virtually every state boasted at least a few miles of track. The hectic pace of railway construction only came to a halt with the outbreak of World War I in 1914, which coincided, too, with the growth of road transportation and the shift of investment away from the iron road to the tarmac. It would be impossible to mention every country where the railways were beginning to be developed at this stage, as the list would be both lengthy and repetitive, but there are several stories worth outlining in some detail as examples.

Latin America, for the most part, had been a late starter, and its rugged topography made construction difficult. However, those mountains and ravines were obstacles for roadbuilding, too, and therefore railways were often the cheapest way of providing access to remote regions. At the time, parts of the continent were highly developed, with economies that were as affluent as those in Europe, and consequently railway building was equally rapid. This was particularly true of the most economically advanced South American state, Argentina, which opened its first line relatively late but would eventually boast a remarkable network of 26,500 miles—in three different gauges—just before the onset of World War II, built largely with British capital and equipment. The country's network was more extensive than that of its far larger and much more populous northern neighbor, Brazil, because Argentina's rulers had a better understanding of the potential of the new technology. Initially, they used the railways to attract settlers to remote but fertile regions; this, in turn, stimulated rapid economic growth and further expansion of the rail system. The Argentine system developed westward out of the ports, creating an extensive network in the hinterland primarily for goods traffic. Given that Argentina had the eighth-largest land area in the world, and yet its population was under 2 million when the first line opened, the rapid growth of the network was a remarkable achievement.

The concession to build the first railway was won by an Englishman, William Bragge, and it was built to the strange gauge of 5 feet, 6 inches, the same as the railways in India, reputedly because equipment destined for the subcontinent had been acquired by Argentina.[4] The inaugural railway, between Buenos Aires and Flores 12 miles away, opened in August 1857 and was the start of what would eventually become the 7,400 miles of the Ferrocarril Oeste de Buenos Aires (Buenos Aires Western Railway). The length of the line was doubled by 1860, but despite attracting considerable patronage, the railway soon fell into financial difficulties and was taken over by the Buenos Aires government. That did not stop its expansion, however. The railway reached Chivilcoy, 100 miles from the capital, in 1866, and branch lines were quickly added. Cattle, the country's main export, were the mainstay of the freight operations, though the railway also carried wine and grain.

Three other large companies, all British-controlled and using the same broad gauge, soon emerged. In the north, a concession was granted to an American, William Wheelwright, who had built the first line in Chile, to create a railway between Rosario, a port on the Paraná River, and Córdoba, a state capital 250 miles away, but the financing came from Britain. Opened in 1870, the purpose of the Ferrocarril Central Argentino (Central Argentine Railway) was not only to transport goods to the port but also to encourage the settlement of the land between the two towns, which, though small, with around 25,000 people each, were the two largest population centers outside of Buenos Aires. The railway proved successful, carrying 35,000 passengers and 35,500 tons of freight in its first year, and attracted large numbers of settlers to the *pampas*. At the opposite end of the country, the Gran Ferrocarril

Sud (Great Southern Railway) was created with the idea of linking the capital with Chascomús 75 miles away. Not only was this railway reliant on capital from Barings Bank, but it was also built by one of the great British contractors, Sir Samuel Peto, who, helped by a local firm, laid the railway with remarkable speed, taking just twenty-one months to complete the line, which opened in December 1865. That railway, too, quickly became profitable; it soon had more than 500,000 passengers per year, which allowed the company to pay generous dividends of 8 percent. The fourth main rail company was the Buenos Aires & Pacific, which completed its first line in 1888. It eventually built up a 3,200-mile network, but despite its name, it never reached anywhere remotely near the ocean. Much of the British capital would remain in the country right through to World War II, earning generous dividends for the shareholders until competition from motor transport began to eat into profits in the interwar period. By the time of the nationalization by Juan Perón in the aftermath of the war, nearly two-thirds of the railway was still owned by British interests and a tenth by the French, with the rest already in state hands.

These early Argentinian railways had an instant effect on the economy, creating an economic boom based on agricultural production, notably cereals, sugar, and wine, which in turn stimulated rapid growth of the network. Beef, too, soon flourished, thanks to the advent of refrigerated ships that permitted the meat to reach European markets in good condition and thus command a high price. The massive Vestey food empire, which included the Blue Star Line and the Argenta chain of butchers' shops, depended on the Argentinian railways and their refrigerated trains to transport beef cheaply to Europe. Beef then stimulated a remarkable boom in the railways that was similar to the earlier ones in Europe and America and was funded by both state and (largely British) private investment. The British lines tended to focus on the ports, while the relatively unprofitable areas further from the sea were left to state enterprise. Even remote areas such as Patagonia began to get railways, and there was considerable duplication. The rate of growth matched what had happened in the railway mania in Britain, but in a country where the population was still a mere 3.4 million at the end of the 1880s. By 1890, there were already 6,000 miles of railway in Argentina, but that figure rose to 10,000 at the turn of the century and double that by the outbreak of World War I.

No other country in Central or South America could match such a phenomenal density of railways. Brazil, for example, with an area three times that of Argentina, had 24,500 miles of railway at its peak between the wars, fewer than Argentina and barely a tenth of the total reached by the United States, whose main landmass[5] is only slightly bigger. Brazil made many of the same mistakes as Australia, which is nearly the same size, by failing to develop the railway as a national, rather than a provincial, resource and therefore ending up with a hodgepodge of gauges and administrations that hampered railway development. The problem, as in Australia, was that responsibility for granting the early railway concessions lay with the provinces[6] rather than the Brazilian government. This led to the establishment of a series of railway systems with different gauges, many of them separated by large tracts of territory without any lines, where the navigable waterways remained the most efficient form of transport, and consequently no national network ever emerged. Railway development was confined to the coastal areas and the immediate hinterland, with only a few tentacles stretching far inland, and none, apart from a connection with Uruguay, stretching into the ten countries with which Brazil shares a border. Moreover, the Brazilian railways were largely developed to serve the needs of specific industries, notably coffee in the south and sugar in the north, rather than to provide a means of traveling around the country. Coffee, indeed, was the spur to the creation of much of the country's rail network. Previously, the beans had been carried from the plantations to ports by mule trains on primitive roads and tracks, which tended to turn into quagmires in the rainy season, resulting in the loss of much of the crop. Brazilian coffee production grew rapidly in the mid-1800s, ensuring that the country had a virtual monopoly on the product at the time, and the need to find better ways of transporting it stimulated the adoption in 1852 of the first law to allow the construction of railways.

The inaugural Brazilian railway was promoted by Baron de Mauá, the country's richest man, who was involved in virtually every early railway project in the vast nation, but who, like many railway pioneers, died in poverty and obscurity. Mauá obtained a concession from the local province for a railway from the port of Mauá, in the bay of Rio de Janeiro, to Petrópolis, dubbed the Imperial City because it was a summer playground for the emperor, who had a summer palace built there, and his coterie of aristocratic friends. The first section of 10 miles, built, like Argentina's first railway, by the Englishman William Bragge, was completed in April 1854 amid much celebration. The significance of the opening in terms of the democratization of society and changing the way people lived was not missed by a reporter for a business paper, who wrote: "The railway also invites the poor, the humble[,] . . . to take a seat in the carriage next to the one which carries the prince, both moved by the same locomotive."[7] The real revolution would come when they traveled in the same carriage! Moreover, in a country that had only just emancipated its slaves and had not yet embarked on industrialization, the railways were the first significant use of advanced technology, unlike in the more developed countries, where various steam-powered machines had already been commonplace before the arrival of the iron road. The railways alleviated concerns among plantation owners about the freeing of the slaves, as rail technology greatly reduced the demand for labor in transporting agricultural produce.

Again, the choice of gauge seems to have been happenstance. The first locomotive, built by a Manchester company, had originally been intended for Ireland but was sent to Brazil and consequently was intended for use on a 5-foot, 3-inch, track.[8] Later, however, when the line became part of the Leopoldina Railway, the track was changed to meter gauge. The line terminated at Fragoso at the foot of the Serra do Mar, a mountain wall just inland from the ocean that cuts off the coast from the plateau, rising, at times, to 8,000 feet. And there the terminus remained for thirty years until a rack railway was carved into the cliffs, finally allowing trains to reach Petrópolis.

In the meantime, other, more significant railways were being built in Brazil, including, notably, a line that would breach the Serra do Mar wall, giving access to the sea from the vast hinterland where most of the world's coffee was being grown. Better connections were urgently needed between Santos, the principal port used to ship coffee, and São Paulo, the largest city in South America, on the plain above. The plantation owners had long pressed for a railway and had even submitted a scheme to the ubiquitous Robert Stephenson in 1839 that had come to nothing, not least because the Brazilian government at the time was opposed to the concept of railways.

It was not until 1856, after the coffee and sugar growers in the region pressed for a line, that a concession was granted for the São Paulo Railway,[9] with the ever-active Baron de Mauá as one of the promoters. The capital needed to build the line inevitably came from Britain, as did the engineering expertise. The mountain wall was only 2,700 feet high at that point, but covered with virgin forests and dotted with several vertical cliffs, it presented a formidable obstacle. Advice was sought from an experienced railway engineer, James Brunlees, who had built the Ulverstone & Lancaster Railway through difficult terrain across Morecambe Bay. Brunlees sent a young acolyte, Daniel Fox, to survey the site, and after months spent climbing through the forest and up the crude paths—during which young Fox's skin was said to have turned white because of the lack of sunlight—he found a way through. The solution, he suggested, was to use cable haulage on four sections, with a steep grade of 1 in 10, separated by level lengths of line that would house the winding engines that hauled the trains up the incline. The inclines had three rails, with the trains climbing and descending the hill on different pairs in order to keep the cables safely separated, and only the middle rail was used for both ascent and descent.

Construction started in May 1860. Even the flat section from Santos to the foot of the Serra posed considerable difficulties. Embankments were required to raise the line above the swamps, and a series of bridges were needed to span the rivers and the arm of the sea that makes Santos an island city. The section that climbed the Serra, through virgin forest and granite rock in an area of heavy rainfall and frequent landslips, is one of the heroic railways in this story, almost defying description. Not only did a very steep mountain have to be crossed, but there were the usual problems of working in a tropical climate in which disease ravaged the workforce. Heavy rains in 1862 wrecked much of the early earthworks and there were the usual complaints about the contractor, whose "bad service," according to the historian of the Brazilian railways, was "a constant factor among the English contractors who came to Brazil to construct the first railways."[10] Just to make things worse, after all the difficulties of construction had been overcome there was an accident at the ceremony marking the opening of the Santos to São Paulo section in September 1865—reminiscent of the tragedy that marred the inauguration of the Liverpool & Manchester Railway thirty-five years before when a prominent politician, William Huskisson, was killed. This time, a locomotive derailed and fell into a ditch, killing the driver and injuring several people, although, fortunately, the president of the province and other VIPs in the open carriages escaped unscathed.

Despite this mishap, the 87-mile line to Jundiaí, beyond São Paulo, was completed in 1867. It was an immediate success, making substantial profits for its promoters by attracting vast quantities of freight as well as large numbers of passengers. It marked the start of a boom in similar railways serving the coffee plantations and in fact was such a success that the line up the Serra was doubled before the century was out to cater to the growing traffic. The expansion of coffee cultivation, which soon accounted for more than half the value of Brazilian exports, and railway construction went hand in hand throughout the southern provinces around São Paulo, and by the 1870s a full-blown railway mania had developed to serve the needs of the burgeoning coffee plantations.

At the same time that the São Paulo Railway was being built, another line up the Serra do Mar was under construction, the second section of the Estrada de Ferro Dom Pedro II. This railway connected Rio de Janeiro with the provinces of Minas Gerais and São Paulo, which, again, was vital to facilitate agricultural exports, mainly coffee. The first section of railway on the coastal plain, a 40-mile line between Rio and Belém (now called Japeri), had been completed in 1858, and then the promoters had to face the task of climbing the Serra. They chose North American rather than British engineers, who decided that the cliff could be climbed without the use of a rack railway, but instead reduced the gradients to a maximum of 1.8 percent through the construction of 3 miles of tunnels to ensure that the trains could climb the slope under their own steam. The longest of these tunnels was a mile and a half long and took seven years to build. As testimony to the vision of the engineers, the line is still in use today with very heavy traffic, including trains that weigh twenty-five times more than originally expected, and it remains the most efficient of the eight railways that were eventually built up the Serra.

The fact that the railways in Brazil were confined to a relatively narrow area did not diminish their importance. Though they were primarily intended for freight, passengers flocked onto them and communities grew up around the freight depots, which soon turned into busy stations. The completion of the railways had a curious side effect, the virtual abandonment of the old roads and mule tracks, which were difficult to use and expensive to maintain. The railways quickly gained a monopoly position with the result that many villages and small towns on those roads were abandoned. The monopoly was so entrenched that one major road, the Unidão-Indùstria, was even ripped up to allow the railway to be laid on its bridges and much of its route. Roads, of course, would get their revenge in hundreds of places around the world in the twentieth century, when the railways would suffer the same fate, their bridges and embankments handed over for use by motor transport.[11] Because of the primitive nature of the existing transportation system in Brazil, the impact of the iron road was, according to the historian of the Brazilian railways, "greater than in Europe or the United States, where many animal traction and stagecoach firms existed before the railways, and which were therefore less sensitive to the changing habits."[12]

The symbiotic relationship between railways and development is illustrated by the pattern of economic growth in Brazil. New railways were concentrated in coffee areas, since the capital was available from profits and the need was pressing, whereas few railways emerged in regions where sugar, once the country's main crop, predominated because by then the sugar industry was struggling in the face of competition. By 1876, Brazil had 1,300 miles of railway, and over half of them were in the southern coffee regions of Rio de Janeiro and São Paulo, whereas the main sugar region, Pernambuco in the northeast, had just 150 miles, including those still under construction. Unlike the southern railways, the railways of the northeast struggled to make an adequate return, which discouraged expansion. Competition from abroad kept sugar prices down and meant that sufficient capital could not be generated to allow for the construction of railways in sugar-production areas. Sugar cane was bulkier, too, than coffee beans, and therefore more expensive to put on rail. The barges that had traditionally been used for sugar transport were thus able to continue operating profitably.

In Cuba, by contrast, where no such waterborne transport was available, sugar did stimulate the creation of an extensive rail network. A narrow island 800 miles long, Cuba was, as mentioned in Chapter 2, one of the earliest countries in the world—and the first in the region— to have a railway. Sugar plantation owners had long tried to find a way of transporting their produce more cheaply, but there were no suitable rivers and the roads were too poor to withstand the tropical climate. Therefore the country soon built up an intricate network of railways almost entirely geared toward servicing the needs of the sugar industry. As with coffee in Brazil, the development of the sugar industry and the railways themselves in Cuba were so interlinked that neither would have happened without the other; it is uncertain, however, which one was the main catalyst for the growth of both industries. The railways were essential for turning sugar into a product that could be exported worldwide, but clearly they would never have been developed without the need to provide transport for the crop.

At the time of the opening of its first railway in 1837, Cuba was blessed with just the right set of circumstances for a significant growth in sugar production: an abundant supply of fertile land, principally in the interior of the country; substantial capital accumulated by healthy profits from existing production; a rising world demand; and an ample supply of cheap imported labor, either slaves or very poorly paid indentured labor working in appalling conditions. The inauguration of the first railway, built to connect the rich valley of Güines with the city of Havana, immediately changed perceptions of how agricultural produce could be transported, as the railway offered a method that was far quicker and more economical than the carts on poor roads used previously. A spate of railway building was promoted through a process that was almost identical every time, stimulated by a group of plantation owners and merchants who joined together to promote a line in order to facilitate the export of their produce. Primarily, however, these local investors were not interested in the railway enterprise itself but only in how these lines would benefit their business. By 1852, nine companies had built a total of 350 miles of railway to serve the sugar market. The pace slowed briefly when production of the crop stagnated; then, as output boomed again, so did railway construction, and the country embarked on a veritable railway mania. Mileage reached a total of just under 800 in 1868, when the rate of growth fell again as a result of economic difficulties caused, inevitably, by drops in the price of sugar.

Given that Cuba was an undeveloped country with barely any industry other than sugar, and was treated parsimoniously by its colonial power, Spain, the density of the railway network it had created by the 1860s was quite remarkable. Indeed, in terms of railways per square mile, it was bettered only by a few developed European nations, but unlike these, the Cuban population benefited little from the railway system. The lines were built exclusively to carry sugar, and even their intersections did not allow for timely changes between trains, because junctions were designed primarily to afford the sugar shippers the opportunity to choose among ports for the export of their produce. Consequently there were long delays in the spartan stations, which were not even blessed with any permanent structure, let alone waiting rooms. Passengers had to endure circuitous routes, too, as the lines were laid out to maximize access to the various large plantations rather than to provide direct links between major towns, and journey times were consequently slow, although still much faster than any horse-drawn transport on the terrible roads. While small towns did emerge at the junctions, they did not turn into major centers of development as was the norm elsewhere. Nor did the railways stimulate the wider economic growth that they brought in their wake in more developed countries. This was due to their very limited focus and the use of entirely imported technology, principally from the United States.

In fact, the economic effect of the railways was quite perverse. By enabling the plantation owners to continue to grow sugar profitably, they helped perpetuate the slave system that might otherwise have collapsed as a result of its economic inefficiency. Moreover, oddly, the railways exacerbated regional differences between prosperous and poor areas. That was because the railways in Cuba were built to tap those boom areas with flourishing sugar estates; no railway promoter aimed to lay down a railway in the hope of kick-starting a backward region. As the government declined to provide subsidies in the less-developed regions, the railways were mainly concentrated in the western half of the island, while in the east just a couple of lines served the few sugar plantations. As the historians of the Cuban railways put it, "The rail-road development of the first decades lacked the long-term perspective that would permit the growth of a national grid."[13] Despite mild support from the Spanish government, a plan to build an east-west line through the spine of the country, which would have provided an efficient transportation network for most of the population, was shelved because there were not enough promoters. A through line was eventually built in the early twentieth century, but even then it became part of the sugar railway network, as the company building it immediately planted cane along its path.

Despite these shortcomings, the importance of the railways in Cuba should not be underestimated, as they were crucial to the development of the sugar industry. In the thirty years after the construction of the first line, sugar production, which had already been the mainstay of the island's economy, increased fivefold thanks to improved transportation. And the railways, rather than improvements in sugar technology, were crucial in stimulating that growth: "The railroad and not the introduction of steam engines in the sugar mills is the first element of the Industrial Revolution that completely altered the productive conditions of Cuba," according to one of the leading scholars of Cuban history.[14] The railways were also profitable because most of them had a monopoly. They were able to charge high rates to make up for the fact that their business was very seasonal and almost exclusively one way, since they took produce to the ports and returned empty. The railways were mostly owned by the groups of local sugar plantation owners who had promoted them in the first place, which partly explains their limited focus. It was only in the late nineteenth century, when the railway companies fell into economic decline as a result of yet another collapse in the price of sugar, that they were taken over by British investors (who until then had played only a minor role in their development) and, to a lesser extent, U.S. companies. The Cuban railway network at its peak was one of the densest in the world, with 5,000 miles of track. Half of the lines used the standard gauge and were employed principally to shift sugar out of the plantations; the other half used narrow-gauge tracks to haul cane on the plantations themselves.

As well as Cuba, several other countries of the Caribbean provide interesting insight into the ubiquity of railways at their peak in the period running up to World War I. Intuitively, most of the islands seem to be unpromising territory for a railway, given their size, but the majority developed networks nonetheless. Sugar was the predominant reason for the construction of most of these rail systems, with even small islands developing railway lines to transport cane to a port. For example, the tiny island of St. Kitts built a 30-mile railway around nearly the whole coast, cutting off a peninsula, to link all the sugar plantations. The even smaller neighboring island of St. Croix had a railway of about 12 miles long, also serving the sugar industry, and at the other end of the scale, the Dominican Republic, which shares the island of Hispaniola with Haiti, built a remarkable series of sugarcane railways that extended for more than 1,000 miles and made use of a variety of gauges, from the tiny 1 foot, 10.5 inches, up to the standard 4 feet, 8.5 inches.

Other railways in the Caribbean served a variety of obscure purposes. In the Bahamas, a mule-hauled railway was built in the 1860s to transport sea salt; later, another was constructed for carrying timber. Dominica, too, built a small timber railway, while Bermuda, in the 1920s, established a 22-mile line solely for the purpose of providing for passengers from the numerous visiting cruise ships. On Trinidad, where a 150-mile network was eventually built, the railway was created partly to serve the unique asphalt extraction industry.

Apart from Cuba, the largest general-purpose railway in the Caribbean was the one developed in Jamaica, a substantial island 145 miles long. The British colony opened its first railway as early as 1845, linking the capital, Kingston, with Spanish Town 13 miles away. One of the first drivers, Isaac Taylor, suffered the ignominy of incurring the island's first speeding ticket when he was fined £2 by the Jamaica Railway Company for having taken his train up to 40 miles per hour, twice the permitted speed, causing panic among the passengers. The inaugural line was the start of a network that extended across the island to Montego Bay by 1894, a distance of 112 miles, and the network eventually reached nearly 200 miles. In addition to freight such as bananas, coconuts, oranges, and, inevitably, sugar cane, substantial numbers of passengers were carried as well. The island also had several freight railways created solely for the transport of minerals, notably bauxite and alumina, from mines to ports.

As elsewhere, there were many projects in the Caribbean that were never completed. Puerto Rico, a Spanish colony until 1898, when it was ceded to the United States, had ambitious plans to build a 283-mile coastal line around the whole island for passengers, but only 100 miles were ever completed. Painfully slow services were provided on it until the railway's closure in 1957.

Aside from Brazil and Argentina, it was in the Andes, remarkably, that most of the significant Latin American railway developments took place. Peru, Chile, and Bolivia all heroically built railways climbing to heights undreamt of elsewhere. Of the thirty-five lines that reached 2 miles (10,560 feet) or more above sea level around the world, all but two[15] were in South America, mostly in Peru or Bolivia. These Andean railways were built primarily to bring the mineral wealth in the mountains down to ports, but they also provided passengers with the most spectacular, and occasionally perilous, rail journeys in the world. They were used by the strangest collection of rail coaches, converted buses, and automobiles ever to run on railway tracks.

A series of railways was built running to the sea from the mountains (mostly unconnected with each other). Chile, a country 2,700 miles long and barely 100 miles wide, did build a north-south line running approximately two-thirds of its length—inevitably not all in the same gauge—and this eventually spawned thirty-five branches, largely to service various mines. Elsewhere the lines ran almost exclusively between the sea and the mountains on a broadly east-west axis. The most spectacular and ambitious railways were built in Peru, and it was Henry Meiggs, who had rescued the Santiago-Valparaiso scheme, Chile's first significant railway (mentioned in Chapter 6), who was the crucial figure in the construction of these remarkable railways. Meiggs—or "don Enrique" as he was known locally—had a close, and not entirely above-board, relationship with the Peruvian government, and when a plan to build a line deep into the Andes was proposed, he was the obvious contractor.

Brian Fawcett, the author of the standard history on the railways of the Andes, has argued that "besides the Central of Peru all other Andean systems are child's play."[16] While that assessment may have a whiff of a rail fan's overenthusiasm, it is unarguable that the conquest of the Peruvian Andes was the most ambitious of all the great railway projects. The first line in Peru, like the inaugural railway in so many countries, linked the capital with a port—in this case, the city of Lima with the port of Callao just 9 miles away. Extending this small section of the Central up the mountains to exploit mineral wealth was a priority for the Peruvian government, which had filled its coffers on the brief guano boom (bird droppings that had accumulated over thousands of years were gathered and exported as fertilizer in the mid-1800s). Meiggs's notion was to follow the route where the llamas went, perilous ledges carved into the sides of the mountains. These old paths did, indeed, offer the easiest route, but the climb up to the highest point of the main line, 15,700 feet above sea level at the Galera Tunnel 100 miles from Lima, was relentless. The line was built through a series of valleys that were wide and gentle at their mouths but became steeper and narrower at the head, where invariably they would give out onto another gorge. To bring the railway up these steep inclines, Meiggs used more than a dozen zigzags with short straight sections at the end where the train could be switched from the lower line to the upper one, the same method developed to climb the Ghats from Bombay twenty years previously (see Chapter 2). It did not make for fast or easy operation, especially when the trains were longer than the level sections at the end of each zig or zag, resulting in complicated shunting maneuvers, but it was the only feasible way to build the railway at a reasonable cost. The locomotives alternately pulled and pushed their loads up the incline, which meant that the rail journey averaged barely 15 miles per hour, but that was still much faster than could be achieved on the old llama trails. Meiggs, however, built inclines that were just about gentle enough, 4.4 percent,[17] to allow the locomotives to climb by simple adhesion rather than racks or cables, which he thought were too expensive and complex. The switchbacks offered passengers a thrill worthy of the best fairground rides, since their only protection was the flimsy buffers that would not prevent a runaway train from plunging into the abyss.

Meiggs used American rolling stock, which was sooner adapted to the particular circumstances of the Andes than British stock. English locomotives were not sufficiently robust to cope either with the steep gradients or the rough track. The civil engineering methods and equipment were largely American, too. A bridge built by the British for

£3,350 (about $16,500 at the time) soon collapsed and was replaced by an American one that proved both serviceable and durable; it cost just £2,000 ($10,000 at the time), weighed half as much as the first bridge, and took just eight days to erect (compared with eight weeks for the British one). The gauge was the standard 4 feet, 8.5 inches, which was remarkably ambitious for a railway in such difficult terrain. It made construction far more expensive than it would have been with a narrower gauge, but it eventually enabled the railway to carry far greater loads.

Meiggs died in mid-project in 1877 at the age of sixty-six (the cause was not recorded). The line had reached a location only 90 miles from the coast, and work had stopped, inevitably because of a lack of money. The death toll of his workers was notoriously high, partly because many were Chileans unused to the height and with no immunity to a virulent local disease, Oroya fever, that affects the liver. Some estimates suggest that as many as 10,000 workers died, principally from disease, but, as ever, no attempt was made to determine the exact toll.

Construction continued in fits and starts, halted by a war with Chile as well as a shortage of cash, but the Peruvian Corporation, a British-owned company that took over much of the country's rail network in 1890, was contracted to complete the project. Work restarted that year but was not completed until 1908, thirty-eight years after the first sod had been ritually turned. After the Galera Tunnel, the highest on any railway in the world, the line descends into the high valleys, but it still remains above 10,000 feet for much of its route. In the early days of the line, passengers accounted for just 6 percent of revenue, and they were seen as a burden by the rail managers, as freight-only railways can be operated far more cheaply and with lower standards of maintenance. Moreover, goods do not complain of bumpy rides and are oblivious to the occasional derailment. Emergency oxygen had to be provided for the passengers, because many invariably suffered from altitude sickness.

Despite Fawcett's suggestion that the other railways in the Andes were far easier to build than the Central, there were several others whose construction stretched the limits of contemporary technology. As well as the transandine (described in Chapter 6), there was the Southern in Peru, another standard-gauge railway, which Meiggs had started building simultaneously with the Central, and which also became part of the Peruvian Corporation. It also took amazing engineering skill to build, and it rose swiftly, though not quite as impressively as the Central, from the Pacific coast far into the Andes. Starting at Mollendo on the coast, after 100 miles it reached the city of Arequipa, at 7,500 feet, then climbed up to Crucero Alto, 14,666 feet above sea level. One branch was built to Puno on Lake Titicaca, the world's highest navigable lake at 12,500 feet above sea level, where a ferry connected with the meter-gauge Bolivian railway at Guaqui, while the main line continued through Cuzco, the Inca capital, eventually descending into the watershed of the Amazon and terminating at 11,000 feet.

Meanwhile, in Bolivia the huge complex of the Antofagasta Railway was being created to exploit the nitrate wealth of the uninhabited southern area bordering Chile and the copper and tin mines of the interior. The first section of the Antofagasta, built to a remarkably narrow gauge of just 2 feet, 6 inches, was initially operated by mules when it opened in 1873. Steam locomotives were introduced three years later and by the end of the decade the railway already stretched 100 miles away from the coast. The Antofagasta Railway was eventually built up into a remarkable network of 1,821 miles, including a main trunk route through the Andes to La Paz, 729 miles from the coast, with various branches to serve mines. It stretched deep into Bolivia and Peru and even later connected with the Argentine rail system. The Antofagasta was built, oddly, in two gauges: Half of its length used the tiny 2-foot, 6-inch, gauge, while the main Bolivian section was meter gauge. The line has no zigzags or rack sections as it runs through slightly easier territory; as with the Australian transcontinental, the main obstacle to both construction and operation was the lack of water for the initial section, which reached up from the Chilean port of Antofagasta. It crosses a desert where it never rains, which means there are no water courses, and this greatly increased the cost of construction. Again, it was British capital that financed most of the construction of the line after the railway company, the Ferrocarril de Antofagasta a Bolivia, was floated on the British stock exchange in 1888.[18] Passengers were never more than an afterthought, although for a time dining and sleeping cars ran on the tiny 2-foot, 6-inch, section, and, along with nitrate, the railway became vital for the transport of copper and several other minerals. It was not until 1928 that the final 180 miles in the dry regions up from the coast were converted to meter gauge, a massive enterprise given the lack of water.

There are several contestants for the prize of the world's toughest railway ever built. Though the Trans-Siberian or the Panama may seem like obvious winners because of the sheer scale of the projects and the difficulty of the enterprise, and the remarkably high lines in the Andes have to be considered, there is another in Latin America that should be up on the podium sharing the champagne, partly because of the sheer madness of the project. Brazil's most remote railway was an enterprise so fraught and seemingly without obvious purpose that it is incredible that it was ever undertaken, let alone started five times before it was completed. The 230-mile Madeira-Mamoré railway was probably the most isolated line in the world. Its story is particularly poignant because it took thirty years to complete, and soon afterward became virtually redundant owing to a change in the pattern of the rubber industry, its principal raison d'être. The Madeira River is navigable from the sea as far as a town called Porto Velho, but access to another 1,000 miles of navigable river upstream in the Amazon jungle is obstructed by a series of rapids. Efforts to tame them through canalization proved unsuccessful, and a company was formed in London in 1872 to build a railway to circumvent this obstacle. The first attempt to build the line foundered soon after work started, and a subsequent expedition in 1874 also made no significant progress, as the workers succumbed to everything from malaria and yellow fever to attacks from alligators and local indigenous tribes.

A third attempt was made by an American contractor, P&T Collins of Philadelphia, which managed to build 4 miles of meter-gauge track and bring in an American Baldwin locomotive before its supply steamer sank in a storm. Not only did the disaster cost the lives of the eighty men on board, but the loss of the ship cut the 1,000 American workers off from civilization; around a third of them succumbed to starvation or disease as they left the site to make their way 2,000 miles downriver to Belém. A fourth effort in this saga was begun in 1883 when a Brazilian team surveyed the route, but the hapless crew had to be helped out after they all fell ill. As one historian of South American railways put it, "for twenty years the jungle grew over the Baldwin locomotive before local political troubles revived the lost cause of the railway."[19]

It was because of the rubber boom at the turn of the century, stimulated, ironically, by the demand for tires from rail's great rival, the automobile, which made the project viable again. Bolivia had ceded some territory to Brazil and in compensation was promised an outlet to the Atlantic Ocean via the river and the Madeira-Mamoré railway. The railway company was reconstituted in 1907, and lured by the promise of high wages, workers came from the Panama Canal and the West Indies to build the line. Over 2,000, however, were to pay for their daring with their lives before, five years later, the line was completed, running between Porto Velho and Guajará Mirim. Amazingly, the old Baldwin engine was recovered from the jungle and put into service along with another eleven locomotives from the same supplier.

Almost as soon as the line opened, however, rubber seeds smuggled out of South America were taken to Malaya, which had a similar climate, and profitable plantations were rapidly established there. This knocked the bottom out of the market for rubber from inaccessible places like the deep Amazon, where the difficulty of transporting the product would inevitably make it far more expensive. Along with the freight trains, there were, for many years, three weekly passenger services. These trains undertook a journey that could never have been very pleasant in the heat, since it took between twelve and eighteen hours to cover the 230 miles, and yet even today a small part of the line has been preserved for tourists.

Like Argentina and Brazil, the biggest country in central America, Mexico, had a railway boom that resulted in a rapid expansion of its system from just 400 miles in 1877 to 12,000 miles by the start of World War I. The spur was a government law enacted by Porfirio Díaz, a despot but a modernizer, which not only gave generous subsidies to the construction of railways, which were seen as a vital catalyst for economic development, but also provided draconian powers to the railway concessionaires to seize land. The first significant railway in Mexico was a transcontinental line from the Gulf of Mexico to the Pacific completed in 1873; its success motivated Díaz to offer concessions for contractors to build a huge network, most notably a line running through the spine of the country to link Mexico with the United States.

Elsewhere in Central America, the networks were mostly sparse. Many were designed to serve the needs of the growers of a single crop, particularly bananas or sugar. They were built in a wide variety of gauges, which proved to be a major obstacle to their development. Minor C. Keith, the nephew of Henry Meiggs, wanted to blend them into a coherent network linking up much of Central America. Keith began as a railway pioneer and struggled to build Costa Rica's first line, but he later became a banana magnate as well, the founder of the United Fruit Company. His aim was to create a system of lines throughout the region; he was defeated, however, by the scale of the task, the difficulty of the terrain, and the political divisions of the region. The dense network of lines in Honduras, for example, was designed solely with the banana crop in mind and did not even reach the principal city, Tegucigalpa, making it one of the few capitals in the world without a railway. Similarly, Guatemala built up a 500-mile 3-foot-gauge network, starting in 1877, which served the banana industry but did not connect its major towns, while El Salvador, beginning slightly later, eventually had around 300 miles of railway that offered few useful services to local people. Like many of the railways of the Caribbean islands, the great majority of these networks are now abandoned, though some special-purpose railways survive to serve a particular mine or an agricultural product.

BACK IN EUROPE, railways were popping up in the most unlikely places. The islands of the Mediterranean were another strange setting for a series of railway networks. The Sicilian rail network was a perfect example of a system built with funds from the government, which saw it principally as a social service and a way of stimulating economic development rather than as a viable enterprise. The first railways were initiated in the early 1860s by the Società Vittorio Emanuele, which planned to build two lines running broadly north to south across the whole island, a total of nearly 400 miles. The motivation for the construction of lines in such an underpopulated and impoverished area as Sicily was the hope that the reunification of Italy would stimulate the same kind of economic development that had followed the recent Civil War in the United States.

The Vittorio Emanuele had funds available for investment, having been forced to sell its French holdings to the Paris-Lyon-Mediterranée, as part of the spoils of war after Piedmont, with the help of Napoleon

III, had defeated the Austrians in Savoy. Eager to keep expanding its railway, the Vittorio Emanuele took up generous government subsidies to embark on its Sicilian scheme. But the plan was far too ambitious, even for the supposedly cash-rich company. Work started in 1863, but after five years just under 100 miles had been completed and the company went broke, leaving, as ever, the government to pick up the pieces. The slow progress was partly a result of malaria, which was then endemic on the island. In the summer up to 80 percent of the workforce was absent at times because of the disease. Its cause was still unknown, and its debilitating effects were not understood by the company. Even those who turned up to work were often weakened by its effects. The state continued the work, and the initial plans were completed by 1872. A rail network of more than 1,500 miles was completed by World War I, far more than a sparsely populated island less than 200 miles long would ever need.

Sardinia, the next biggest island in the Mediterranean, built up an extensive railway system far more quickly than Sicily did, again with the help of generous state subsidies. It, too, was a malarial island, with stretches on the western side that were so little populated because of the disease that the navy used the area for target practice. And yet a standard-gauge railway running up the coast for 260 miles, virtually the whole length of the island from Cagliari in the south to Golfo Aranci on the northeast coast, was constructed with remarkable speed. It took just a dozen years from the start of construction in 1863 to complete the line, even though there were the usual delays caused by lack of money. The line, which was primarily intended for passengers, though there was some freight, was built by an Englishman, Benjamin Piercy, who designed a mansion for himself at Macomer in the center of the island, 100 miles from Cagliari. The author of a history of Mediterranean railways likens it to "a Victorian country house of the 1880s [that] would blend without remark into stockbroker Surrey or, better still, the outskirts of Sheffield or Leeds."[20] An extensive system of narrow-gauge railways, which ran to nearly 1,000 miles at its peak, provided connections to every corner of the island, albeit at a grindingly slow pace. The terrain was frequently mountainous, and the railways had been built cheaply in order to keep down costs. The train on the 142-mile trip between Cagliari and Arbatax on the east coast took twelve hours when steam-hauled, and even the diesel railcars in operation in recent times make for a seven-hour haul. The ride is punctuated by stops at remote stations where shepherds board from a rugged countryside without a house in sight, accompanied by the occasional goat.

Three other Mediterranean islands—Corsica, Cyprus, and Majorca—developed mainline routes linking their major towns, but, with few branch lines, their railways were on a much smaller scale than those of the two large Italian possessions. Malta only managed a small suburban railway in Valletta. Just 7 miles long, it ran from 1883 to 1931, apparently providing "the tribulations offered [by] contemporary British railways [to] the season-ticket holder of London or Manchester."[21]

THE FAR EAST was also something of a latecomer to the railway age. By and large, the railways arrived there in the last quarter of the nineteenth century, reaching, for example, Japan in 1872, China in 1883, and both Malaya and Vietnam in 1885. Japan, which had only recently given up its centuries-long isolation, embraced railways just eighteen years after it was forcibly opened up following the arrival of the U.S. Navy in 1853. The U.S. mission had been led by Commodore Matthew Perry, who had brought a model railway with him. It caused a sensation in a society that previously had made very little use of wheeled vehicles. Despite this American influence, it was the British who were responsible for both the financing and the construction of the early lines. An Englishman bearing the grand name of Horatio Nelson Lay raised £1 million ($5 million) from the London money markets to build the first line between Tokyo and Yokohama, and British engineers were sent out to undertake the survey and organize the construction. They found a land that was ill-suited to railway construction. The country was mountainous, and the lines would be threatened by both earthquakes and landslides. Inevitably, costs were higher than anticipated.

Nevertheless, the initial section of the line was completed in 1872, and the importance of the event can be gauged from the fact that at the opening ceremony, in a break with tradition, the emperor appeared for the first time in full state regalia in front of foreigners. Further progress on the railway was slowed not only by the difficult geological conditions but also by shortages of government money and by a violent civil war, which stopped all work between 1876 and 1882. Japan was a very different place from the country it is today, and the early train services reflected the pace of life of a society that had been isolated from the rest of the world for many centuries. When the line between Kobe and Tokyo was completed, trains took about a day to travel the 350 miles on a winding track.[22] There were about fifty intermediate stops, making for an average speed of just 15 miles per hour. As an early traveler, B. D. Timins, fondly described it, Japan through her railways offered a "new and surprising sensation to the jaded globetrotter—that of being unable to hurry."[23] For the peasants in third class, the lengthy journey had to be endured in stifling hot weather on seats lacking upholstery, but in first class Timins found "the seats ran round all four sides of the carriage [with] in the centre a diminutive table, upon which reposes a tea service properly furnished with all necessary requisites for brewing a cup of green tea *à la Japonaise*." He loved, too, the *bento* lunch boxes on sale at every station and was skeptical that the Japanese would ever seek to change this gentle pace of life, explaining that people arrived at the station several hours or even the night before the train was due, eschewing all notion of "speed, haste or flurry," which were entirely foreign to their nature. Ah, the pitfalls of prediction!

Despite the stuttering start and the slow service, the idea of the railways had been sown, and with the government short of cash, several private companies sprang up in the 1890s. By the turn of the century, Japan was well on the way to obtaining the heavily used network for which it is famous today. Curiously, even though most of the early locomotives were British, the gauge was 3 feet, 6 inches, the size used in New Zealand and parts of Australia.

Like Japan, China was slow to take advantage of the new technology, largely because the development of the railways became bound up with nationalist politics. At the time of the opening of the first railway, parts of the country were occupied by European and Japanese "missions," which were effectively an occupying force that undermined the weak Chinese administration. It was these foreign interests that were keen on building railways in order to exploit the country's resources, but they were met with fierce local resistance. Indeed, the very issue of modernization was so fraught that the country's first railway between Shanghai[24] and the port of Woosung, completed in 1876, was ripped up the following year. The Chinese administration was angered that it had been built by foreign interests, despite the fact that control of the railway was handed over to China in an effort to encourage its acceptance. Nowhere else in the world did the advent of the railways stimulate such controversy and opposition. But that was inevitable given that for centuries the country had been ruled by an isolationist and conservative regime with a strong xenophobic element that had resisted any kind of change. It was as if every possible argument against railway construction that had been raised in other countries coalesced in China. This slowed the pace of expansion in a country whose size should have made it a natural for early development.

The opposition was led by high officials, the mandarins, whose objections chimed with popular feeling. In particular, would the railways displace thousands of coolies from existing jobs and cause unrest? The core Luddite argument was elegantly put by one Chinese official, Chang Tzu-mu: "The peasants in the fields of the south, the miners in the mountains of the north, and those who haul the cars and man the boats—these number many millions of people. They bear the calluses of their struggle to survive. If machines are suddenly introduced, they will lose their livelihood. Won't they join together to cause disturbances?"[25] Or, as another writer put it, thousands of people would "end up starving in the ditches [or] gather as outlaws in the forests."[26] These officials spoke with the awareness that several recent uprisings, notably the majestically named Red Turban Rebellion of 1854, had been started by transport workers who lost their jobs as a result of changes in government policy or technical innovation.

Some objections were rooted in fears of change, such as the official who was worried that "when one uses coal with such profligacy, coalfields soon disappear."[27] Or the fear that railway property would be a tempting target for thieves. However, there remained well-founded economic objections to the development of the railways in China, which still stimulate debate today. Given the widespread poverty, would the Chinese be able to afford the price of rail travel? Would the concentration on developing railways damage other efforts to modernize the economy? With the railways' undoubted need for subsidies and the desire of foreign companies to tailor the railway to their narrow interests rather than national considerations, the fundamental question of whether the railways would be good for Chinese interests remained.

As a result of these fears, the Chinese refused foreign investment in their railroads, which resulted in lamentably slow development. The first line built after the destruction of the Woosung was completed in 1881, but for the next decade and a half little progress was made, with a mere 18 miles per year being completed until 1895. But with foreigners then being allowed back to invest, and the Chinese authorities realizing, following the Boxer Rebellion at the turn of the century, that railways could be used to carry troops quickly around the country, the pace increased. A railway boom developed, with 350 miles being completed annually until the revolution of 1911 and the outbreak of World War I. The war then slowed progress. The most populous country in the world had at last embraced the railway age, though only reluctantly. Total mileage by 1911 was just 6,000 miles. China was way behind other countries. As Nicholas Faith has pointed out, China's mileage was "half the figure for India, [which was] smaller and itself not oversupplied with railways."[28]

New Zealand used the same gauge as Japan—3 feet, 6 inches—but presents an unlikely contrast. For a time it had more railway mileage per head than any other country;[29] the railways were used by the government not just as a way of stimulating economic development, or even as a way of cementing the country together, but rather as a means of creating the nation. New Zealand, a country of barely 500,000 people, including the native Maoris, made a desultory and late start to railway construction, with only a few short lines, comprising just 46 miles, completed by 1870.

Then a politician came up with a plan and kept to it. Julius Vogel, the colonial secretary, published a plan to reawaken the "colonizing spirit" by embarking on a massive publicly driven program of railway construction. His idea was to use the railways to fuse the colony's far-flung and isolated settlements into a unified, prosperous nation. The cornerstone of this development drive was a commitment to building 1,000 miles of railway over the next nine years. To keep costs to a minimum, the railway would be built on the American model using light rails, with sharp bends and stiff gradients. The builders would also plan for low line speeds. Although some of the earlier lines had been built to the generous Irish gauge, Vogel insisted that the New Zealand system would use a 3-foot, 6-inch, gauge, also in the interest of keeping costs down. It was a project with a strong socialist and nationalist flavor. Vogel hoped that the construction, funded by public borrowing, would create employment for the indigenous Maori population, stimulating integration, and thus spearhead a peaceful European conquest of the Maori heartland. The settlers attracted by the railway would, he hoped, soon outnumber the Maoris. But there was an inherent threat in this policy, as Vogel later acknowledged, saying: "The Public Works Policy seemed to the Government the sole alternative to a war of extermination with the natives."[30]

The program was well-timed. With the British railway network almost complete, contractors and laborers were looking for work and therefore were willing to travel to the far end of the Pacific in search of riches. Inevitably there were delays caused by all the usual problems—construction difficulties, manpower shortages, disputes over finances and methods—but within five years 316 miles had been built and the program was in full swing. Vogel had kickstarted New Zealand into the railway age and it never looked back. Neill Atkinson, the author of a history of New Zealand's railways, explained that "in the 1870s, iron and steel rails welded cities and towns to their hinterlands, connected farms, forests and mines to markets and ports, and widened New Zealanders' physical and cultural horizons. By establishing reliable overland communication between the original coastal 'islands' of European settlement and 'opening up' previously inaccessible inland districts, railways promoted centralization and standardization and helped forge a straggle of disparate provinces into a modern nation."[31]

This was very much a government enterprise. A remarkable 48 percent of the government's entire public works budget was spent on railway construction between 1870 and 1929, more than on roads, telegraphs, immigration, tourism, defense, mining, and much else put together. The railways were not expected to make a profit, as they were perceived as a public enterprise serving the interests of the country as a whole.

It was, though, not all good news. The railways provided access to the Maori heartland of the central North Island, hastening the destruction of the environment by transforming virgin forests into British grasslands: "The chief beneficiaries of the rail revolution were the Pakeha [European] inhabitants of the main centres, the towns that flourished along the rail routes, and the freshly opened hinterlands."[32]

For the Maoris, as for the Indians of America and the Aborigines of Australia, the railways were an instrument that hastened the destruction of their traditional way of life.

By the end of the century, then, apart from large swathes of Africa and the Middle East, a few islands, and the occasional laggard, such as Albania, the railways were ubiquitous. Every significant country had joined the railway age, and wherever lines had been built, the changes they brought in their wake were profound and irreversible, as we see in the next chapter.

NINE

The Railway Revolution

I T IS ALMOST IMPOSSIBLE TO EXAGGERATE THE PROFOUND IMPACT OF the railways. They transformed the agricultural economies, which had prevailed since mankind emerged from the caves, into the industrial age. By the latter part of the nineteenth century, in the more advanced European countries and in flourishing economies elsewhere such as the United States, South Africa, and Argentina, the railways were omnipresent. Every town of substantial size was on the railway map, as were many villages and hamlets, even if, in many cases, their connection was through a narrow-gauge branch line built on the cheap and served by infrequent trains. There was barely an aspect of life in the later Victorian years left unaffected by this remarkable invention, which stimulated changes that were later entrenched and intensified, rather than initiated, by the arrival of the automobile. It is almost easier to list aspects of life left unchanged. The railways were developed before those other life-changing aspects of technology—before electricity for lighting, gas for cooking, the telephone, the typewriter, and the bicycle—and therefore their creation marked the dawn of the modern age.

The Chinese mandarins who objected to the establishment of a railway network in their country may have been misguided, but their instincts about its profound impact were sound. Their arguments were rooted in the history of a country with a strong isolationist tradition and their opposition was influenced by narrow national interests, but that does not mean their fears could be dismissed as reactionary nonsense.

Although it is undeniable that the railways stimulated economic development and social progress, they also disrupted existing ways of life that had survived for centuries, even millennia, and this process was bound to generate significant numbers of losers as well as winners. While broadly most people's lives improved when the railway arrived in their town or village, there were exceptions; inevitably, some lost out as their business or trade disappeared or they became the victims of unfair treatment at the hands of monopolistic railway companies.

It was not only in China that the arrival of the iron road was vehemently opposed. In Mexico, the Catholic Church was deeply skeptical of its worth. In 1856 railway promoter Manuel Escandón built a 3-mile railway between Mexico City and the shrine of the Virgin of Guadalupe to win over ecclesiastical support and counter the impression that the railways were a corrupting influence. He had to transport the locomotive several hundred miles by cart, and all this effort was just to run a railway on Sundays and the twelfth of every month (the day of the Virgin).

In England, romantics such as William Wordsworth and John Ruskin were horrified at the incursion of the railway into the countryside. Wordsworth opposed, in particular, the construction of a railway to Windermere, in the center of the Lake District. He was an early environmental campaigner, but rather than setting up a protest camp, he wrote a series of poems and letters to the *Morning Post* in which he suggested that there was no need for a railway because there were no quarries, industries, or substantial agriculture to warrant it. Then, when it was pointed out that the railway would open up the Lake District to tourists who could enjoy the stunning countryside, he responded sniffily, claiming the working class did not have the capacity to enjoy the beauty and "character of seclusion and retirement" offered by the Lake District: "A vivid perception of romantic scenery is neither inherent in mankind, nor a necessary consequence of a comprehensive education."[1] Then he deployed that old argument of the early tourist, which of course has a grain of truth in it, that the masses would destroy the undiscovered beauty of what they had come to visit.

Given the unpopularity and intellectual weakness of such arguments, the opponents, for the most part, were brushed aside or shouted down. The power and savagery of the steam locomotive is echoed in the sheer brute force of the railway itself. At its peak, nothing could stand in its way. By the outbreak of World War I, just eighty-four years after the opening of the Liverpool & Manchester, the world's mileage had reached 630,000 and was still rising, with the car and the truck only just beginning to emerge as competitors.

In the main, the railway promoters and their sponsors in government won the argument because the success of the early railways plainly demonstrated their beneficial effect and their ability to generate profits, and, quite simply, because railways were seen as exciting in the same way that cars would be perceived a century later. As a result, the speed and extent of the changes across the world over the space of just two generations, between 1830, the year of the completion of the Liverpool & Manchester Railway, and the turn of the century, had no parallel in history. This was no accident. As mentioned in Chapter 1, the preconditions for this change had been set by the Industrial Revolution of the eighteenth century, which meant that in 1830, unlike, say, fifty years before, the world was ready for the railways and the changes they would bring in their wake. Railways obviously needed steam engines, but they needed capital, too, and it was the availability of both that paved the way for their invention and rapid spread. The crucial innovation, the placing of a steam engine on purpose-built tracks, was made possible by a host of technical—and, indeed, social and financial—changes triggered by the Industrial Revolution. The simplicity of this invention was the crucial factor that ensured the technology could be readily imitated and developed. Thanks to their adaptability, the railways provided an organized form of power in the shape of a mobile and efficient source of energy unlike anything before—as witnessed in the variety of sizes and gauges of the early railways and locomotives. Of course, the much earlier invention of the wheel had made societies more mobile, but only in a limited way; it allowed for relatively short journeys, and for the most part people's horizons remained restricted to a few miles around their homes. The railways operated on a far larger scale.

The other key innovations—a flanged wheel running on a metal track, together with the traction provided by steam-powered locomotives—allowed the carriage of loads that were ten or more times heavier than anything previously hauled by man or beast. Moreover, passengers could be carried in unprecedented numbers, the equivalent of dozens of stagecoaches on one train. As a historian of the American railroads put it, the difference "between a primitive railroad and a typical horse or ox drawn wagon on the national road was . . . almost beyond calculating."[2] This complete reworking of the equation between cost and distance quite simply transformed the economics and social geography of the way people lived.

It was the sheer versatility of the invention that ensured its success. The promoters of the Liverpool & Manchester had built their line with the transport of freight in mind, but they soon found that they made far more money from passengers than from carrying goods. Right from the beginning, railways were built for a variety of purposes: There were mineral lines carrying coal or ore to a port, suburban railways designed exclusively for passengers, and, often as the inaugural line, railways linking the capital with a port. Then a variety of other purposes were found for this ubiquitous invention. Railways were built to subdue colonies or indigenous populations, to transport armies, to bypass unnavigable stretches of river, to conquer territory, and frequently, to unite countries. Many were built largely on speculative grounds, with the promoters having little idea how they would recoup their money, while others were sure-fire investments tapping an obvious market. The popularity of railway development was driven partly by the fact that trains became fashionable. While there was always, as we have seen, the odd opponent, arising from entrenched interests or fear of technology, most people wanted a railway in their backyard—or at least near it. Passengers and goods were transported on railways for a fantastic variety of purposes. Everything from mail-order goods to milk was carried, while schoolchildren were ferried to school, suburban residents to work, and spectators to sports events. Every conceivable activity could be helped or generated by the railways. The historian of the New Zealand railways encapsulated it thus: "Rail travel accelerated the formation of regional and national organizations, the distribution of newspapers and magazines, and touring by local and international cultural performers."[3]

In 1830, few people other than sailors and soldiers had traveled beyond the confines of their parish, and their access to goods, particularly fresh food, was strictly limited by local availability. Once the railway gave them access to the outside world, it was those who had *not* traveled who found themselves in a minority. Villagers could, for the first time, travel to the local market town cheaply and quickly, and even further beyond to the nearest major city or even the capital. In return, urban dwellers could visit the countryside and, in particular, the seaside in a way that previously was not possible. One immediate and universally beneficial effect was on diets, particularly those of people living in towns. Almost overnight, with the opening of a line, fresh food became available from the countryside, which helped both the farmers, who had a far wider market for their produce, and the consumers in the towns. Take, for example, the good dairy folk of Orange County in upstate New York, who before the completion of the New York & Erie Rail Road in 1841 sent their produce to the city as salted butter. Until then, New York's milk supply had come from evil-smelling, swill-fed cows kept in basements. The railway's innovative local agent persuaded farmers to put milk in barrels on the trains, which quickly found a market among New Yorkers because it was both better and cheaper than the milk they'd had before. And when it got so hot in the summer that the milk turned sour before it reached the city, one of the farmers devised a way of keeping the inside of the barrels cool with ice in a pipe. With that innovation the flow of milk was assured, and the farmers of Orange County saved themselves the trouble of having to churn it into butter.

London lost its cows a few years later when an enterprising dairyman named George Barham, the founder of Express Dairies, took advantage of an outbreak of cattle plague among the poorly housed London cows to organize the transport of milk to the capital and gradually rid the city of these urban herds.[4] That not only improved the supply of milk for Londoners, but greatly improved the quality of the air in the city, which had a distinctly agricultural whiff until the cows were sent off to the fields. Bread, too, could be sent by train. The first Swiss railway line, as noted in Chapter 5, was named after the *Brötli*, the bread rolls that could now be provided fresh on people's breakfast tables thanks to transport by train. More sophisticated dishes, too, were now within reach of ordinary people. In Britain fresh fish was brought by overnight trains to inland towns and became part of the urban diet for the first time. As a result, fish and chips, once the sole preserve of seaside towns, became the national dish.

There were countless other examples of the railways improving not only people's diets but their very ability to obtain food. France, for example, had periodically suffered famines as a result of adverse weather conditions right up to the 1840s, but once the railways began reaching the most rural parts of the country, food could easily be sent to districts suffering shortages. Moreover, it was at a price people could afford. The price of food became more uniform because the expense of transporting it represented only a very small proportion of its total value, and the railways allowed these movements to take place quickly, ensuring it remained fresh. The consumption of fruit and vegetables by the French urban masses doubled in the second half of the nineteenth century almost solely as a result of the railways. Nicholas Faith noted, somewhat lugubriously, that "'only' half a million Chinese died in a famine in 1920/21 whereas twenty-five times as many had perished in a similar disaster fifty years earlier, before the railways had reached the regions involved."[5]

If the railways put food into people's stomachs, they also opened up their horizons; by accident rather than by design, they became a powerful force for social progress. Railways may have been designed by elites, whether Liverpudlian merchants, colonial adventurers, American carpetbaggers, or Mexican generals, who naturally exploited them for their own personal benefit, but their creation also helped large sections of the lower social classes in often unintended and unplanned ways. As mentioned in the previous chapter, the rapid expansion of the Mexican railways in the 1880s was stimulated by the authoritarian regime of Porfirio Díaz, who wanted to foster foreign investment and shore up his own power base. Although these aims were partly achieved, the railways had a much greater, and largely favorable, effect on the population in the regions they crossed. The vulnerability of railways to attack or even moderate sabotage makes it difficult to build and maintain them without the support of the local population, and railway companies, particularly private ones, realized they had to win over the support of the indigenous population. In Mexico the railways were forced to recognize the rights of local people, despite the central government's attempts to ride roughshod over them. The author of a social history of Mexican railways explains: "To win the support of the recalcitrant, railroads figured not only as agents of modernization and progress, but more important, as a right of citizenship to which every Mexican was entitled."[6]

By their very presence and size, the railways challenged entrenched local power structures in rural areas that previously exploited the population without fear of outside interference. The railways provided an alternative power source giving local residents greater ability to negotiate with federal authorities. Moreover, railway legislation itself had to be framed in a way that enabled local people to benefit from the iron road, so that, for example, freight rates were cheaper for subsistence goods on which the peasants depended, third-class passenger fares were set at a level they could afford, and strict safety standards were imposed for both construction and operation.

Even in a practical sense, many peasants in Mexico benefited from the creation of the iron road, which they used as a thoroughfare. The railroad was far better than the roads that had preceded it, which then, unfortunately, as in Brazil, deteriorated further through neglect as traffic migrated to the railways. The railway forded rivers and canyons, cut through mountains, and, wherever possible, took the shortest route, unlike the old roads, which were circuitous as well as badly maintained. There were other unintended benefits too: The railway was used as the route for water pipes, and the station buildings served as meeting places, as they were often the only permanent structure in the village.

There were, of course, examples to the contrary. The railways were for the most part monopolies, and without tight regulation they could easily exploit their position. For example, without regulation, most of the benefits of providing farmers with new markets for their produce could accrue to the railway company rather than to the people growing the produce, a form of exploitation that was particularly prevalent in the United States, where the economic importance of the railways was paramount. This phenomenon was brilliantly recounted in Frank Norris's novel *The Octopus*,[7] where the farmers of the San Joaquin Valley in central California were held to ransom at the turn of the century by the powerful Pacific & Southwestern Rail Road when it raised the rate for carrying their wheat so sharply that their land became worthless. The tenants of the railroad's own land were given preferential rates, an example of the widespread discriminatory schemes used by American railroads that gave favored customers far cheaper rates, a practice known as "rebating." The railroad had ensured that through bribes and subterfuge it controlled the state legislature in Sacramento, which meant that the farmers' pleas for the government to intercede on their behalf against the railways fell on deaf ears.

That was not the only form of corruption engaged in by railway companies. In countries where subsidies were available for the construction of railways, huge amounts of money were purloined by corrupt promoters and operators. That was almost inevitable, given the size of the contracts needed for the construction of a new line. As mentioned in Chapter 1, George Hudson, the "Railway King" who persuaded investors to part with their capital but diverted the money to pay dividends on his other railways, was the first major fraudster to be exposed, but there were to be countless others. Most losses occurred when promoters who were either fraudulent, stupid, or simply overoptimistic obtained huge sums of money for lines that were never completed, so that investors lost all their cash. Investors were most vulnerable during the railway manias that raged through different countries at various times; they were swept up in the rush simply because everyone else seemed to think it was a good idea. These dishonest aspects of the railway bubbles simply fit into the history of similar scandals, from the Dutch tulip mania of 1637 to the recent banking crisis. There's no shortage of elegantly embellished but completely valueless railway company share certificates, still adorning living-room walls, that date from the various railway manias of the nineteenth century.

However, it was when governments became involved that unbelievably huge sums could be purloined by corrupt promoters, and the world center for such activity was the United States, where several later scams dwarfed even the dodgy dealings outlined in Chapter 6 during the construction of the first transcontinental. At least the transcontinental was built, unlike many lines that sucked up millions of taxpayers' and investors' money without ever coming to fruition. The rapid growth of the railways following the end of the Civil War in 1865 and into the beginning of the twentieth century was matched by a series of scandals on an unprecedented scale, precisely because the amount of money going into the railways offered so much opportunity for corrupt dealings. Fraud was particularly rife in the southern states, where northerners moved in to "reconstruct" the railroads damaged or destroyed by the war but frequently pocketed the state grants while neglecting to build the lines. In North Carolina, for example, in 1868–1869, $28 million was given in state bonds to thirteen railroads but fewer than 100 miles were built, barely a tenth of what should have been possible with that amount. Most of the money went into the pockets of a "railroad ring" of promoters and "spent on gambling junkets in New York City and bribery in North Carolina."[8] There was even a bar in the statehouse that dispensed free wine, liquor, and cigars paid for with railroad money. In neighboring Georgia, almost $40 million was "authorized for thirty-seven companies, nearly all of which had little more than a paper existence."[9]

In the northern and eastern states the corruption established itself on an even grander scale. One historical account reported that "Vanderbilt, Drew, Gould [three railway magnates] and others were making millions and tens of millions in manipulating rail systems extending for thousands of miles. . . . Watered stock, stockmarket rigging, corrupt and dishonest management, rate wars, rebating and labor violence were all part of the railroad picture in the postwar generation."[10] It was, of course, not only in America that railways supported by public funds were built for dubious reasons. Pork barrel politics resulted in the construction of innumerable railways that could not be justified economically in many countries, such as, for example, in France, where a vast number of branch lines sprang up in the 1860s and 1870s as a result of lobbying by local municipalities to be connected with the rail system.

Given that the railways could, almost at will, wreck people's lives or endow them with fortunes, the fear of the big railway monopoly was widespread. It was not surprising, therefore, that demands for government intervention, stimulated by antipathy toward the railways' monopoly position and corrupt practices, were universal. This feeling was neatly summed up by the views of a Berwick fish trader in the 1890s, who said: "What we want is to have our fish carried at *half* present rates. We don't care a —— whether it pays the railways or not. Railways ought to be made to carry for the good of the country or they should be taken over by the Government."[11] In fact, railways around the world were increasingly subjected to regulation toward the end of the nineteenth century as their power over economic life attracted public criticism and government action, and, in several cases, nationalization. Although throughout the century, state railways were greatly outnumbered by those in private hands, following the government takeover of several large systems, including those in Italy and Prussia, by the outbreak of World War I, barely half the world's railways remained in private hands.

Despite these examples of exploitation, the railways were a democratizing force. Indeed, they had been opposed for precisely that reason by some of Europe's autocratic rulers. The Austrian emperor Francis II resisted the establishment of a railway network as he felt it would foment rebellion, but as we saw in Chapter 5, most intelligent rulers soon realized that the military potential of the railways, both against internal and external foes, made them work to their advantage. The fears of the absolute monarchs about the effects of the railway were well-founded. The shift in Europe from authoritarian to democratic politics in the nineteenth century went hand in hand with the spread of the iron road. The railways not only afforded people the ability to travel, but stimulated economic development that ensured that more of the population qualified for the suffrage or were in a position to press for the widening of democracy. Dr. Thomas Arnold, the headmaster of the Rugby School in England, was in no doubt that the railways had put an end to feudalism. He commented, when seeing a train thunder past the school: "To see it and think that feudality has gone for ever. It is so great a blessing to think that any one evil is really extinct."[12] In France, the railways were perceived as creating that revolutionary dream of *liberté, equalité, fraternité*. A French utopian writer, Walter Pecqueur, said, rather optimistically,: "By causing all classes of society to travel together . . . the railways quite prodigiously advance the reign of truly fraternal social relations and do more for the sentiments of equality than the most exalted sermons of the tribunes of democracy. . . . The railways most generally provide a continuous lesson in equality and fraternity."[13]

These early enthusiasts for the railways were a tad over-optimistic. Railways may have been democratizing, but they did not bring about universal peace or the end of conflict. As we have seen, they were also powerful unifiers of countries such as Italy, Belgium, and Canada, helping to create nations and cement boundaries over which great wars would be fought in the twentieth century. In this respect, while the railways opened up vast swathes of often deserted country, they were, paradoxically, a centralizing force, deepening the power of federal governments. Although the railways helped to stimulate the preconditions for democracy, they did not, by themselves, ensure its creation. After all, they allowed for rapid troop movement and therefore enabled ruling elites to crush uprisings by groups fighting for regional interests or for greater democracy. While in general they were a liberalizing influence, their very flexibility meant they could be used for a variety of ends. Railways made fighting wars more efficient and therefore more deadly, and as we shall see in the next chapter, World War I could not have been fought on such a massive scale without them.

In a sense, too, the railways mirrored the class system rather than breaking it down. Although they did offer poorer people their first opportunity to travel, they created different standards of comfort for their passengers according to their ability to pay, entrenching preexisting differences. Though people of different classes traveled in the same trains hauled by the same locomotives, the provision of three or even four classes of carriage was commonplace. The poor had, on occasion, their own trains, such as the parliamentary services guaranteed by legalization in Britain, but many express services in several countries were only available to first- or second-class passengers. This meant slower journeys for the less well-off.

Reputedly only Norway can claim to have never had different classes on its railway network. Even in America, where, as previously mentioned, the practice was to have open carriages with a central aisle, a class system soon developed, with "coaches" for most travelers and "parlour cars" for the rich; notably, until the 1960s, when the Civil Rights Act was passed, there was separate and inferior accommodation for black people on American trains. For the seriously rich, or antisocial, there remained in many countries well into the twentieth century the opportunity to charter one's own train. According to Nicholas Faith, "Winston Churchill was considered frightfully extravagant when he hired [a train] for election purposes in 1910,"[14] but this hardly matched the hauteur of Auguste Belmont, a New York banker who had a special car of his own on the subway that he had helped to finance. Sherlock Holmes, too, regularly hired his own carriage, but, more modestly, he rode with everyone else when he took the London Underground, which abandoned its class system before World War I. The early lines in the Underground system had three classes, but the deep tubes had none, and that classlessness later spread through the network.

The railways not only helped continue the class system by defining new types of discrimination but also led to the massive expansion of the working class. They enabled the migration of huge swathes of the population from land-based employment or subsidence farming to paid work for capitalist businesses, not only for the railway companies themselves but for the huge supply chain that grew up to service them as well as a wide range of other industries needing large amounts of labor. The railways not only allowed people to live further away from work, but also stimulated the development of the financing system on which capitalism's growth depended. Indeed, the expansion and spread of capitalism paralleled that of the railway. The construction of the railways was the biggest enterprise ever undertaken—certainly since the Egyptian pyramids and the Roman roads and aqueducts, and arguably far greater than any of those—and to support this huge construction program, a banking and financing system had to be created. Although capital, as we have seen, was raised locally in many countries, overall this process of raising funds to build railways was dominated right up to World War I by British investors who, by then, owned 113 railways in twenty-nine countries. These were valued at £1.6 billion (perhaps £80 billion, or $140 billion, in 2009 money), of which a quarter was in colonial railways and £617 million (approximately $50 billion in 2009) in the United States, where the railways accounted for a staggering four-fifths of British investment at the time. Many of these railways were also built using British contractors, who would bring over a few experienced engineers (whose life expectancy, as we have seen, was sadly low, whether in Africa, India, or South America) and then employ largely local labor (though sometimes these workers were imported from a third country, such as China or India) to build the lines.

Britain may have been the leading investor, but France, the United States, and Germany also channeled considerable sums of capital into overseas railways, and the railways were a massive stimulus to all these economies. Most of this investment consisted of bonds, which paid a fixed amount on the capital invested, rather than shares, which depended on the performance of the company and consequently were riskier. Even for bondholders it was by no means a safe investment because, as we have seen, the cost of constructing railways tended to be underestimated, and many investors lost all their money. A whole section of the finance industry developed as a consequence of the need to protect these investments, adding to the City of London's importance in the U.K. economy.

Railways provided a far more enduring boost to the economy than the construction of roads or even canals had previously. Once a road was built, apart from the occasional patching up, it could be largely left alone. In contrast, railways not only needed continual maintenance, such as regular patrols checking that the track was safe, but also required a sizable organization to operate them involving thousands of workers.[15] The human impact of this process swept right through all the social classes. As a historian of the German railways put it, "one must attempt to conceive of this remarkable expansion in terms of the people involved: the hundreds of thousands of individuals—executives, employees, workers, miners—whose lives were thrust into the maelstrom of industry, all of whom needed to be housed, clothed and nourished."[16]

The impact of the railways varied across the world, influenced particularly by the level of economic development. The experience of the United States, for example, was very different from that of Europe, as was neatly described by one railway historian: "The European countries developed their railroads; the American railroads developed their country."[17] Many of the spin-offs were indirect and consequently have been largely ignored by economists. For example, railway construction created a sudden and substantial demand for professional expertise, in effect stimulating the establishment of a raft of professions. Indeed, the railways helped to establish the very idea of "profession," as the economic historian Terry Gourvish suggested: "It is not an exaggeration to say that the [rail] industry played a key role in encouraging the growth of occupational professionalism based on specialized work. Engineering, law, accountancy, and surveying all received an important stimulus."[18] The effect was international, stimulating cross-border trade. Coal from the northeast coast of Britain was exported to the Baltic states, which in turn sent back pine, which was deemed to be the best material for railroad ties.

The railways opened up the world to millions of people who benefited from economic progress far beyond simply the ability to access resources for the first time. The Mexican railways, as the author of a book about their social history stressed, stimulated local capitalism since they "prompted the arrival of multiple suppliers, thereby breaking any monopolies or market strangleholds, and provided smallholders affordable access to markets beyond their local community."[19] In Russia, the village usurer was made redundant by the local peasants' newly discovered ability to turn their produce into cash, either by selling it on the spot or by taking it to a market town on the train. The new stations "swarmed with a mass of small traders, exporters and commission merchants, all buying grain, hemp, hides, lard, sheepskin, down and bristles—in a word everything bound for either the domestic or the foreign market."[20] In short, the railways were empowering. They were a grand capitalist enterprise that in turn stimulated petty capitalism.

One of the odd aspects of the rail industry is that even some of the failures, lines that never came close to paying their way, proved to be of enormous benefit to the economy of a country. The Grand Trunk Railway in Canada, which seemed to be one transcontinental line too many and consequently was a financial basket case, is a case in point. According to a historian of the line, "for eighty years, the Grand Trunk Railway was subject to recurrent financial crises and was a failure as a commercial enterprise [but] the railway made a significant contribution to the economic development of Canada."[21] In the language of economists, this is because railways generate "externalities," benefits that cannot be captured through the ticket office or freight receipts. Most obviously, the construction of a railway leads to an increase in land prices on its route, but, with the occasional exception, the companies building the lines rarely benefited from this uplift. Even a railway that had no economic purpose in mind, such as the line through Senegal in what was then part of French West Africa, built by the colonial power in order to impose itself on the population, helped to stimulate development because it allowed "the rubber, the cereals and the peanuts from a rich hinterland to reach the Senegal river and transported them towards the interior manufactures produced on the coast, such as textiles, foodstuffs and machinery"[22]—a story that was repeated across the world.

But it was also more than just a matter of cheaper trade. There are revisionist studies[23] suggesting that a detailed analysis of the price of transport shows that the railways did not cut costs as dramatically as claimed, and therefore were not as economically or socially important as has been stated by their supporters. This is bunkum, a failure of imagination on a grand scale. It is misleading simply to try to add up the narrow economic effects of the creation of the railways and assume that the sum of the parts adds up to the whole. The railways did so much more than merely reduce the price of carrying coal and other minerals. Making transport cheaper was merely one small part of their effect. First, there was the construction of the lines themselves, which represented an enormous investment and stimulated a whole series of other industries to supply them with equipment and materials, from locomotive manufacturers and ironworks to providers of signaling equipment and station buildings. In the beginning, of course, it was the coal, iron, and steel industries that the railways did most to stimulate. Coal's relationship with the railways was symbiotic, since they both carried and consumed vast quantities of the black gold. Coal had been the stimulus for the creation of the first railways, and the railways made it possible to exploit mines that previously would have been uneconomic. In turn, the railways reduced the price of coal, making it possible for town dwellers to heat their homes, which further increased profitable traffic for the railways. The German experience neatly encapsulates the importance of the railways. Before 1848, when there were few railways, the German mines had produced just 3 million tons of coal; by 1870, after the construction of a substantial rail network, production had increased tenfold, "and a rising percentage of that freight was simultaneously carried and consumed by trains."[24] Coal continued to play an important role in the economics of many railways right into the twenty-first century, long after steam locomotives had disappeared; for example, rail freight on the U.K. network has grown considerably since 2000, largely as a result of increases in imported coal.

The railways, as Nicholas Faith put it, "acted as midwives"[25] to countless new enterprises. He cited the transformation of Birmingham, Alabama, into a major industrial center because the Louisville & Nashville Railroad provided favorable freight rates to enable the iron ore deposits at Red Mountain to be exploited. Another somewhat bizarre example, which illustrates how technologies first require and then outgrow the demand for railways, was ice. Starting in the 1860s, ice, mostly from a particularly suitable pond in New England,[26] was transported in huge quantities to warmer climes on the eastern seaboard of the United States, and later, with the completion of the transcontinental, from even further afield. It was a massive business that at its height provided two-thirds of a ton of ice for every person living in a major U.S. city. The trade declined in the 1890s with the development of cooling equipment, but that innovation, in turn, spawned another commercial opportunity for the railroads, the transport of perishables such as fruit and fish in refrigerated cars. The example of Argentinian beef, cited earlier, is just one demonstration of this effect.

The list, therefore, is endless. The railways carried everything—from newspapers to nuts, minerals to mail—at a cost that was invariably cheaper than the alternative, even when monopolistic railways exploited their position. Britain's *Railway News* in 1864 mentioned trains arriving at a London goods yard every ten minutes in the morning laden with "Manchester packs and bales, Liverpool cotton, American provisions, Worcester gloves, Kidderminster carpets, Birmingham and Staffordshire hardware, crates of pottery from North Staffordshire and cloth from Huddersfield, Leeds, Bradford and other Yorkshire towns."[27] Think of the most obscure industry possible and then discover how the railways changed it. Take wine, for example. The boom in the Argentinian wine industry in the late nineteenth century was centered around Mendoza, the result of the arrival of Italian immigrants brought in by the railways. They modernized the vineyards and exported their large amounts of produce on the railway, moaning, as ever, about excessive freight rates. In Europe, the landlocked vineyards of Chianti developed their worldwide reputation thanks to access to wider markets through the arrival of the railway, while in many regions the taste improved, because, as Nicholas Faith pointed out, "in pre-railway days, many wines tasted decidedly resinous because they had been carried on muleback in hog skins painted with pitch—a treatment and taste which survive today only in that appalling beverage, retsina."[28] There was a downside too. By creating expanded markets, similar to today's globalization, the railways removed the protectionism that inefficient enterprises had enjoyed as a result of high transport costs. Whole areas that produced crops that were inferior to those in other regions were forced out of business. As Faith noted, "rail borne grain flooded from the American and later Canadian prairies, then from the Argentine pampas, causing a major agricultural slump throughout Europe."[29]

Besides reducing transport costs, trains speeded up journey times, allowing traders and businesses to reduce the level of stocks they held, and thus releasing cash for investment. It was, in a sense, the start of the "just in time" concept that became fashionable toward the end of the twentieth century. For livestock animals, though, the railways were bad news. Instead of being led on a cheerful amble toward market and slaughter, they were now killed at the stockyard and dispatched as carcasses to the new meat markets springing up in major towns. For the farmers, this was more profitable, since previously animals had lost weight when driven to market.

The reduction of transport costs enhanced the advantages of specialization for towns concentrating on the production of a particular product. As Michael Robbins pointed out, "the Victorian child's game which showed steel and shipbuilding at Barrow, jute at Dundee, straw hats at Luton, boots and shoes at Northampton, cutlery at Sheffield, was not seriously misrepresenting the facts."[30] This phenomenon was international. The Swiss developed their specialty in watches and precision engineering because the railways made their agricultural products, grown in difficult mountainous conditions, uneconomic; Denmark flooded the British market with butter and bacon products.

The changes were worldwide. America, as we have seen, was shaped and virtually created by the railway; as a U.S. rail historian said, "railroads have played a decisive role in nearly every major movement in our history. They were important in hastening the rise of the Atlantic seaboard metropolises, in peopling and supplying the west, in attracting and transporting immigrants, in shaping the enterprise of trappers, cowboys, miners and farmers."[31] It is ironic that a description of the railways in the United States, a country that has all but shunned them today, should best demonstrate their sheer ubiquity in their heyday:

Few Americans lived more than a short buggy or hack ride from a depot where a through train paused. . . . A well disciplined world of affairs was built on the trains' regular comings and goings. Salesmen, vaudeville performers, commercial travelers set their itineraries a month or even a season in advance. Businessmen arranged their appointments to avoid interfering with the regular twice a day mail delivery that made it possible to mail a letter from Boston to New York at the end of the day and be handed a reply before 10 am two days later. . . . Farmers paused in their fields to set their pocket watches as the express roared through. . . . Bicycles, velocipedes, live billy goats (the goat drawn wagon had a brief vogue among upper middle class children) and countless other treasures, destined for birthday parties or Christmas trees, were deftly offloaded from the express cars.[32]

While there is a touch of romanticism about this, the railway undoubtedly was just as central to people's lives as the automobile is today, and possibly more so, since it represented such a radical change from the past, whereas the adoption of the automobile was more gradual and prolonged.

Even time itself was standardized by the railways. Until their invention, towns had their own time, determined by their longitude, and therefore Plymouth in Devon, southwestern England, for example, was twenty minutes earlier than London. This hardly mattered when it took days to travel such distances, but when the railways began to create networks of lines with interconnecting trains, it became essential to standardize time. In 1847, the Railway Clearing House in Great Britain adopted Greenwich Mean Time as "railway time"; it was used by most railways and eventually became the universal standard. The problem was far greater in the United States, where distances of 3,000 rather than 300 miles had to be managed. As a historian of the American railroads suggested, "the matter of time on American railroads, and indeed in all American life, was pretty much chaos."[33] He cited the case of Buffalo in New York State, where the railroad station boasted three clocks: one on New York City time, used by the New York Central; a second set to the time in Columbus, Ohio, for the Michigan Southern and other regional railroads; and the third showing the local time. In Pittsburgh, a big railroad center, there were no fewer than six clocks showing different times, and transcontinental travelers might have to change their watches twenty times during their journey. Illinois and Michigan both had twenty-seven local times, while Wisconsin had thirty-eight. The complexities and confusion for both railway timetablers and passengers were legion, and yet it was not until November 18, 1883, that standardized time was adopted in the United States.[34]

There were no riots on the streets in America as there had been in Europe in the 1500s when the Gregorian calendar supplanted the Julian and eleven days were lost, but there were a few protesters arguing that the "immutable laws of God" had been changed. The *Indianapolis Sentinel* caught the mood in expressing the ambivalence felt about this change with gentle irony: "The sun is no longer boss of the job. People, 55,000,000 of them, must eat, sleep and work as well as travel by railroad time."[35] Oddly, the timing of the change was at noon, which involved the railroads in hugely complex scheduling, and there were fears that a disaster might be caused by the confusion over old and new time. In fact, although a few passengers missed their trains and many others arrived to the station far too early, there were no accidents. Probably very few Americans know today that their time zones are the product of the railroads' need for coordinated timekeeping.

Although many early passengers rode the rails for business purposes, the railways were quick to catch on to the fact that leisure travel could provide a huge new market. It is barely an exaggeration to say that the railways created the tourist industry. Until their arrival, there had been the odd steamship trip for pleasure seekers, but that was a very limited phenomenon. The rich, of course, could travel relatively long distances in their carriages, and the occasional family might take a stagecoach to Brighton for the day or, especially in Germany, have a day off for a group ramble, a *Fussreise* (literally, "foot journey"). The curious or the idle rich went on the Grand Tour around Europe, but it was the railways that democratized leisure and turned it into a mass enterprise. It was in Britain, the birthplace of the railways, that this phenomenon began. Within days of the opening of the Liverpool & Manchester in September 1830, a train was being used to carry day-trippers from Manchester to the Liverpool Charity Festival. In May the following year, the first excursion train took 150 members of Manchester's Bennett Street Sunday School to Liverpool for an outing. Soon whole trains were being chartered for trips to the races or to church bazaars, or even, in the case of the Bodmin & Wadebridge Railway in Cornwall, to a public execution in 1840. It was Thomas Cook who famously built up a huge business on the idea of running special train services. He started modestly by taking 570 teetotalers from Leicester to a temperance fete at Loughborough. It may have been a journey of merely 11 miles, but Cook had discovered a winning formula, publicizing his trips extensively, and was soon organizing railway tours around Europe (still for the affluent, unlike today's InterRail travelers). He even, by 1872, offered an around-the-world trip taking 212 days, at a cost of £210 (around $1,000).

Most railway companies soon developed their own excursion services, and while the "bucket and spade" specials to Scarborough or Blackpool were largely the preserve of the northern working classes, the more affluent soon developed a taste for resorts further afield. The Côte d'Azur, or Riviera, as the British called it, on the French Mediterranean coast soon became the holiday destination of choice. Cannes had been a sleepy fishing village when it was "discovered" by a few English travelers in the 1830s, but after the arrival of the first trains in 1863, it became a booming resort. Other towns along the railway line, such as Antibes, Nice, and the spa town of Menton on the Italian border, all soon sported hotels, and the British upper classes began to move to them wholesale for the pleasant summers and, later, the mild winters. Monte Carlo, the gamblers' paradise, which is in the principality of Monaco and therefore exempt from the French ban on casinos, owes its affluence to the railway, which brought in the high-rollers in ever-increasing numbers to fill the burgeoning number of hotels. Queen Victoria overcame her dislike of trains to visit Nice annually in the late 1890s, travelling in a special coach kept at Calais on the French side of the Channel. Her son, the future Edward VII, favored Cannes, while regarding the Riviera in general as "a country of good company where everyone meets up, rather like a garden party."[36] It was not only seaside resorts that attracted vacationers. Other playgrounds of the rich included spa towns such as Baden-Baden in Germany or Saratoga Springs in New York State, and seasonal hideaways, such as the numerous hill stations in India, became far more accessible and flourished thanks to the railway.

The railway was crucial to the creation of the grandest tourist development of the nineteenth century. Seeing the success of the French Riviera, an enterprising American oil man, Henry Flagler, the partner of John D. Rockefeller who created Standard Oil, sought to create an equivalent in the United States. He chose Florida, at the time a desperately poor state described as a "wilderness of waterless sand and underbrush." The weather, though, was wonderful, and Flagler decided to invest in a series of resorts that he developed along the coast. First he built the massive 540-room Ponce de León hotel in St. Augustine, and then the Royal Poinciana at Palm Beach, which was twice as big. This was followed by a third hotel, the Royal Palm at Biscayne Bay, and eventually another in Miami, which at the time was not even incorporated as a town. All these resorts were linked by the railway, which he knew was essential to bringing in affluent guests from the East Coast.

The Florida East Coast Railway[37] was an ambitious enterprise, formed by taking over a series of older lines and adding several new sections, including one of the most ambitious railway projects in the world. This was a 128-mile line reaching far out into the Atlantic Ocean over rocky islands and stretches of sea to connect Key West with the mainland. The line, popularly known as the "railroad that goes to sea" or, more prosaically, "the Overseas Railroad," involved crossing forty-two stretches of sea, including a 7-mile section, and the construction of 17 miles of viaducts and bridges and 20 miles of filled causeways. Three severe hurricanes, each worse than the previous one, delayed completion of the project, but eventually the line was opened in 1912, just in time for Flagler to ride a train triumphantly into Key West before his death a year later at the age of eighty-three. Key West was actually nearer Havana than it was to Miami, and the main passenger train, scheduled for four hours but normally taking six or seven because of the difficult conditions, was called the *Havana Special*. Not surprisingly given the arduous nature of the journey, the tourists never flocked to Key West in the way they did to the other resorts, and despite the fact that it was the southernmost deep-sea port on the Atlantic seaboard, and consequently nearest to the booming Panama Canal, there were never very many freight trains.

The construction of Flagler's magnificent creation, which cost $27 million and several hundred lives, was largely perceived as an old rich man's folly, yet it proved, in an unexpected way, to be the savior of the Keys' economy. Although Flagler would have been saddened by the destruction of the railway by the Labor Day hurricane of September 1935, only twenty-three years after its opening, he would have been heartened that the "Eighth Wonder of the World" became the base infrastructure for the road to the Keys. Indeed, even before the great storm, occasional cars had started bumping along the tracks, since train services had been reduced to just one per day. After the storm the railroad directors gave up trying to keep the railway open, allowing vehicles to use the track without fear of a disastrous collision. The route survives today as the southernmost section of U.S. Highway 1, running to a series of exotic resorts in addition to the port and naval base.

Along with enabling the development of mass tourism, the railways stimulated more short-term leisure activity, such as attendance at sporting events. In Great Britain, it was racing that initially attracted mass crowds arriving by train, but the visit of the first Australian cricket team in 1878 and the creation of the English Football League in 1888 generated huge numbers of excursion trains. By enabling far more sports fans to attend matches, the railways played a role in the development of professional sports, as matches began to attract enough paying spectators to make them profitable. The English soccer cup finals at Crystal Palace, for example, were attended by more than 100,000 spectators by the end of the nineteenth century, virtually all of whom arrived by rail.

The railway proved to be a boon, too, to more spiritual activities. Lourdes owes its renown as a place of pilgrimage to its early connection with the railways. As Nick Faith suggested, "Bernadette Soubirous' visions [in 1858] could not have been better timed. She saw them just as the (Jewish) Pereires were deciding the route of the railway line south of Bordeaux."[38] Lourdes duly obtained a direct connection with Bordeaux and even Paris, and as a result 80,000 pilgrims descended on the little town annually over the next few years, establishing its reputation as a source of healing.

The railways became, in the words of the railway historian Jack Simmons, "unequalled movers of people in large numbers at a good steady speed,"[39] enabling people to live further from their work than ever before and thus creating the very concept of commuting. Although a few workers were starting to commute in London by rail when the first railway designed exclusively for passenger traffic, the London & Greenwich, opened fully in 1838, it was not until the 1860s, when local lines spread out of cities and dormitory suburbs sprang up, that the habit really caught on. The railways began offering season tickets at vastly discounted rates. In Britain, legislation requiring companies to provide workmen's trains with cheap fares stimulated a massive growth in commuting for all classes.[40] The early lines of the London Underground were heavily used by commuters and were soon running numerous early-morning services for workers. Operated by small steam engines hauling sparsely furnished trains with hard seats and little lighting, commuter services offered few comforts except that they allowed people to live further from the overcrowded and often unsanitary town centers.

Other countries were slow to follow the example of the Underground, with the first major subway systems, in New York and Paris, not appearing until the early 1900s. In the 1880s, New York, in fact, had gone the opposite way, building a system of elevated railways in the middle of the grid-patterned streets. Though these were very efficient at transporting people, they were a nightmare environmentally. The coaches were hauled by steam locomotives at intervals of barely a minute for most of the day, and the noise of the trains clattering along the viaducts, reverberating from the tall office buildings on the streets, was deafening. They were fantastic mass people movers, carrying huge numbers, but were unpopular because of the effect on the street environment. They soon made way for the subway, which allowed most of the elevated railways to be torn down, much to the relief of New Yorkers.

In Chicago, however, the elevated railway created in the 1890s survives today, supported by an extensive network of suburban lines. The Illinois Central, which built most of the lines, was probably being just a tad optimistic in portraying "the very perfect system of Suburban Trains now run by this Company [in which] seated in their comfortable cars, the business man, after a hard day's work in the city, finds the homeward ride along the beautiful Lake Front a pleasure rather than a task, as he waits for no drawbridges, railroad crossings or other annoyances"[41]—except, of course, the numerous stops that these trains invariably made. The Illinois Central deliberately set out to attract suburban dwellers, and the railways, therefore, were responsible for urban sprawl a couple of generations before the automobile made it possible for every town, rather than only those blessed with a suburban railway network, to expand in that way. Commuting railways were transforming life in big cities all around the world, greatly extending the area in which workers could live. This changed the nature of cities, stimulating the creation of central business districts that could be served by train, with high office blocks densely concentrated in an area of expensive land and urban sprawl to outer areas where land was cheaper.

While in Europe the railways were built into existing cities, in the United States the central terminals became the focus of development. Albro Martin, a historian of the American railroads, has even suggested that the railroads created the very idea of downtowns in U.S. cities: "The new terminals brought to the American city hotels, theaters, office buildings, and department stores whose size and magnificence proclaimed a giant step forward in American sophistication and the standard of living. Downtown was a center of commercial, mercantile and financial activity during the day."[42]

The impact of the railways was not, of course, always beneficial. Their speed, combined with the lack of safety features, created a level and frequency of disaster that previously had been caused only by natural events or maritime accidents. Wherever there were railways, there were accidents, and their spectacular and deadly nature attracted considerable criticism of the railway companies.

At first, accidents were not very serious, as speeds were low, and the opportunity for collisions was limited, as there were not many trains. In the 1830s, there were a few accidents in Britain and the United States, but with death tolls in single figures. The first indication of the potential for major catastrophes came with the 1841 disaster near Versailles on the Paris-St. Germain Railway, after a locomotive broke an axle and derailed, killing more than fifty people, as described in Chapter 2.

As the railways grew and trains were improved, so that they could go faster, the number and extent of disasters also increased. The railway authorities were learning through a process of trial and error, and safety standards largely grew out of the lessons learned from accidents, but often the process was slow owing to the resistance of the companies to spend money on reducing the risk. The litany of the causes of accidents include those that still occasionally lead to disasters today, such as derailments resulting from faulty track, or collisions due to poor working practices, bad weather, or road vehicles crossing the tracks, but there were numerous others that have subsequently been remedied, such as the absence of adequate braking or signaling systems, exploding boilers and other mechanical failures, and collapsed bridges. Rear-end collisions were a particularly common type of accident in the days before the introduction of signaling systems, as trains were dispatched at time intervals, and a breakdown could easily lead to a later service smashing into the back of the stricken carriages. Brake failures could result in runaway trains, leading to very high-speed collisions or derailments on curves, with particularly deadly results. Once a train had suffered a mishap, there was the ever-present danger of fire caused by the hot coals from the engine igniting the wooden structures of the carriages. This is what happened in Britain's worst accident, which occurred at Quintinshill near Gretna Green in 1915, when 227 people, mostly troops on their way to war, perished. Improving the accident record was not simply a matter of adopting technological improvements, but also of creating a culture of safety among railway staff and managers, as sheer sloppiness or incompetence was a cause of many early crashes. So was tiredness, as railway workers were expected to put in very long hours, frequently without extra pay for overtime and with little supervision.

Although the notion of learning from disasters may suggest that the railways were remiss, it has to be remembered that this was an entirely new industry operating at speeds far greater than anything before, which inevitably led to unforeseen hazards. There had been doomsayers who suggested that the very notion of such travel would lead to breathlessness, headaches, or even death. While these Cassandras were obviously misguided, their predictions of disaster were not entirely off the mark; the wide variety of causes of early accidents demonstrated the difficulty for railway operators, who could not have foreseen all the potential risks. However, by and large, each railway system became safer as the companies gained experience through investigations into the causes of every accident.

This was by no means a smooth process, as railway companies were often unwilling to learn from these disasters because of the cost implications. Consequently, independent regulatory authorities emerged in most countries. The public outcry over accidents forced governments to create these bodies and put pressure on the rail companies to reduce the number of accidents. Basic braking and signaling systems were only introduced after considerable pressure was exerted on the companies by both the public and government authorities, not to mention countless deaths, even in developed countries such as the United Kingdom and the United States, with their extensive railway networks. In less developed countries, the process was far slower, as witnessed by the long list of huge disasters on the Indian subcontinent, with tolls often in the hundreds.

The railroads in the United States were particularly disaster-prone in the mid–nineteenth century because the lines had been built so cheaply, and as traffic intensified, the disasters multiplied. In a period of just a couple of weeks in the spring of 1853, two major accidents, with death tolls of twenty-one and forty-six, respectively, highlighted the inadequacies of the railroads' safety protocols. After the Civil War had ended in 1865, there was renewed public outcry over a series of seventeen fatal accidents in the single month of August that killed a total of eighty-eight people. The U.S. railroads did not use any external signaling on the lengthy sections of single line but relied on a dispatch system that was dependent on communication between the controllers, a method still widely used today, and the result of any error was, all too often, a cornfield meet.

In the United Kingdom, it took a particularly deadly disaster, the Armagh collision in 1889, in which eighty-eight people, mostly schoolchildren, were killed when the train carrying them rolled back down a hill into the path of another, to convince the government to become involved in regulating railroad safety. Following this, passenger trains had to be equipped with continuous automatic brakes and signaling systems were implemented to keep them apart. It was in America that the technology for safer braking had been developed—by Westinghouse—but it was slow to be adopted universally. In the United Kingdom, the implementation of a fail-safe system was delayed by arguments over what sort of brake—vacuum or air—should be used. Even in the United States, despite some railroads fitting continuous brake systems throughout the train, it was not until the 1890s, when federal law specified the adoption of the air brake, that real progress in rail safety was made.

Although railway accidents in the nineteenth century were front-page news, as they are today, train travel was nevertheless relatively safe compared to other methods of transportation. A rare early study of the rate of railway accidents found that in France, between the introduction of the railways in 1832 and 1856, 642 people were killed, which translates into a rate of about one fatality for every 2 million journeys, which suggests that traveling by train was seven times safer than traveling by stagecoach.[43] France did, later, have what was the worst accident ever to occur in Europe, and possibly the world, when an estimated eight hundred soldiers were killed in an overladen train that lost control descending from the Mont Cenis Tunnel in 1917. Indeed, in the long litany of catastrophic rail accidents across the world, it is noticeable that a high proportion occur to specials, such as excursion trips or troop trains, partly because these are not in the normal schedules, and partly because they were often carrying far more people than was normal in old rolling stock. In Russia, however, where trains were very accident-prone throughout their early history, not even luxury was a guarantee of safety. In October 1888, the Russian Imperial Train jumped the tracks at Borki, 400 miles east of Moscow, because of poor maintenance, killing sixteen members of the tsar's court. The tsar himself was in the dining car and escaped unhurt, but he ordered a much more solid version of the Imperial Train to replace the wrecked coaches. This accident was an exception, though. Such VIP trains were generally treated with far more caution by the railway authorities than the regular trains. When Queen Victoria went by train, the track was checked for possible bombs placed by Fenian activists, suggesting that the terrorist threat to trains is almost as old as the railways themselves.

Away from the newspaper headlines, railway workers suffered in far greater numbers than passengers, with thousands losing their lives every year in accidents that were drearily similar and attracting none of the attention of the passenger train disasters. Because of the absence of interest and publicity, only very gradual improvements were made to the working conditions on the railway, and consequently the job was as dangerous as far more obviously perilous professions, such as fishing or construction. To take one example, Russian statistics show 184 employee deaths and just 9 passenger fatalities in 1874, a year in which 25 million journeys were made on the 11,000 miles of track.[44]

The most dangerous job on the railway was that of shunter, who worked in railway yards surrounded by moving locomotives and wagons. His job was to uncouple wagons, usually when they were in motion, and quickly change manual points so that the wagons went onto different tracks. He ran beside the wagons with a shunting pole, using its hook to lift the three-link couplings, and he could only stop a wagon by forcing down the handbrake. One false step and he was dead. All large stations had shunters, who would stand at the head of a train between the buffers as a locomotive bore down on them, relying on the skill of the driver and fireman to avoid being crushed. Minor accidents were commonplace, and rare was the shunter who retained all his fingers for the whole course of his career.

The death toll from tragedies was not the only negative effect of the railways. While they spread prosperity, they also led to massive movements of people, and in heavily populated countries this increased the speed at which diseases spread. Therefore, although the railways may have staved off starvation, they undoubtedly intensified epidemics, especially in India, where huge pilgrimages involving hundreds of thousands of people were a regular event.

The trains, too, had far more devastating consequences for the environment than merely bringing too many people to disturb Wordsworth's favorite views. Their environmental impact caused numerous others to protest as well, though it was not so much the introduction of the railways themselves that caused problems—since, after all, they used up far less land than roads and blended easily into the countryside—but the economic development that ensued. Virgin forests could be cleared in months and the timber transported economically, mines could be dug in previously inaccessible places, and industrial towns would spring up where barely hamlets existed before. The railways themselves caused the most damage not in unspoiled rural areas but in towns and cities where the multi-track lines cut huge swathes through built-up areas that were inhabited, invariably, by the poor. By dividing towns into two distinct sections with a ribbon of railway that was difficult to cross, they would often create an undesirable part of town that was "on the wrong side of the tracks." But the advantages so outweighed the negatives that much of this was forgotten as the railways spread their tentacles ever further around the world. For most journeys, train travel had become the only method of getting around, the first choice for rich and poor alike. Moreover, as the next chapter shows, with the networks largely complete, the railways hoped that by improving services and facilities they would win over the public.

TEN

Getting Better All the Time

T HE SPREAD OF THE RAILWAYS ACROSS THE GLOBE MEANT THAT RAIL travel became routine for a significant proportion of the world's population. By the early years of the twentieth century, millions were traveling by train daily and the numbers were growing annually but it took a long time before the experience could be described as comfortable, let alone pleasurable. The literature of the railway tends to focus on the top end of the market, the express and luxury trains enjoyed by the rich and the profligate, and there are plenty of books detailing these prestigious services, but few explaining what it was like for the poor ordinary Joes and Juanitas to travel by rail. For the most part, the experience of railway travel was far more banal than described in these books; certainly, for the first few decades, it was mostly downright uncomfortable.

The first trains were utilitarian affairs as the railway companies concentrated on engineering rather than on providing for customers, and the improvement in conditions on trains was consequently gradual. The Liverpool & Manchester, as with many other aspects of the railway, set the tone, along with several of the inaugural railways in other countries. In particular, there was little regard for the poorer passengers. Third-class carriages, referred to by some of their users as "pig class," were open to the elements and had hard benches—some did not even offer that amenity—and the second-class accommodation was little better, apart from a canopy giving modest protection from the elements. Luggage and sometimes people were carried on the roof, but as soon as trains speeded up this practice was discontinued.

The first-class passengers enjoyed rather more comfort. Their carriages had three compartments, essentially stagecoach bodies joined together, and each one accommodated six people, who, according to a contemporary account in slightly obtuse language, were "defended from the annoyance of pressure by arms, like those of a chair, with which each seat is provided."[1] Seats were numbered and reserved, so that no one was left behind "for want of room," but the journey would have still been very bumpy as the carriages only had four wheels and only basic springs. The fare in first class for a 62-mile return journey on the Liverpool & Manchester was 7 shillings (35p), just over a penny farthing (1¼d or 0.5p) a mile. Second class cost 4s (20p), still not cheap for workers who generally earned less than £1 (around $5 at the time) per week. Even for first-class passengers, there was no lighting or heating, nor toilets or refreshments.

The discomforts, however, pale into insignificance compared with the fear with which many people approached train travel. The most frightening aspect was not so much traveling at 20 or even 30 miles per hour, far faster than a galloping horse, but the prospect of meeting a train on a set of rails less than 5 feet away, hurtling in the other direction at the same speed. However, the advantages of the new form of transport, and its improvement over the expensive and slow stagecoaches, were so great that people overcame their concerns and the discomfort. Virtually everyone flocked onto the iron road.

Other lines followed the Liverpool & Manchester's practice of accommodating third-class passengers in open wagons until the 1840s, when one of the earliest rail accidents, at Sonning, in Berkshire, on Christmas Eve 1841, caused the death of several people, who were hurled out of the open wagons. The accident put a stop to the practice, but even then, facilities remained basic. The *Pictorial Times* described the Bristol & Gloucester Railway's "parliamentary train," the service that had to be provided by the railway companies to comply with William Gladstone's 1844 legislation specifying that on every route there had to be at least one daily train offering fares of no more than 1d (0.4p) per mile and speeds of at least 12 miles per hour, including stops: "[There is] no night lamp, one door only on each side. A sight of the country is confined to the passengers who are fortunate enough to get near the door. No provision is made for the admission of air in bad weather when the doors and windows are closed. The badly lighted and ventilated carriage carries fifty-four passengers."[2]

French trains were similar, with three classes and the same stage-coach design for first-class passengers, and other early starters in Europe, such as Belgium and Bavaria, adopted similar models. There was just one compensation for the third-class travelers: According to Wolfgang Schivelbusch, the author of the book *Railway Journey*, which analyzed the effect of the railway on people's lives in the nineteenth century, they had a lot more fun than their better-off peers in the more expensive coaches. He cited a German traveler in the late nineteenth century who reported how, "while traveling alone or with people with whom it was impossible to start a conversation, [I] envied the travelers of the third and fourth class, from whose heavily populated carriages merry conversation and laughter rang all the way into the boredom of my isolation cell."[3]

As we saw in Chapter 4, American carriage design was different from British models because of the length of journeys and the far harsher weather conditions, with the result that even poorer passengers were offered reasonable amenities right from the start. In the early days, there were no classes, and on the Baltimore & Ohio, which started running in 1831, all the carriages had open platforms at each end and an entrance in the middle, a style that would be universally adopted throughout the United States. Center aisles and backs for the seats soon appeared, as did heating, which was far easier to provide in large open carriages than in compartment trains. Ventilation was ever a problem, the choice being between opening the windows to fresh air laced with soot and sparks or putting up with a stale atmosphere.

Women faced particular problems on trains. While being able to travel easily was a liberating experience, freeing them from the need to be shepherded by men, it also had its risks. In the United States no particular provision was made for them, but in Europe, several countries, notably the Germanic-speaking states, Russia, and Italy, created women-only compartments but also discouraged them from traveling alone at night. In India women faced harassment from railway officials and fellow passengers, and their treatment was the subject of widespread protests by the native population against the railway authorities. Indeed, more widely, the treatment of third-class passengers, who were exclusively Indian, caused complaints throughout the period of the British Raj. The doubts about whether Indians would use the railway were instantly proved groundless as the local population flocked to use them, frequently traveling long distances across the country, notably for pilgrimages. But their treatment was appalling. They traveled in cramped, unlit carriages that invariably stank because there were no toilet facilities. There was no drinking water at stations, the vendors licensed by the railway authorities exploited their monopoly by charging exorbitant prices for their food, and women and children had no one to help them with their luggage. The worst conditions were endured by pilgrims who were dumped into freight wagons normally used to transport livestock because of the shortage of third-class rolling stock. The wagons had neither windows nor amenities and were policed by officious conductors who gave no succor to the pitiful passengers. The wooden wagons were bad enough, but the iron ones, which had no insulation, were insufferable, becoming intensely hot or cold depending on the outside temperature.

As early as 1866, the British Indian Association of the North West Province petitioned the British government about conditions on trains for the "poor passengers." The petition called for proper shelter at stations, restaurant facilities, medical staff, and, most of all, elimination of the "unfailing bad treatment of Native passengers of all classes and grades."[4] The lack of toilet facilities in either second or third class was a real hardship for all passengers but especially women, as was the lack of any lighting, which exposed them to harassment and theft. Worst of all, though, was the attitude of the railway officers, belonging to "a low class of Europeans," who treated even those more affluent Indians able to afford second-class travel with rudeness and contempt.

What is cruelly striking about the situation on the Indian railways is that despite this litany of complaints, the authorities did nothing to change the situation until the dying days of the Raj. The same complaints expressed in the 1860s were made seventy years later by Mahatma Gandhi, and the treatment of the native population by the railway authorities contributed to the hostility toward the colonial power that his popularity embodied. The author of a history of Indian railways suggested that it was not so much the lack of facilities that angered the Indians, but rather "the ill-treatment [that] smacked of racism and . . . was deeply resented by Indians."[5]

Nowhere in the world was the contrast so great between first and third class as in India—though Russia, with the authorities' long-entrenched disdain of the masses, came close. First class in India was a comfortable and pleasant experience. A Frenchman, a Monsieur Rousselet, who spent the mid-1860s in the subcontinent, recounted how he traveled "an immense distance with comparatively little fatigue—sleeping at night on a comfortable little bed and walking down in my carriage during the day."[6] Though dining cars were a twentieth-century innovation in India, from the early days there was a well-organized system of ordering meals from an attendant who would then telegraph the details to the station ahead, ensuring that a freshly cooked meal would be delivered at the next stop.

Although conditions on Indian trains barely improved over time, in Europe there was a marked improvement in the final three decades of the nineteenth century. Oddly, it was royalty that, unwittingly, came to the rescue of ordinary rail passengers, followed by an American, George Mortimer Pullman. More comfortable coaches with additional wheels giving a better ride and improved facilities were first developed for royal travelers and then introduced more widely. Queen Victoria was a reluctant train traveler, and the rail companies had to work hard to tempt her on board. The Great Western built a special state carriage for her in 1840, with eight wheels rather than the standard four, but even then it was two years before she ventured onto it for the short trip between Slough (near Windsor) to Paddington. That year, the dowager Queen Adelaide had a special carriage built for her use on the London & Birmingham, which was soon equipped with a heating system, but it would not be until 1852 that this began to be introduced for ordinary passengers. The first lavatory was provided for Queen Victoria on the royal Great Western coach in 1850, though the plumbing was crude and the royal excreta dropped onto the tracks just as that of her subjects did when, in the 1870s, this facility was at last provided for them. In the rest of Europe, too, these state carriages were eventually followed by better accommodation for the greater mass of travelers.

Because of the sheer size of the United States and the length of train journeys, American railroad companies had to address the problem of feeding their passengers and providing both toilet and sleeping facilities far earlier than such things were attempted in Europe. The carriages also had to be better sprung because the poor track quality would have made long rides on four-wheel coaches unbearably uncomfortable and tiring. By the 1850s, the "bogie coach," a long carriage supported at both ends by a set of four wheels, which were able to swivel rather than being held rigid like the wheels on older trains, had become standard. The extra sets of wheels not only improved the quality of the ride, but spread the weight of the carriage in order to reduce impact on the track. As the coaches were longer and open plan, fitting a lavatory at each end, usually segregated by sex, was an easy matter. Clerestory roofs, which had an internal lining to insulate the carriages from extremes of hot and cold, quickly became the norm as well.

The Baltimore & Ohio developed the first dining cars in 1848, but for many years most trains still had scheduled meal stops where food of usually very poor quality was on offer. Right from the beginning, basic arrangements had been made for people to sleep on trains, but these consisted of little more than providing a blanket or a pillow, and the ride was, in the words of an English visitor, "like sleeping on a runaway horse."[7] The earliest purpose-built cars were introduced by the New York & Erie Railroad in 1843, when beds were made up with the help of iron rods linking opposite seats, but since the cushions contained horsehair cloth that penetrated all but the thickest clothing, only the drunk and thick-skinned would have enjoyed much beauty sleep.

There were a few advertised sleeper trains in the 1850s in the United States and even in India, but it was not until the arrival of George Pullman that the railways began to meet people's basic need for a proper night's sleep on a train. Just as the influence of George Stephenson in the early days of train travel is difficult to overemphasize, so is George Pullman's role in spreading the idea of luxury travel for long-distance travelers. Pullman, one of the few people whose name became synonymous with his product, was already renowned for moving buildings without damaging them when he turned his attention to rail travel. Most famously, in 1855 he had organized the raising of Tremont House hotel, then Chicago's tallest building with four floors. The hotel owners had thought that the building would have to be demolished when local authorities decided the streets had to be raised to clear the local swamps, but instead Pullman put the whole building on 5,000 jacks and raised it inch by inch—for a total of 6 feet—while chambermaids continued to make the beds and waiters served lunches to the guests.

There was, therefore, a certain logic in Pullman deciding to develop the concept of hotels on wheels. He first converted two carriages of the Chicago & Alton Railroad in 1859 to provide upper and lower sleeping berths. When they were put into service between Chicago and Bloomington, Illinois, the passengers were somewhat bemused. They were reluctant even to take off their boots before settling on their berths, which, unlike those that would be introduced later in Europe, were set out lengthwise and did not have individual compartments, a style that endured in the United States. Nevertheless, the service proved popular. But the Civil War interrupted Pullman's efforts to promote his ideas, and it was only in 1863 that he was able to build the *Pioneer*, which was far grander and more luxurious than any previous railroad car. It was also the most expensive, at $20,000, five times the average cost of a railroad car in the 1860s. The extra expense was partly for the "interior finish of polished black walnut, candles set in elaborate chandeliers, pure linen bedding and marble washstands."[8] The *Pioneer* was so heavy that the railroad companies were reluctant to use it for fear it would damage their track. However, when it was selected to be part of the cortège for Abraham Lincoln's funeral in 1865, the Chicago & Alton had no choice but to adapt its permanent way to take the heavy train, whatever the expense. Other railroads soon followed suit, and soon, being able to withstand the weight of Pullman cars became the basic standard for railroad tracks. Within a couple of years Pullman had built a fleet of forty-eight coaches, all of which were given individual names.

The sleeping car naturally led to Pullman's next innovation—a "hotel" car that combined sleeping accommodation with a restaurant, where a steak could be purchased for just 60 cents, on the Great Western Railway of Canada. Then, on the Chicago & Alton, he introduced full-length dining cars, which soon developed a reputation for excellent cuisine and rapidly spread to other railroads. There followed the luxurious day saloon with light refreshments on tap and a relaxed hotel *salon* atmosphere. These facilities became all the more necessary as the transcontinental routes opened, requiring six days at best to cross America; Pullman responded by introducing new cars and additional luxuries specifically designed for them. Technical developments, however, were still at times lagging behind the need for luxury: Steam heating from the locomotive, which was cleaner than the stoves used previously, only became available in the 1880s, and it was not until the following decade that electric lighting began to replace smelly kerosene.

Another crucial development for these long journeys was the introduction of an inter-car vestibule, protected by flexible pleated shields, since previously walking between coaches had been a hazardous task, especially for ladies in long dresses, as it involved stepping over the couplings from verandah to verandah. Ladies, however, got something of a raw deal. The editor of the Baedeker guide noted that in the open-plan sleeping carriages, men were allocated twice the space given to women in which to dress and make their toilets. The company pleaded that there were far more men to accommodate in these unsegregated carriages, but Baedeker's scribe was not appeased: "It is considered tolerable that they [women] should lie with the legs of a strange, disrobing man dangling within a foot of their noses,"[9] he snorted. He was critical, too, of the slowness of branch line travel and the overall standards on the American railroads, which were poor compared with those of the local steamboats, which he likened to "floating palaces."

Although the poorer travelers could get across the United States for under $30 in the 1880s—such a bargain that there were complaints that hobos were no longer finding it necessary to jump freight trains—Pullman passengers had to pay a stiff premium to enjoy the luxurious facilities. Given that standards on most trains were still very basic, there was no shortage of people prepared to pay extra for the privilege of Pullman travel, especially on long journeys, and his company thrived. Pullman's strength was in providing high standards and being meticulous about maintaining them, in contrast to most railway companies of his day, which had little understanding of the concept of passenger service. His early trains, as Geoffrey Freeman Allen suggested, "were an incitement to gluttony."[10] The hotel cars on the Chicago-Omaha route in the 1870s offered no fewer than fifteen seafood dishes, together with thirty-seven meat entrées, including a prodigious selection of game. It would be interesting to know how often the waiters had to say, "Sorry, that's off today," though it would have been rarely, if Pullman had anything to do with it. He wrote a remarkably detailed primer on how the waiting staff should treat their customers, setting out, for example, twelve steps on how to serve a beer, starting with asking what type was required, going through a detailed list of the order in which to pour the bottles, and then, finally, removing the tray. And that was just for serving beer!

Pullman was asked to produce ever more luxurious carriages, costing up to $500,000 each, for the truly rich who would not deign to mix even with those who merely could afford a journey on Pullman's expresses. Sparing no expense, he created what were called "mansions on rails" that were taken from private siding to private siding. They were fitted with whatever their owners requested, regardless of cost or weight: marble baths, hidden safes, Venetian mirrors, and even, for J. P. Morgan, an open fireplace burning balsam logs. The railroad baron Jay Gould had a set of four carriages, one of which was occupied by a cow that was felt to be essential because it could provide fresh milk with precisely the right butterfat content not to upset the sickly magnate's delicate digestive system. These carriages were attached to a scheduled service train at a small cost, or even for free, since the railway companies were grateful to be chosen to transport these masters of the universe on their tracks.

Pullman, who was fiercely anti-union, died in 1897, his health having been weakened by the strain of a particularly bitter and rancorous labor dispute in 1894, the worst ever in U.S. railroad history. It had involved 125,000 workers on twenty-nine railroads and had culminated in pitched battles that resulted in the death of thirteen strikers. Pullman was, in fact, an old-fashioned paternalist in the Rowntree or Cadbury mould, creating a company town with pleasant housing and good amenities for the members of his workforce, who, in return, were expected to conform to his rules, which included a ban on public speeches, town meetings, and independent newspapers and a requirement to keep their homes clean. He died as one of the richest men in America, but because of his intransigence during the strike, one of the most hated, with the result that his lead-lined coffin had to be covered in cement because of fears that his grave would be desecrated by unionists.

Of course, there were rival sleeping-car providers in the United States, notably Webster Wagner, who built up a fleet of seven hundred and, incidentally, was killed when one of his cars was destroyed in a rear-end collision. None of the competitors, however, could match Pullman, and he kept his lead by buying many of his rivals out, including one called Rip Van Winkel. Eventually, in 1927, the Pullman company established a monopoly in the United States, by which time the company had long been established in Europe.

In Britain there had been no pressing need for sleeping cars until the Anglo-Scottish trains started running in the 1850s. The first attempt to provide for reclining passengers was a strange "bed carriage" provided in 1838 by the Grand Junction, which ran between Birmingham and Manchester. Rather like the early American services, these were compartments that ran lengthwise. Passengers rode with their midriffs supported by extra cushions in the aisle provided by stewards, and, bizarrely, their feet would tuck into a specially created cubbyhole in the opposite partition.

Remarkably little progress, however, had been made by the time Pullman arrived on the scene. The far-sighted James Allport, the long-serving manager of the Midland Railway, crossed the Atlantic in 1872 to observe Pullman's trains in service and brought the concept back to the United Kingdom, initially providing restaurant cars with a table service, and later sleeping carriages. Interestingly, Allport's moving restaurants fell foul of the strict British licensing laws; to evade them a complicated charade was devised that was not unlike the procedure for obtaining food on Indian trains. Orders were taken on departure, then a manifest was thrown out at a handy signal box and information relayed up the line so that the drinks could be picked up at the first station. Since technically the refreshments had been bought at the station buffet rather than en route, this was akin to bringing one's own bottle to an unlicensed restaurant and therefore remained within the law. Allport also provided better service for poorer customers in 1875 by abolishing second class and providing the carriages for third-class passengers, who previously had to content themselves with hard benches. This move, which appalled rival railways, marked the belated start of a process of gradual improvement for all passengers on British railways.

Pullman offered the height of luxury travel in Britain. He provided a series of extremely well-appointed dining cars that were either attached to existing trains or made up entire trains, such as the prestigious *Brighton Belle* and the *Golden Arrow*, and the term "Pullman" became synonymous with the best in luxury train travel.[11] Pullman's sleeping cars, however, did not prosper in Britain, as the railway companies provided their own in response to his attempts to introduce them.

The popularity of Pullmans spread to many places around the world. After the transandine railway opened in 1910, the arduous thirty-eight-hour journey from coast to coast could be made somewhat more comfortably by traveling on the Pullman service. The first part of the journey from Buenos Aires to Mendoza, in the Andean foothills, took just under twenty-four hours, arriving at 6:10 A.M., and then passengers could choose either the ordinary train, leaving at 7 A.M., or, half an hour later at an extra cost of just over £1 on the £10 fare (say, the equivalent of around £500, or $800, today) a Pullman, which offered a dining car and comfortable "day" coaches for the twelve-hour journey across the Andes.

In Europe, however, another pioneer in train design soon proved more successful than Pullman. Pullman's open-plan sleeping cars were not as popular in Europe as they were elsewhere, especially with the French, who disliked being surrounded by strangers while they slept. George Nagelmackers, the Belgian creator of the Compagnie Internationale des Wagons-Lits, teamed up with Colonel William Mann, an American manufacturer of "boudoir" cars, which were sleepers with individual compartments and doors, a design that proved to be more acceptable to the European public. Nagelmackers's real breakthrough, however, was to create through trains using these cars across frontiers between various European capitals, establishing, in effect, an international train service across the continent.

His most famous train was, of course, the *Orient Express* but he developed numerous other services that traversed Europe and made cross-border travel far easier than before. In 1872, Nagelmackers ran his first sleeping car from Ostend to Brindisi on the tip of the Italian heel, oblivious to the Franco-Prussian War, which had just ended. This service was part of the *Indian Mail*, the mysterious, ghostly even, train that traveled overnight through France to connect with ships destined for India. Its origins stretched back to 1839, when carriers took mail overland by road through France to Marseille to avoid the long boat trip via the Bay of Biscay. Starting in 1855, the *Indian Mail* ran at speeds of up to 50 miles per hour at night, without stopping, and was given priority over all other services. Because of its stealthy nighttime progress, a host of myths grew up around the train, suggesting that it was carrying gold or jewels, or even errant maharajahs. The truth was more prosaic, as it was primarily used for mail and urgent freight, taking advantage of the high speed of rail travel compared with steamers. With the opening of the Mont Cenis Tunnel in 1871, Brindisi became the port of embarkation for India, but not for passengers until the Compagnie Internationale des Wagons-Lits trains were introduced. The Brindisi service later became the *Peninsular Express* and enabled British travelers to leave London on a Friday at 3:15 P.M. and arrive, via Paris Gare du Nord, at Brindisi at 4 P.M. on the following Sunday. From there they could take the steamer for India through the Suez Canal. The mail was then taken in a second train; the service survived until the outbreak of World War II.

The East and the Balkans were opening up with the weakening of the Ottoman Empire following its defeat at the hands of the Russians in 1878, and a train service linking them with the rest of Europe was an obvious next step. Nagelmackers needed all his well-honed skills as a supreme deal-maker and negotiator to thrash out arrangements with eight mistrustful railway administrations across Europe, a process that took several years and required finding solutions to problems ranging from technical details of locomotive couplings to regulations concerning the security of wine lockers. Eventually, with much press interest but, initially, few passengers, the *Orient Express* was launched. The train was rather less grand than its name implied, consisting of just three coaches, two sleepers, and a diner sandwiched between a couple of baggage vans, which departed from the Gare de l'Est in Paris on the first journey to Constantinople (Istanbul) via Vienna, Budapest, and Bucharest on June 5, 1883.

Moreover, the line was not complete. At Giurgi, on the Romanian side of the Danube, a local train took the passengers to the Black Sea coast; from there they sailed to their destination in an Austrian Lloyd steamer. The London *Times* man, bearing the extraordinary name of Henri Stefan Opper de Blowitz, was happy; he reported that he could shave despite the train's speed, which was hardly surprising given that "*Express*" was somewhat of a misnomer: The *Orient Express* crossed France and Germany at 45 miles per hour, but thereafter progress was slow, 30 miles per hour in Hungary and just 20 in Romania. With stops, the timetable allowed eighty hours for the 1,800 miles, and that was not counting the all-too-frequent unscheduled halts on the eastern sections. From the beginning, a high standard of comfort was assured. The first coaches accommodated twenty people each and had a *salon* for the ladies and a smoking room for the gentlemen with red plush armchairs. A connection from London and Ostend on the Belgian Channel coast, the *Ostend Vienna Express*, was added in 1894, by which time ten hours had been shaved off the Paris-Constantinople journey time.

A service linking Calais with the Mediterranean, the *Train Bleu*, soon followed, using William Mann's boudoir coaches and exploiting what a contemporary railway history suggested was "the most profitable bit of traffic in the world."[12] It proved immensely popular, especially among the British, for whom the train was largely designed. Another overnight service linked Paris with Berlin. Indeed, by the outbreak of World War I, there were, in fact, a huge variety of "Orient" and other expresses, or just assorted connecting carriages with similar names run by Nagelmackers's company, allowing passengers to reach such far-flung cities as Copenhagen and Moscow, usually with a connecting service to part of the main *Orient Express* service. When the Simplon Tunnel opened in 1906, the *Direct Orient Express* was launched, which went through Lausanne, Milan, Venice, Zagreb, Belgrade (where branches from Athens and Munich linked up with the train), and Sofia, taking fifty-three hours. Initially the *Orient Express* was only a weekly service, but as patronage built up daily services were operated on the western section between Paris and Vienna and twice-weekly services further east.

The literature on the *Orient Express*, with its intrigue and murders, suggests that this was a train solely for the rich, as there was only one class of accommodation, which cost 300 francs for a single journey from Paris to Constantinople, around two weeks' wages for a manual worker of the day. In fact, second- and third-class carriages attached to the train were used extensively for shorter journeys by poorer people traveling in cramped conditions, especially in the eastern countries. In the comfortable *couchettes*, there were indeed diplomats, affluent traders, aristocrats, and even royalty, who were served attentively by a vast array of well-trained staff. The corps of sleeping-car attendants had to be masters of various languages, shield their charges from all kinds of enervating frontier formalities, be prepared to administer to their medical needs, and, above all, be discreet. The stories of train attendants hiring prostitutes for bored gentlemen and even bishops by telegraphing the station ahead, and the intrigues concerning spies and diplomats, were undoubtedly true, but probably not as frequent as the novelists would have us believe. Certainly, there is no record of a murder on the *Orient Express*, though there was the mysterious death in the 1950s of a U.S. official who fell from a train. Nagelmackers set out detailed rules to maintain high standards, requiring the attendants to be smart at all times. On special occasions waiters were dressed in blue silk breeches and buckled shoes, while the locomotive crew were expected to wear white coats, hardly a sensible choice of color. The passengers, too, were expected to look their best, with evening dress *de rigueur* for the seven-course dinner service.

Away from this world of luxury, however, train travel, even for the aristocracy, remained spartan. The Baroness de Stoeckl,[13] a regular traveler down to the Côte d'Azur from London, described a journey in 1886 that showed that on the routine trains, little had changed. She was content until she arrived in Calais, where "there was a wild rush for the buffet as in those days there were no restaurant cars on the trains." Everyone ordered the same—*potage, demi-poulet, pommes purées, un demi de vin rouge*—only to find the train officials hustling them out by shouting "five minutes to departure" long before the *poulet* could be finished. After paying their bills and grabbing their coats, the passengers "[ran] like maniacs along the platform, maids with anxious faces pointing to the various compartments, the porters shouting '*prenez vos places.*'" There were no lavatories or corridors, but the baroness reported mysteriously that "most people took with them a most useful domestic utensil, the emptying of which necessitated the frequent lowering of the window." And this was in first class!

After changing trains in Paris, the baroness reported, things were slightly better. The compartments "consisted of three large seats with padded backs, these had a handle which one pulled and down came the bed." There were, though, no blankets or sheets, only pillows that could be hired for a franc, and the only heating was from *chaufferettes*, footwarmers that were changed at various stations during the night by men in blue blouses who made a tremendous kafuffle by removing the old ones, which were hurled noisily onto the platform, and replacing them with steaming hot new ones. To keep warm, the baroness "had a travelling rug, but one never undressed, it was not considered safe in case of accidents." Her breakfast was unsatisfying, too. At Toulon, "a chipped cup of coffee with a *croissant*, already moist from the overflow on the saucer (*plus ça change!*) would be thrust through the window; that was all, yet one lived through it."

The baroness may have been a tad unlucky or simply too demanding, but there is little doubt that she reflected the general experience of train travel for most people. Fortunately, the conditions improved greatly in the last quarter of the nineteenth century throughout Europe. The railway companies became less concerned with expanding their networks and were thus able to invest in existing services. They provided better lighting and heating, and cushioned seats became standard; a key improvement was to have a corridor linking compartments, which gave passengers access to dining cars and toilets. The Nord company in France first introduced this as an external passageway open to the elements, a fairly hazardous innovation, but soon the corridor was brought inside to provide an internal connection, and this became standard in large swathes of Europe.

Countries where the railway came late seemed to lag well behind in provision for passengers. A journalist traveling on New Zealand's first major railway in 1864 reported how he and fellow passengers were "politely requested by the guard to leave the carriage and help to push the carriage and engine to the summit of the bank . . . and on returning to our seats, the guard promptly collected 2s 6d [12.5p] from us as our fares!"[14] The accommodation on the trains was a throwback to the Europe of thirty years before: "The standard carriages of the 1870s were tiny four and six wheel boxes with rigid axles, longitudinal (lengthways) bench seats and gloomy colza-oil lamps. Heating, toilets and passageways between carriages were non-existent."[15] There were only two classes, but despite paying 50 percent extra, first-class passengers were in the same carriages as the second-class ones, separated only by a partition, and benefited only "from horsehair cushions, coir floor mats, brass spittoons and the 'quality' of one's travelling companions."[16] An English visitor described it as barely up to the standard of second class back home but far slower, since the trains averaged only 20 miles per hour.

In New Zealand, where the railway system had been built on the cheap, it was hardly surprising that speeds were low, but in Europe there was no such excuse for the failure to improve journey times in the middle years of the nineteenth century. It is a reflection of the mores of the age that the introduction of luxury on trains for those who could afford it preceded speed, which would have benefited all passengers.[17] Trains had so speeded up communication that neither the passengers nor the rail companies initially showed much interest in making them go any faster. The French were the first to take an interest in speed, running trains in the early 1840s that did not stop at every station and calling them "expresses," a word that soon crossed the Channel. In Great Britain, the early trains averaged 20 miles per hour, but in 1852 the Great Western started running an express service between Oxford and London that was timetabled at 55 miles per hour, and one from London to Exeter, a distance of nearly 200 miles, that averaged a commendable 43 miles per hour.

These fast trains were, however, very much the exception, and it was not really until the last quarter of the nineteenth century that much effort was focused on reducing journey times. By then there was no great technical barrier, since steam locomotives were now far more developed than previously and able to run at high speeds for long periods. Thanks to the widespread use of steel, steam locomotives were able to haul heavier loads, go faster, and climb steeper gradients. The locomotive industry was maturing and able to respond to a variety of needs, whether for more powerful, more robust, or faster engines. With better locomotives readily available, it was more a question of the railway companies being willing to provide faster services, given that the extra cost was considerable, than having the technology to go faster. The track had to be in better condition, gradients had to be reduced, and the sharpness of bends lessened. There is always a trade-off between keeping costs down and raising line speeds, and parsimonious railway managers tended to err on the side of caution.

A survey by a pair of contemporary timetable analysts, E. Foxwell and T. C. Farrer, provides a detailed snapshot of timings across the world in 1889.[18] They were only interested in "expresses," which they defined, with some generosity, as any train averaging 29 or more miles per hour, except, oddly, in Britain or America, where the cut-off was 40 miles per hour. This requirement immediately ruled out large swathes of the world; there were no such services in South America or India, although they did find a few in Australia. Leaving aside their rather jingoistic bias toward British services, they praised Holland for being "blessed with British locomotives" and estimated that France, with a few trains averaging 43 miles per hour, had the "best set of expresses on the Continent," but they were critical of the lamentable standards on the great Paris-Lyon-Méditerranée line. In Germany 35 miles per hour was very rare, and in Austria-Hungary there were even fewer fast services. They found that there was only one express daily linking Milan with Venice and noted that for passengers from London, the connection involved a six-hour wait. Services in Sweden were reputed to be "very poor," but Denmark attracted praise, as did Hungary, for allowing the *Orient Express* to average 32 miles per hour, faster than it could go in Austria.

In the United States, they found a mixed bag of services, which was hardly surprising since there were now 200,000 miles of line. They found that the fastest U.S. train was the one between Washington and Baltimore, covering 40 miles in forty-five minutes, an average of 53 miles per hour. Virtually all the fast services were in the East, with the Pennsylvania Railroad having by far the greatest number of fast services, with almost half the total. Foxwell and Farrer were disappointed, however, at the poor service offered by many famous trains, even those with grand names such as "Flyer" and "Limited" that indicated they did not stop at all stations. The *Chicago and St. Louis Limited*, for example, took nearly thirteen hours to run between Jersey City and Buffalo at an average of 31 miles per hour. One cause of low average speeds was that trains had to slow down while crossing towns. The lines ran directly through the center of many towns, and there were perilous crossings over main streets without the protection of fences, thanks to the American habit of eschewing expensive bridges and tunnels. Over the next decade, however, many American railroads made heroic efforts to speed up their services, often in response to competition from each other.

By the start of the twentieth century there had been a big improvement in many countries. France still had the best trains, with twenty daily expresses now averaging 56 or more miles per hour, and had created a whole series of international trains running from the Channel ports to a variety of places beyond its borders, such as Vienna, Warsaw, and Milan, aimed largely at rich traveling Brits. By then, express times in both Britain and Germany averaged more than 50 miles per hour, but the norm for most passengers was still the slow train that might spend a few minutes stopped at every station while having wagons or carriages added or removed. The best trains offered services that were not so different from today's timings. In 1888, for example, the standard scheduled trains between Rotterdam and Amsterdam, a distance of 54 miles, took seventy minutes, barely different from today's timing of around sixty minutes with two stops.[19] On several routes, both in the United States and Britain, races developed between competing railways on particular journeys, such as New York–Chicago (which, in the early 1900s, was served by no fewer than eighteen possible routes), New York–Boston, and London–Scotland, which spurred on companies to reduce journey times.

Not only did train services speed up in this period, but the frequency of trains also increased greatly as railways began to cater to the growing demand and realized that providing extra services on existing tracks was generally a profitable exercise. Fares were coming down, too. A rare comparison of the cost of travel between 1889 and 1900 by W. M. Acworth, the author of a classic history of the railways,[20] showed that both European and British fares reduced over that period, though interestingly it was the more affluent travelers who benefited most. The cost, for example, of traveling between London and Milford on the Welsh coast, a distance of 280 miles, was reduced in that period from £2 7s 9d in first class and £1 2s 8½d in third to £2 3s 6d and £1 1s 8½d, respectively (the exchange rate at the time was around $5 to the pound and there were 20 shillings in a pound). In France, reductions were even sharper, and geared more toward poorer passengers, with the 540-mile journey between Paris and Marseille costing £3 17s 4d for first class and £1 14s for third in 1900, a reduction of 9 percent and 27 percent, respectively, from the 1889 fares. Interestingly, the comparison suggests that overall fares were broadly similar in Prussia and Britain at the turn of the century, averaging around 2d (0.83p) per mile in first class and around half that in third, while in France and Italy rates per mile were around 25 percent cheaper. In Prussia prices had risen because of the widespread introduction of "concertina" trains with corridors for which an extra charge was made. In truth, such comparisons are extremely difficult to make, since there were no overall statistics available, a fact Acworth lamented. As a more recent writer put it, when discussing whether the conditions on trains in the late nineteenth century were better in France or Germany, "so many coaches travelling countless kilometers over such a long period of time under a wide variety of conditions do not permit a secure global judgment of that sort."[21]

Acworth, in his first edition, published in 1889, pointed out that the difference in fares between Britain and other countries was more marked for first class, but that only accounted for 3 percent of travelers. He noted that as "nine Englishmen out of ten go third," the key point was the condition of these carriages, and he concluded that the British ones were "incomparably superior."[22] Although he still felt this was the case a decade later, he had to concede that some key express trains in France, and, particularly, in the United States, had speeded up at a far greater rate than in Britain. He concluded that "the best expresses in the world are no longer, as they were ten years ago, English: they are French and American."[23] Indeed, Acworth realized it was more than that. This was the stage in railway development where America had seized the initiative, where its technology was beginning to be the first choice for railways around the world: "Twenty years ago foreign railwaymen in search of new ideas and improved methods came, as a matter of course, to England. Today the intelligent foreigner thinks there is nothing new in English railway matters to be studied and he accordingly betakes himself to America."[24]

It would be wrong to draw widespread conclusions about the generality of the travel experience on these respective railways from the performance of a few expresses; however, there is no doubt that services were improving in every way during this crucial period before the advent of the automobile and the truck. In fact, by the early 1900s, the Americans' claim that they had the best trains in the world—as one historian put it, "the superiority of American [rail] passenger travel to any in the rest of the world was beginning to be taken for granted"—was probably true by virtue of the fact that they had such a wide range of high standard services. Whereas European railways could generally boast a few excellent prestige expresses, the regular intercity services in

America were probably far superior to those existing at that time anywhere else in the world. The efforts of American railroads from the
early 1890s to speed up timetables had paid off, and the widespread extension of "limited" trains greatly improved the service by enabling
long-distance travelers to avoid having to change trains. Rival railroads
began cooperating to provide through services, and a process of consolidation started to develop that would see the emergence of a handful of
huge railroad companies in the early twentieth century. These trains
would be of flexible length, with carriages from branch lines being attached on the way and, most important, Pullmans being connected at
night to give travelers a good sleep. The services did more than just afford Americans the opportunity to travel between major cities. Rather,
they were instrumental in stimulating growth during this key stage of
American industrialization: "It was a period of integration and unification into one great national economy. The railroad was the chief agent
by which this was being accomplished."[25] These vastly improved rail
services, most of which were established by the early 1900s, would set
the pattern for the next fifty years, forming the backbone of the U.S.
transport system, especially in the eastern half of the country.

Travel to the American west was still a lengthy affair, as the trains
had to cross hundreds of miles of desert in often intolerable heat—air
conditioning did not become commonplace on trains until the 1930s—
and this was a great hindrance to the economic progress of the region.
Robert Louis Stevenson, who crossed the United States in 1879, described an interminable journey[26] during which he was crammed into a
compartment with two fellow travelers as they traversed a featureless
landscape: The "train toiled . . . like a snail," he said, punctuated only
by stops for meals that were "rarely less than twenty minutes," and
they slept on uncomfortable straw mattresses laid on stiff boards. The
best part was when he transferred from the coaches of the Union Pacific
to those of the Central Pacific, which "were nearly twice as high and so
proportionately airier" and had better designed seats that obviated the
need for the uncomfortable bed boards. The scenery through Nevada,
though, was equally mindnumbing for poor Stevenson, who was not in
the best of health: "We travelled all day through deserts of alkali and
sand, horrible to man, and bare sage-brush country that seemed little
kindlier," he wrote.

The famous London *Times* war correspondent, William Russell, who had reported on the Crimean War, described a similar nine-day journey[27] a few years later that was equally tedious and so hot that the blinds had to be shut, which "deprived ourselves of the opportunity of what was called 'seeing the country,'" but an occasional glance proved more than sufficient. Russell, however, seems to have enjoyed far better food than Stevenson, including, much to his surprise, fresh fish and a "liberal and unfailing" supply of ice, though the train crew had to labor hard to maneuver the ice out of the special cold larder, using grappling irons. The ennui induced by these long journeys was exploited not only by traders of all kinds, who congregated to sell their wares at stations, but also by card hustlers, who made their living from cheating bored passengers. A poster showing a hand of five aces was displayed on some trains with the warning: "Card sharks can beat any honest player. If you do not play cards with strangers, you are safe. If you do, you invite possible robbery."[28]

Whereas train services took a good half-century to show substantial improvement, stations had quickly become public demonstrations of the power of the railway companies, which were eager to erect temples to steam that dwarfed neighboring buildings. As with everything else that had to do with the railways, the early pioneers had to work out what was needed from first principles. The Liverpool & Manchester had quickly understood the need for some sort of building at its termini, Liverpool Road in Manchester and Crown Street in Liverpool. Both had simple structures—single boarding points (not really yet a platform) and a room for the sale of tickets and waiting. Even the notion of platforms and buffer stops had to be conceived and developed. Brunel, for example, favored impractical stations that were only accessed from one side, causing much delay as trains had to cross each other's paths and inconvenience to passengers. As mentioned in Chapter 2, the strange habit of holding passengers in the waiting room and only allowing them on the platform a few minutes before departure developed in continental Europe, but this practice was not imitated elsewhere.

There was a marked contrast between termini in big cities, which were often built on a scale that was only matched by the local cathedral, and the huge number of stations in smaller towns and villages that were often little more than a hut to keep off the rain, if that. The basic needs of a station were soon established: a booking office, a waiting room, preferably segregated for the sexes, toilet facilities, and a room where the oil lamps could be kept and replenished. Refreshment rooms were a later addition, though they became universal in sizable stations, and telegraph offices became essential, too. Retailing chains, such as W. H. Smith in Britain (which opened its first station newsstand at Euston in 1848) and Hachette (which followed four years later) in France, set up shops at station platforms to sell newspapers and books to rail passengers, but their wares were bought only by the more affluent travelers, not least because most of the poorer ones were illiterate.

Even the most modest stations rapidly developed into the focus of their local communities, attracting not only passengers but greeters and traders, becoming the place of hellos and goodbyes, happiness and sorrow, where soldiers went off to war never to return, or brides came from distant parts never to leave. The importance of the local country stations was greatest in the years that straddled the turn of the century. Whether it was in the United States or Britain, Japan or Mexico, the daily scenes of the movement of people and goods were similar. Everything that the village or little town required came by train.

O. S. Nock may have taken a slightly romantic view of the railways when he described a British scene in the late Victorian era, but he was accurately describing a scene that was replicated across the world, differing only in details and the architecture of the station:

> There would be a small goods yard, a shed, a coal stack and simple equipment to deal with any special local traffic. The station master and all his men would be well-known personalities in the village and in those days it was a prestige job to be on the railway, in however humble a capacity. . . . The station master, in his immaculate array of gold braid, has a rose in his buttonhole; the veteran ticket collector, unofficial head gardener of the station (the station narrowly missed out on the best kept station competition), sports a carnation.[29]

Nock could not resist the nostalgia as he described how a squire and his lady arrived from a London outing and were shepherded off in the carriage that had been waiting for them, while a trader's cart was in the goods shed picking up coal. In Britain, these country stations generally

consisted of one or two cottage-style buildings; elsewhere they also reflected the vernacular: dark varnished wooden chalets in Switzerland
and Sweden, mini-châteaux in France, and in Russia, every country station was wooden and "roofed with thin sheet iron painted red-
brown . . . [with] in the front of the building: a clock, a bell and a large
thermometer."[30] Improvisation was often the order of the day, with an
oak tree serving as a ticket office at Moreton-on-Lugg, on the Shrews-
bury & Hereford Railway, as late as the 1860s. In the United States, undercapitalized railroads tended not to spend much on stations and
where possible used an existing inn or general store to handle ticket
sales and deal with freight.

It was the great termini, though, built with seemingly little consideration of expense or modesty, that best illustrate the importance and
power of the railways in their pomp. The first station with claims to
grandeur was inevitably in Britain. From its opening in 1837, Euston
Station, the terminus of the London & Birmingham Railway, boasted a
200-foot engine shed and a 70-foot-high Doric arch with columns that
was purely decorative.[31] The addition a dozen years later of the classical
Great Hall, designed by Philip Hardwick, with its double staircase and
coffered ceiling, confirmed Euston's status as by far the grandest of the
early stations. Writing a year after the completion of the Great Hall, Dr.
Dionysus Lardner expressed the awe widely felt at the growing power
of the railway companies: "It is impossible to regard the vast buildings . . . without feelings of inexpressible astonishment at the magnitude
of the capital and boldness of the enterprise."[32] The first continental
cathedral was the Romanesque Gare de l'Est in Paris opened in 1849,
which, with its enormous curved windows, was considered the most
stylish station of the age.

The elegant train sheds with their enormous spans, which needed to
be high enough to allow smoke from the steam locomotives to dissipate,
characterize these large stations, and as the technology to build these
huge structures became more sophisticated, their size increased. The
most famous and biggest at the time was the William Barlow train shed
at St. Pancras completed in 1868, which, combined with the Gothic
Midland hotel, opened a few years later, was arguably the world's most
impressive nineteenth-century station. It was a deliberate show of
strength by the Midland Railway, which had reached London later than

its rivals. Indeed, nothing illustrates the crazy competitive spirit among railways better than the juxtaposition of the fantastic Gothic folly of St. Pancras next to the functional, but magisterially elegant, King's Cross, built sixteen years previously. The disunity of the British railway system was neatly encapsulated by the fact that the clock atop the Italianate tower on King's Cross rarely agreed with the one on its Gothic counterpart across Midland Road. Another London terminus, Victoria, housed two railway companies separated by a wall, which made transfer between the lines of the two companies difficult.

This competitive spirit extended to the principal European nations, which, during the second half of the century, seemed to be trying to outdo one another in building huge termini. Most of them survive to this day as the focal points of their cities. In Paris, the Gare de l'Est itself had a rival a few hundred yards away, the Gare du Nord, completed a few years later and based around a triumphal arch that boasted a series of huge sculptures, which represented international destinations that could be reached from the station. The Germans tended to opt for *Hauptbahnhöfe*, single central stations that brought together all the local rail services. Frankfurt's station, completed in 1888, incorporated three sets of lines serving different regions, a practical expression of new-found German unity. The triple shed covering these lines stretched out 610 feet and was 549 feet wide, more than twice the size of Barlow's St. Pancras.

These vast stations were an eclectic collection, borrowing from every conceivable style but cleverly adapting classical designs to the modern needs of a railway station. In Rome, the design of the Termini station owed much to the Gare de l'Est, whereas Milan's original Stazione Centrale was based on the Louvre. Amsterdam's Centraal harks back to the city's maritime past, while in Tours, twin train sheds were projected forward, joining onto an extrovert art nouveau facade. Antwerp's station, built in the dying years of the nineteenth century, was a celebration of the arts and crafts movement. But perhaps the most exuberant of all the stations in Europe was the Gare Quai d'Orsay, the Paris terminus of the Paris-Orléans company completed in 1900 to serve the International Exhibition marking the new century. The French government demanded that its architecture be of the highest quality because it was next to prestigious government buildings. The Beaux-Arts design was by Victor Laloux, who was also responsible for the station at Tours.

There was no need for a huge train shed, as the trains were brought from the previous terminus at Austerlitz by electric locomotives. This was a very early use of electrification and resulted from fears that the connecting tunnels from the old terminus would fill with smoke and choke the passengers. The Gare Quai d'Orsay was the gateway for visiting heads of state, but it survived as a mainline station only until 1939 because its platforms were too short, and was closed entirely in 1958. In the late 1970s the building became the site for an arts museum.

In Asia and America most designs were borrowed from Europe, though the most ostentatious station, Victoria terminus in Bombay, was a wonderful and crazy blend of the European and the Oriental, a style termed "Gothic Saracen" that combined domes with spires and blended an eclectic mix of "Middle English, Venetian, Romanesque and Orientalist features."[33] The era of great stations in the United States occurred somewhat later than in Europe, though it was not long before every major American city acquired one. The train shed for the first New York Grand Central, completed in 1871, was based on Barlow's design, as was that of Jersey City Station. Chicago's Grand Central had stained-glass windows, marble floors, and a massive fireplace, and Union Station in St. Louis, opened in 1894—America's biggest station—was built in a style "variously described as Romanesque, Norman Revival and Chateau."[34] The busiest in America, and at the time in the world, was the art-nouveau-style South Station in Boston, terminus for the Old Colony, Boston & Albany lines, which served 50 million passengers in 1900, the year after it opened.

The symbolic value of these station buildings as a naked expression of the power of nations, as well as of the companies themselves, was demonstrated by the unnecessary reconstruction of Metz Station in 1905 by the Germans, who had taken Alsace-Lorraine as the spoils of their victory in the Franco-Prussian War. The old French station was demolished to make way for a new building whose squat, Romanesque style contrasted with the Beaux-Arts idiom that characterized contemporary French designs and was a clear expression of German triumphalism. It was only in the early years of the twentieth century that a new vernacular of stations was developed; at last designs moved away from the classical, with buildings such as Eliel Saarinen's Helsinki, an early art-deco temple, which also had a deeply political undercurrent because it was an expression of Finnish nationalism at a time when the country was still part of Russia (although, with its hard lines and stylized figures, the station's rather brutalist architecture would not have been out of place in Stalin's Russia). Oddly, although these stations were normally second in importance for their cities only to town halls, the surrounding areas frequently became run-down as the cheap hotels required by travelers would attract prostitutes and the stations themselves became honeypots for hustlers and thieves.

The big stations, like the new prestigious trains, reflected the leading position of the railways for an all too brief period of history running up to World War I. They were the dominant enterprises of their day, and they took advantage of their strong position to consolidate their power by joining together to form large enterprises. However, that was partly a defensive move too. Many railways struggled to make money even before the arrival of competition from the road, and joining together made them more viable. And in most countries, the state was ever present to pick up the pieces of failing railways by nationalizing a service that was seen as essential. Although both Britain and the United States had numerous private railway companies, their networks were increasingly dominated by a few big enterprises. In France there were just half a dozen regionalized companies, still for the time being in private hands, whereas three of the former German states, Prussia, Bavaria, and Saxony, had large state-owned systems with private feeder lines. The Austrian Südbahn and the Hungarian Magyar Allamvasutak (MAV) were also large private railways, while Russia effectively had a single unified system. Elsewhere, too, there was a trend toward consolidation and integration. The railways, though they mostly did not realize it, were about to face their biggest battle: competition from the car and the truck, which would undermine their position after half a century of virtual monopoly. First, there was a world war to be fought, and life on the rails would never be quite the same again.

The *Rail Zeppelin*, an experimental railcar designed by the German aircraft engineer Franz Kruckenberg in 1929. Like many attempts over the years to produce radical alternatives to conventional trains, it proved a failure.

The *Fliegender Hamburger* (Flying Hamburger) high-speed diesel service, one of several similar trains launched between Berlin and Hamburg in 1932. They were all abandoned when Germany went to war in 1939 because of the shortage of fuel.

(Left) Rail-mounted guns were used in several wars but with limited success because of their inflexibility and the problem of recoil.

(Right) Soviet prisoners captured by the Germans, in 1941. The railways were used extensively in the Second World War to carry prisoners, as well as Jews and other concentration camp victims.

This railway bridge over the Rhine was attacked in 1944. Railways were the constant target of bomber aircraft but proved remarkably difficult to destroy and relatively easy to repair.

Facing competition from the car, the American railroads of the late 1940s did their utmost to retain passengers. The lower section of this air-conditioned train, built by General Motors, was designed to be enjoyed by groups of friends or families who could also take advantage of the observation car above.

(Left) Japan built the world's first high-speed railways but their construction encountered surprisingly strong opposition from residents concerned about noise and vibration.

(Right) Is it a train? These futuristic units, called Italo, are built by Alstom and are expected to come into service in 2011. They will travel at speeds of up to 225 mph between Italy's main cities, reducing journey times by up to a third.

(Left) In the post-war period, competition from motor vehicles and aviation caused thousands of miles of track, and in some countries even entire networks, to be abandoned.

(Right) China was a relative latecomer to the railways. Its first line was only completed in the 1870s, but it is now building more high-speed lines than any other country.

Railways have been built in every conceivable type of terrain. Here, a train crosses the world's largest salt flats between Uyuni and El Avaroa on the Bolivian border with China 12,000 feet up on the Altiplano.

ELEVEN

Changing Trains

WHILE ALL SUPPOSED "GOLDEN AGES" OF THE RAILWAYS HAD their flaws, at its height the rail network was a commanding and impressive system. Although many writers suggest the golden age was in the years between the turn of the century and the start of World War I, during the interwar period the railways were still, for the most part, either expanding or at their peak mileage.

The significance of the role of the railways in World War I is impossible to exaggerate. They were crucial not only for the mobilization of troops but also for the movement of equipment, ammunition, and supplies. During the conflict thousands of miles of temporary railways were created, and they were the core of the logistical operation throughout the war both at the front and behind it. The mobilization of the German and Russian armies by rail was such a complicated business that many historians have argued that once it was set in motion, war became inevitable. Indeed, the extremely complex and detailed plans had been prepared over years by military officers who had worked out a precise schedule to deliver troops to the front over a two-week period at the outbreak of war. The Germans had prepared the Schlieffen Plan, which was dependent on a rapid mobilization by rail set out in a sixteen-day timetable. The scheme envisaged the invasion of France through Belgium culminating in a decisive defeat of the French army on the Western Front before the Russian army could attack from the east. The Russians, in turn, had a similar plan to mobilize and knew that once they had started to implement their plan, the Germans would respond by sending their troops west. Tsar Nicholas II was aware of this domino effect, and after the assassination of the Archduke Franz Ferdinand in Sarajevo on June 28, 1914, he hesitated before launching the mobilization plan. He even canceled his orders once, but finally he gave the go-ahead, effectively starting the war a month after the Sarajevo incident.

In fact, the inevitability of war once mobilization had started was probably more the result of the inflexible military mind than a reflection of reality.[1] Reversing the military build-up would have been difficult because of the problem of finding food, both for the horses and then, though less pressing, for the men, but as John Westwood, a historian of the role of railways in war, has explained, "armies in trains are very easily handled and there would have been little difficulty in using the railway telegraph to stop all trains until further notice,"[2] thus giving the rulers the opportunity to negotiate a diplomatic solution to the Serbian crisis. It was not, therefore, as conventional wisdom has it, that the war became inevitable once the mobilization by rail had started, but rather, "the military staffs' conviction that it was impossible"[3] to stop.

If the outbreak of the war cannot be blamed on the railways, there is no doubt that a conflict on the scale and level of devastation of World War I could not have happened without the railways replenishing supplies and troops at the front. During the sixteen-day period when the Schlieffen Plan was being implemented, a train crossed the Hohenzollern Bridge over the Rhine on average every ten minutes day and night. On the other side, the French mobilization scheme, called Plan XVII, successfully delivered 1 million men and 400,000 horses to the front within two weeks of war having been declared. This highlighted a major mistake made by both sides, the emphasis on the use of horses in an era when modern battle tactics and equipment, particularly machine guns, made cavalries largely ineffective. The military seemed to have learned little from the disaster of the Charge of the Light Brigade over sixty years before. The huge movement of horses by Germans, Russians, and French slowed down the progress of their armies, as a cavalry division of 4,000 men required as many trains as an infantry force four times bigger.

Every country involved in the war requisitioned all or part of its railways for military use. Both France and Russia, for example, split their railways into two, with those in the war zone coming under military control and the rest remaining as before. In Britain, the railways were effectively nationalized as soon as the war broke out and placed under the control of a Railway Executive Committee working to the requirements of the war cabinet. The war may not have been fought on British territory, but the South Eastern & Chatham Railway extended its operations into France; later, both the track and the railway equipment from closed branch lines were exported wholesale to France, where the track was installed by the British Army's Railway Operating Division, which was eventually built up to a force of 75,000 men. Remarkably, even in the United States, where nationalization is taboo, the railways were taken over in all but name by the government under the aegis of the newly created United States Railroad Administration. They were then reorganized under military imperatives, which made them far more efficient: They were obliged to send freight by the shortest route, to accept tickets by rival railways, and to standardize their equipment, including locomotives. The government even breached its own antitrust legislation in order to impose order on the railways.

Since the importance of railways was recognized by all sides, an early tactic of retreating armies was to destroy the tracks in the territory they vacated. Surprisingly, given that such sabotage had occurred as far back as the American Civil War, the Germans had relied on the Belgians leaving their railways intact as they fled the invasion of their country, but instead tunnels and bridges had been blown up, delaying the German advance. Troops were often forced to march 30 miles ahead of the last remaining railhead. In contrast, the French maintained rail supply lines right up to the front. This scenario would be repeated throughout the war. While an advancing army would find that the railways had been destroyed by the retreating enemy, the supply lines of the defenders would remain intact. Indeed, as the defenders remained in position, narrow-gauge railways could be built right up to the range of the enemy's guns to reduce the need for horse-drawn transport.

World War I was unique in that it broke out at a particular stage in technological development when railways were well established but motor transport still in its infancy, and aircraft were largely confined to reconnaissance duties. By giving the defending side a built-in advantage, this unfortunate timing contributed to the inertia of the war on the

Western Front, most of which was fought within a boundary of just half a dozen miles. The railways reduced the likelihood of decisive battles in another crucial way, too. Given that air reconnaissance was available to both armies, the need for railway supply lines meant that surprise attacks became impossible. Once evidence of a military buildup, such as the construction of new narrow-gauge railways or intense railway activity to create munitions dumps, was spotted, the defending side would be able to prepare for the assault. The precise targets of offensive operations were chosen with the railway situation in mind, and the first sign of an offensive might well be the firing of the large rail-mounted guns. According to Westwood, "not only did they [the attackers] try to launch their offensives in places where their own troops were well supplied with standard-gauge lines, but they aimed their attacks at areas where the enemy's railway communications were weak or, better still, could easily be cut at an early stage of the advance."[4] And after a successful advance, the first troops to be sent in would be the railway engineers, who had the task of repairing the lines blown up by the retreating defenders.

Even in the quiet periods between offensives, the railways were in perpetual motion. The scale of the railway operation needed to maintain the troops at the front can be gauged from the fact that, at the beginning of the war, it was calculated that for every 100,000 soldiers, a train of 50 wagons would be needed every day to bring in food for both men and horses as well as ammunition and other supplies. By the end of the war, as the battle tactics became more sophisticated, this requirement had tripled to around 150 wagons. The supplies brought in by these trains would be deposited at a railhead, and most were then transferred to specially built narrow-gauge railways that brought supplies as near as possible to the front. These narrow-gauge railways, mostly 60 centimeters (just under 2 feet) became an essential part of the war effort, of far greater importance than had been anticipated by the army commanders at the outset. For example, at the height of the German Verdun offensive, the French had a six-track narrow-gauge railway to the front. Their freight wagons could carry around 10 tons, enough to transport most loads, as well as troops, including stretcher cases. They were particularly useful in winter and in the later stages of the war when many access roads to the front became impassable. The tracks were laid with incredible speed by the armies' railway engineering corps; as a result, not surprisingly, derailments were frequent, but they were easily remedied with the help of a few nearby soldiers. With no signaling, operating procedures were informal, and so were the maintenance arrangements: When locomotives ran out of water, they sucked in a new supply with a flexible pipe "water lifter" from the nearest flooded shell hole.

The worst-performing railways in the war were those of tsarist Russia, and during the October 1917 Revolution the Russian railway system almost collapsed. In building the Trans-Siberian, the Russians had concentrated their railway resources in the east, leaving only a sparse network connecting it with the rest of Europe. The lines leading to the western battle zones were thus immediately overloaded, and there was constant tension between the railway administration and the military, with little clear direction from the incompetent tsarist regime. Supplies to the front were not prioritized and bottlenecks built up throughout the system, severely weakening the war effort. The civil war that resulted from the October Revolution all but wrecked the railways. Coal ran out and trains had to run on newly cut timber. (My father, a White Russian, who fled by rail from Kiev to Odessa after the revolution recalled that his train had to stop several times while the able-bodied men were sent into the woods to cut down trees to fuel the locomotive; the train was also stabled in sidings at various points on the route to allow armored trains through.) By 1921, two-thirds of the stock of 19,000 locomotives was rusting in yards, and it took a decade for the victorious Communists to get the system properly functioning again.

The involvement of the railways in the war was not confined to the Western Front. Most famously, Lawrence of Arabia[5] waged a clever guerrilla campaign against the Turks in Arabia.[6] He led a group of irregular Arab soldiers in a series of attacks on the Hejaz railway with the aim of tying down large numbers of Turkish troops who were then not available to fight on other fronts.

Passenger numbers in the countries involved in the war soared, even in far-off America, resulting in much conflict between military and civilian users. In France this caused serious unrest as *permissionnaires*, soldiers on leave, flooded onto the railways to get away from the front. Once granted leave, the soldiers were supposed to use special trains, but these were invariably full or canceled; unwilling to waste precious leave waiting at train stations, they flooded onto normal service trains, causing chaos. They expected to be treated with a certain degree of respect, having risked their lives to save their country, and were not averse to throwing out all the civilian passengers in order to get home quickly. As the war ground on, the authorities, aware of this problem, gradually put on more and more extra services and created special stations, with amenities such as cinemas, canteens, and information booths, to cater to this huge flow of desperate men. The rowdy behavior of the *permissionnaires* nevertheless continued to cause problems throughout the war. In Russia, the disorder was even worse, with countless stories of soldiers who had been suddenly released from military discipline smashing up coaches, throwing ordinary passengers off the train, and robbing all and sundry.

The most famous rail passenger during the conflict was, of course, the revolutionary V. I. Lenin, who had been forced into exile in Switzerland by the tsar. In April 1917, the German Kaiser, keen to sow chaos and disorder in its Russian enemy, allowed the revolutionary leader to travel in a sealed train from Switzerland through Germany and eventually, via Stockholm and Helsinki, to Petrograd (St. Petersburg). On his arrival, Lenin gave a speech that sparked off the Bolshevik Revolution. He made use of the trains several times in the ensuing months, at one point fleeing back to Finland disguised as a fireman, but the Kaiser's tactic proved successful, as the Russian army collapsed following the Communist uprising. It was, of course, not enough to give the Germans victory in the war, which, fittingly, was ended on a railway coach. Maréchal Foch had based himself and his staff throughout the war in three railway carriages of the *Orient Express*, which of course did not operate in the war though the Germans attempted to run their own version, and ordered another one, number 2419, to be equipped as an office, in which the Armistice was signed on November 11, 1918.[7]

Despite the rapid development of motor transport during the war, the railways remained dominant in its aftermath and were still a booming industry. In 1920, there were 645,000 miles of railway around the world, a figure that would increase to 785,000 by the outbreak of World War II. Of course, the railways were beginning to face the competition that would, eventually, lead to their decline, but neither road transportation nor aviation was sufficiently developed to prevent the railways from retaining their advantage on most types of journeys throughout this period. World War I left many of the world's railways in a state of disarray, short of equipment, overused, and with a huge backlog of needed maintenance. Even though the United States Railroad Administration had run the railways far more efficiently than the private companies, and, incidentally, paid its workers generously, the railways were returned to full private control in March 1920, and America went back to its traditional emphasis on competition rather than cooperation and unification. The lessons were swiftly forgotten and the efficiencies lost as the railways went back to their old bad habits, with roundabout routes for freight, the proliferation of ticket offices, and the construction of nonstandard locomotives in small, and therefore expensive, batches. In Britain, however, the war permanently changed the structure of the railways, with all but some minor ones being consolidated into four large companies. In Canada, too, there was consolidation, as the financial plight of three of its largest railways led to the creation of the publicly owned Canadian National Railways.

In the countries where the conflict had taken place, not only had large sections of line been damaged, but their whole railway systems had been allowed to become run down, leaving enormous repair bills for which the state mostly picked up the tab, either directly or indirectly, through subsidies. Inevitably, given the disintegration of the Austro-Hungarian Empire and Germany's loss of some of its territory, after the war there was a major restructuring of the European rail network. Austria suffered the most. Before the war, its privately owned railway had been the third-largest system in Europe, with 28,750 miles; by the start of World War II, when the railways were taken over by the German Reichsbahn, it was a state-controlled network amounting to just 4,150 miles. The plight of the Austrian system highlighted the inflexibility of railway networks. Prior to the war, the emphasis had been broadly on a north-south axis with Vienna at the center, but now the much diminished country was mostly situated on an east-west axis that was ill-served by its existing railway network.

Germany had to hand over 3,000 miles of line to Poland—which its army engineers had mostly converted from the wider Russian gauge in great haste during the war—as well as 1,200 in Alsace-Lorraine, including the infamous Metz Station, and even more than 100 miles to both

Belgium and Denmark. The rump of the German railways, still divided among seven states, was finally unified into one system, the government-owned Reichsbahn, fulfilling Bismarck's dream in circumstances that would have had him turning in his grave. Indeed, the Reichsbahn was one of the principal vehicles through which German reparations were made. The various Allied nations claimed no fewer than 5,000 locomotives, 15,000 coaches, and 135,000 wagons for those lost in the conflict, and these had to be delivered in perfect order. A team of Allied inspectors ensured that these were manufactured to the latest design and even instituted special trials at sites near the border to put the locomotives through their paces. Not surprisingly, the Reichsbahn started life in a state of virtual collapse; remarkably, however, by the mid-1930s Germany would be a world leader in railway technology, notably with diesel trains.

The spread of both diesel and electric-powered trains really took off in the interwar period, but electrification had its roots in the nineteenth century. Electric power had a number of advantages, being cleaner, more efficient, and ultimately cheaper once the technology had been properly developed. The first demonstration of an electric train had taken place in the 1840s, but at that point a technical barrier to electrification remained: Engines were not yet powerful and reliable enough to be used on a railway's main line. By the turn of the century, Britain already had more than 400 miles of electrified lines, mostly tramways, but also the pioneering Liverpool Overhead Railway. But progress in Britain was slow compared with the advances in several European countries, notably Germany; Switzerland, where the preponderance of tunnels and steep inclines made electrification particularly desirable; and even the United States, which used electricity on a few lines around New York City because the authorities insisted that steam locomotives were unsuitable for the tunnels under the Hudson River.

Switzerland showed the huge advantage of electric traction on the St. Gotthard line where it was introduced in 1920. Whereas two steam engines struggled to climb up the gradients pulling a 200-ton load at 20 miles per hour, just one electric locomotive could do the job with a 300-ton train at 30 miles per hour. That year the Swiss also completed what is probably the most beautiful of the railways through the Alps, the Lötschberg line (the Bern-Lötschberg-Simplon) running from Spiez on the Thun Lake through the Bernese Oberland to the Rhône Valley. It was built by private enterprise, with support from the Canton of Bern, and from the outset used a system of electrification that later became standard throughout Switzerland. The trip along the 50-mile line, from the savage, desolate Lonza Gorge through the summit tunnel and on to the verdant valley of the Rhône, is one of the great railway journeys of the world; in recent years, a new tunnel and rail line have been built as part of the country's plan to take trucks off the alpine roads.

With the success of the Swiss adoption of electricity, as well as the shortage of coal after the war and the increasing availability of hydro-electricity, both France and Italy drew up ambitious programs to electrify several mainline routes. French plans were largely held up by technical problems and the cost of implementation, but Italy emerged from the interwar period with the greatest proportion of electrified lines among European countries. The driving force was the dictatorship of Benito Mussolini, who saw electric trains as epitomizing a modern image to boost Italy's standing in the world. As to whether Mussolini actually made the trains run on time, the evidence is not clear cut. At the end of World War I, the Italian railways were in a poor state and rules were extremely lax. The journey from Naples to Milan, the spine of the Italian rail network, took seventeen hours, and trains rarely maintained even that sluggish timetable. Improvements had started to be made even before the Fascists came to power in 1922, but Mussolini invested heavily in the railways.

According to the railway historian Geoffrey Freeman Allen, "under Mussolini the country's intercity passenger services were transformed not only by acceleration, which generally halved the end to end times of the best trains but by massive expansion of train frequencies." This was achieved not only by electrification but also by the elimination of curves—the legacy of the original railway builders' shortage of capital—halving the timing between Naples and Milan to eight hours. The Italian railways also introduced faster *Direttissima* trains. Mussolini's reputation as the man who made the trains run on time was based on a special press trip staged in July 1939, days before the outbreak of the war. A three-car electric unit traveled from Florence to Milan at an average speed of 102 miles per hour for the nearly 200-mile trip, despite a long section of gradient through the Apennines, a world record that would stand until the launch of the Japanese high-speed train a quarter of a century later (see Chapter 13). Mussolini therefore brought about the same kind of improvements to the railways that had been introduced earlier in other countries—but whether his trains actually ran on time according to these new improved timetables is unclear.

The alternative way of modernizing railways was to use diesel locomotives (also sometimes called railcars) with engines either in compartments at the end of a coach or underneath the floor. Early attempts to use gasoline engines were soon supplanted by the adoption of diesel, which is far more economical than gasoline, especially as the amount of power required increases. Diesel engines—which use compression rather than a spark to ignite the fuel—were first invented by a Dr. Rudolf Diesel in 1897. Owing to technical difficulties, however, it was not until the 1920s that this type of traction was seriously considered as a cheaper alternative to steam locomotives. In 1932, the Reichsbahn announced plans to use a new high-speed diesel train, the *Fliegende Hamburger* (Flying Hamburger) service, between Berlin and Hamburg, traveling at an average speed of 75 miles per hour for the 178-mile journey. The train, looking rather like a Zeppelin because it had been tested in the wind tunnels at the airship works, was introduced in May 1933 and demonstrated the advantages of the new technology. It was a revolutionary moment in rail history, since for the first time a train was required to maintain speeds of over 100 miles per hour in order to keep to the timetable. The interwar period is characterized by dictators using the railways to promote the modernity and futuristic aspects of their regimes, and this train's accelerated development was part of Hitler's propaganda exercise to demonstrate the greatness of his "thousand year Reich." The success of the service led the Reichsbahn to launch an even faster service between Berlin and Cologne two years later, averaging 82 miles per hour, and a similar *Fliegende Frankfurter* service soon followed.

These German high-speed diesels were killed off by World War II and did not return to service, but similar trains were developed elsewhere. It was in the United States that they were most widely adopted, transforming long-distance travel in the face of growing competition and ensuring that, between the wars, America led the way in passenger rail travel. The Americans had also conducted a few experiments with gasoline engines for their locomotives, such as those operated starting in 1922 on the Chicago, Burlington & Quincy, which were little more than buses on rails, and the rather more sophisticated Blue Bird service, a three-car train running between the Twin Cities (St. Paul and Minneapolis) and Rochester, Minnesota, introduced in 1929, but it was diesel that was to revolutionize rail travel in the United States.

After the end of World War I, the U.S. railroads had continued their policy of running prestigious trains on their main routes with better timings and luxurious facilities. The most famous and illustrious was the New York Central's *Twentieth Century Limited*, launched, as befits its name, just after the start of the century in 1902 on the 960-mile route between New York and Chicago. This was luxury of the highest order, with a dozen staff to cater to just forty-two people, who were comfortably accommodated in the inaugural train's five carriages. It was not cheap, as passengers had to pay a supplemental fee, but they were rewarded with a choice of sleeping accommodations ranging from the luxurious to the palatial, a gentleman's club car with a wine bar, a barber's shop, a secretary to take down letters, and, for the ladies, an observation car where they could retire while their spouses discussed business. Dinner invariably started with oysters, then went on to soup; a choice of fish, chicken, ribs of beef, or goose; and finally both cheese and dessert. The *Twentieth Century Limited* gave the world the expression "red-carpet treatment," since at both New York and Chicago the soles of its passengers' feet never had to touch the pavement of the platform as a carpet, embossed with the company's insignia, was rolled out just before the train arrived. The travelers were so illustrious that the railroad, eager to boost its reputation, issued their names to the daily press, rather in the same way that sailing lists of transatlantic liners were published by newspapers on both sides of the ocean.

By the 1920s, the *Twentieth Century Limited* service had greatly expanded. A fleet of 122 coaches and 24 locomotives was allocated solely to the *Century*. At its peak in January 1929, just before the great stock-market crash, seven almost identical trains left from Chicago carrying a total of 822 passengers. This prestigious service was a vital part of the economics of the New York Central, earning annual revenues of $10 million. Every day the president of the company was handed a statement with all the details of the *Century*'s operation, including passenger numbers and timings. For less affluent passengers, there was its rival, the *Pennsylvania Special*, launched the same day by the Pennsylvania Railroad; this line later became the *Broadway Limited*, which provided a similar, though not quite as luxurious, service covering the distance in the same time, twenty hours. These two trains would contest the New York–Chicago route for the next fifty years until airlines made inroads into their business after World War II.

Other U.S. railroads soon emulated these two rivals, launching similar services, such as, in 1911, the Santa Fe's *De Luxe*, which was limited to just sixty passengers and required a supplement of $25. It left Chicago at 8 P.M. on Tuesday night and reached Los Angeles at 9 A.M. on Friday, a sixty-one-hour trip. All these trains sought to distinguish themselves from their rivals in some novel way. The Santa Fe presented every passenger with "a pigskin wallet embossed with the train's title in gilt and at the Californian border pages would swarm aboard with corsages for each lady."[8] The Great Northern *Oriental Limited*'s unique selling point was a famous 5 P.M. ritual when a steward, bearing a silver tea service and followed by a retinue of uniformed maids carrying sandwiches and *patisseries* (nothing as vulgar as "pastries"), would travel down the length of the train handing out treats. The Florida East Coast's *Florida Special* boasted a string quartet, which played among potted plants for pre-dinner entertainment, and later featured swimsuit modeling, to the delight of the male passengers. The *North Coast Limited* carried an electrician specifically charged with looking after the innovative light bulbs; it also had one of the first observation cars, a feature that would soon be universally used. And the *Sunshine Special*, one of the few international trains in North America, which linked St. Louis with Mexico City, had special lounges and dining cars in Spanish style, complete with all-American soda fountains. Most of these trains started as once per week services but soon built up to higher frequencies. The *Sunset Limited*, for example, operated the 2,500-mile trip between New Orleans and San Francisco, which took seventy-two hours and became daily just before the outbreak of World War I. North of the U.S. border, starting in 1920, there was the flagship train of the newly created Canadian National Railways, which ran the 2,900 miles between Montreal and Vancouver, a trip taking four and a half days. This line introduced an on-board radio service, which, as a by-product, led to the creation of the Canadian Broadcasting Corporation, since it required the establishment of eleven broadcasting stations along the route.

For their affluent customers, the railroads provided a fantastic door-to-door service, as neatly described by Albro Martin: "The affluent midwestern family whose son was leaving shortly for Yale College phoned the railway depot the day before and had the expressman call in his van for the lad's trunks and suitcases. Father handed the man the railroad ticket, and he duly punched the square indicating that the traveler's baggage—150 pounds were allowed on a full fare ticket—had been checked to destination."[9] And that destination was not the arrival station, but the student's rooms at the college. All this, however, somewhat covered up the fact that these trains ran pretty slowly, and the U.S. railroads, still largely in a monopoly position in the 1920s, made little effort to improve timings.

It was the advent of diesel traction in the early 1930s, together with the belated realization that the automobile, the bus, and the airplane were beginning to challenge their hegemony, that spurred the railways to improve these prestigious services. Buses had begun to eat into the intercity market in the eastern states, and crucially, for the longer distances, the airplane was beginning to be seen as an alternative to the train. The first scheduled commercial domestic flight in the United States was in April 1927 between Boston and New York, and others quickly followed suit. Though diesel locomotives had been introduced by various U.S. railroads by the end of the 1920s, they were confined to shunting and light duties, as the powerful diesel engines were thought to be too heavy to be economic.

The technical breakthrough was the development of a far more efficient engine by General Motors using alloys that gave a far better power-to-weight ratio. At the Chicago Fair of 1932, the new engine caught the eye of the boss of the Chicago, Burlington & Quincy Railroad, who used it to revolutionize rail travel in America. In May 1934, a streamlined train dubbed the *Pioneer Zephyr* ran the 1,000-mile journey from Denver to Chicago at a record-breaking average speed of just under 78 miles per hour, giving the same performance as the German diesels but over a much longer distance. Not only was the train far faster than its predecessors, but it was amazingly cheap to run, as the fuel cost for the trip was reported to be just $16. The inaugural *Zephyr* was only three cars long and able to carry just seventy-two passengers, but soon longer trains were introduced. In a demonstration run, a similar six-car diesel smashed the coast-to-coast record in October 1934, traveling from Los Angeles to New York in fifty-seven hours. This knocked more than fourteen hours off the previous record, which had been set as far back as 1906—a gap that showed all too graphically that the railroads had not considered speeding up their timings to be a priority. It went into service for the Union Pacific as the *City of Portland* running between Chicago and Portland, Oregon, taking forty hours, compared with fifty-eight for the fastest steam service.

From these beginnings, a family of "Zephyrs" and "Cities" emerged on these two railroads with longer and more powerful trains, such as the *Denver Zephyr*, launched in 1936, which had twelve cars offering an overnight sleeper service between Chicago and Denver. The Atchison Topeka & Santa Fe soon joined in with its *Super Chief* between Chicago and Los Angeles, which, rather uniquely, was able to travel on its own tracks all the way for the forty-hour journey, a full half a day faster than the previous steam service. And so on. America enjoyed a brief heyday of the railways, which continued throughout the war but, as we shall see in the next chapter, would be cut short as competition intensified and the nation took to planes and automobiles. Although the memory of these great trains remains, thanks partly to the companies' advertising and public-relations campaigns, which left a legacy of fabulous posters, most rail journeys, both in America and Europe, remained a mundane experience on systems whose perennial problem was underinvestment and a failure to modernize.

Ominously for the railways, even while these modern long-distance intercity services were being introduced, which the U.S. railroads saw as essential for their survival, the number of people using the railroads was falling. In 1920 the U.S. rail network carried 1.2 billion[10] passengers, its all-time peak, and by the bottom of the economic slump in 1933 that figure had already fallen by two-thirds, to just 435 million. These figures exclude commuters, but there the decline was only more marked. Closures were becoming commonplace almost as soon as World War I ended. The railways failed to foresee this collapse in traffic or to do much to defend their interests. Even as late as 1916, a U.S. suburban railway manager dismissed the threat, saying, "The fad of automobile riding will gradually wear off and time will soon be here when a very large part of the people cease to think of automobile rides."[11] Just how wrong he was can be demonstrated by the fact that America's suburban network of 15,600 miles lost all but 3,000 miles in the interwar period. While the railroad companies were blind to the challenge of the car, the major auto manufacturers were far cannier. They even played a role in this decline by encouraging towns to rip up their tramway systems to make way for their products, as highlighted brilliantly in the 1988 film *Who Framed Roger Rabbit?*

This decline made it all the more important for railways to respond by adopting modern methods. Yet, in this difficult transitional period for the railways, there were still doubts as to whether diesel or steam technology represented the future. Although the introduction of the early diesel services proved popular, the romance of powerful steam engines remained enticing, and technological improvements appeared to suggest that steam still represented the future for rail. Certainly steam engines, helped by improvements, put up a good fight before being outdone by the two superior and more efficient technologies of diesel and electric traction. Although various radical changes to the basic steam engine were tried, such as in Russian locomotives using a combination of both diesel and steam power, the future of the technology seemed to lie more in attempts to refine the traditional design than in radically changing it. The best locomotive engineer of the interwar period was the Frenchman André Chapelon, whose increases to the flow of steam through the engine led to improvements in efficiency of up to 50 percent. His efforts were not always welcomed by the engineering establishment, however, since they exposed the inadequacies of his rivals' engines, and consequently his improvements were not adopted as widely as they should have been.

The most visible expression of steam's adaptability was the development of more powerful and more streamlined, elegant locomotives that were used to haul a series of prestigious trains designed to improve the image of the railways. Modern steam might seem like an oxymoron today, but it was very much a feature of the 1930s, in particular with the introduction of these locomotives in several countries, including the United States and Australia. But it was in Britain that they made the most progress, and it was the rivalry between the London & North Eastern Railway and the London, Midland & Scottish, on the routes between London and Scotland, that stimulated the contest.

The two rivals had long tried to outdo each other by providing extra amenities on their prestigious flagship trains, the LNER's *Flying Scotsman* and the *Royal Scot* of the LMS. The LMS introduced leather settees and armchairs, but it was soon eclipsed by the LNER with its Louis XIV–style armchairs and hand-painted decor, later followed by a hairdressing salon, headphone sets to listen to the radio or recorded music hosted by the world's first mobile DJ, and cinema coaches showing newsreels, cartoons, and travelogues for an extra shilling. When the provision of these luxuries failed to attract more passengers, the companies at last resorted to the idea of trying to improve route times, which had not decreased much since the turn of the century, and two of the great steam-locomotive engineers of the age went head to head into battle. For the LNER, Nigel Gresley produced a class of locomotives, A3 Pacifics,[12] which ran up to 108 miles per hour in a test, and in 1935 LNER inaugurated the highly successful *Silver Jubilee* service between London and Newcastle that took just four hours on a route that previously had taken five. The LMS responded with the Princess Coronation Pacifics developed by William Stanier, which ran between London and Glasgow in just six and a half hours, a saving of one hour and forty-five minutes. It was the LNER, however, that triumphed with its streamlined locomotives, the A4 Pacifics, one of which, the *Mallard*, achieved the remarkable speed of 126 miles per hour on a test run in July 1938.

In America, there was a similar contest, which was given added spice by the fact that there were three players, two steam locomotives and one diesel competing for the lucrative traffic between Chicago and the twin cities. The Chicago North Western triggered the battle in January 1935 by launching its *400*, a steam train that covered the 400 miles in 400 minutes,[13] which required considerable periods of running at 100 miles per hour. The Burlington responded with the diesel *Twin Zephyr*, and the Chicago, Milwaukee & St. Paul Pacific with the steam-hauled *Hiawatha*. Although the two steam services reduced the previous best times for the run, it was the *Zephyr* that triumphed and ran the fastest services, albeit by only a small margin. In France, streamlined locomotives were used to haul the fastest train of the 1930s, the *Aérody-*

namique, which consisted of a maximum of just three carriages between Paris and Marseille, undertaking the 536-mile journey in just ten hours (today, the nonstop Train à Grande Vitesse (TGV) covers the distance in under three hours!).

The 1930s were also the heyday of the Orient Express services, which had been restored in 1919 and now consisted of a variety of trains using that name. There was the *Simplon Orient Express* running from Calais and Paris through to Istanbul via Milan, Zagreb, Belgrade, and Sofia, with some coaches being routed to Athens; the plain *Orient Express*, which ran from Paris to Istanbul by a more northerly route via Vienna and Bucharest; and the *Arlberg Orient Express* to Budapest via Innsbruck and Vienna. It was a five-day delay in a snowstorm to a westbound *Orient Express* at Çerkezköy in Turkey that inspired Agatha Christie's *Murder on the Orient Express*. Prestigious steam trains were not confined to Europe. By the outbreak of World War II, the Compagnie Internationale des Wagons-Lit et du Tourisme, which ran the *Orient Express* and many Pullmans in Europe, owned a staggering 2,260 cars operating in more than twenty-five countries, ranging from Senegal to Syria, and just before the war the company even ran trains as far as Baghdad. For example, in Egypt, the train between Cairo and Luxor was named *Star of Egypt* in 1929 and provided with Pullman carriages to serve the growing number of tourists visiting the Valley of the Kings.

Although steam-engine technology did not stand still, there is no doubt that too many of the world's railways held on to their traditional form of traction for far too long. Britain, which made little effort to adopt the new technologies, was particularly guilty of this. In 1933, the Great Western Railway began to run diesel railcars on local services, and both the London Midland & Scottish and the London & North Eastern Railway electrified a few suburban lines, but only the Southern Railway made considerable progress between the wars, converting much of its network in a rolling program using a third rail system. No trunk main line in Britain adopted the far more efficient overhead electrification until the West Coast did so in the 1960s.[14]

In the main, those railways that converted early did so as a result of special circumstances—such as steep gradients, access to cheap fuel or electricity, or a shortage of coal—since railway managers rarely took a long-term strategic view, failing to realize that change was inevitable. In the Andes, for example, the high-altitude railways acquired a strange collection of converted gasoline and diesel road vehicles that would wind their way up the mountains more like a traffic jam than a "train." The transandine itself was partly electrified in the 1920s because of its steep gradients, using electric locomotives imported from Switzerland. Wherever there was abundant cheap hydroelectricity, the economics of electrification became overwhelming, such as in Sweden, where the 280-mile Lapland Railway, which carried iron ore, was electrified in 1923, and the Stockholm-Gothenburg line followed suit two years later.

Morocco, then a French colony, was also a surprising early beneficiary of electrification. The North African country had come late to the railways; before the war, there had only been a few military railways there to provide transport for the Foreign Legion. This changed in 1922 when the French governor, Maréchal Lyautey, decided to develop the infrastructure of the Maghreb and embarked on a rapid railway-building program. Thanks to the cheap hydroelectricity available from the Atlas Mountains, large parts of the network used electric traction, including the *Marrakech Express* made famous by the eponymous 1969 Crosby, Stills & Nash song. It actually was a pioneering railway, as it was one of the first in the world to use regenerative braking—in other words, the electricity generated by the trains braking down the 1 in 66 hills could be put back as power into the system. Morocco therefore developed from scratch one of the most modern rail networks in the world.

Such examples only reinforce the point that in an ideal world many more railways would have adopted the new forms of traction far faster. Not only was there built-in resistance to change, however; both diesel and electric technologies also were slow to be adopted because the railways had huge amounts of capital tied up in existing technologies. Making the change would have required a leap of faith because it involved major investment at a time when the railways were having to battle against competition from other modes of transportation for the first time in their history. Electrification has a particularly high initial cost because of the expense of installing the lineside equipment and so, at first, it was only advantageous in countries where there was cheap hydroelectricity. Diesel used oil, which in many countries was less available than coal. Ironically, Britain was to be grateful that it had not converted more railways to diesel since, in World War II, fuel was scarce but home-produced coal was abundant.

But that is hardly a justification for the original failing, which was a missed opportunity, not just in Britain but elsewhere. Steam locomotives were cheap to build, and though maintenance was frequent and quite expensive, the technology was well known and didn't require fabulously expensive workshops—any old shed would do! The great locomotives produced by the likes of Gresley and Chapelon seemed to be the prelude to even more improvements for steam technology, but in fact they were a dead end, because the unfortunate truth was that steam locomotive development had more or less reached its peak by the late 1930s. Automobile engineering, on the other hand, improved rapidly after World War II as engines became much more powerful and economical. So did airplanes. In 1935, the DC3 carried twenty-one passengers over 1,000 miles at 165 miles per hour. By 1965 the DC9 was carrying 120 passengers over 1,000 miles at 565 miles per hour using less than a quarter of the fuel per passenger of the DC3.

In retrospect it was tragic that the railways kept steam traction for so long. Had more railways converted earlier to modern forms of traction, they would have proved more viable, and many lines that eventually closed for economic reasons might have been saved. This was particularly true of branch lines, where a diesel railcar could operate at a fraction of the cost of a little steam engine and offer a far better service, since no shunting or turning around was necessary because there was a driver's cab at both ends. Had the diesel railcar been more generally deployed earlier, and had the unions accepted driver-only operation sooner, then some lines would have remained viable and possibly not been slated for closure at all.

It was not only diesel- and electric-powered trains that were developed in the fruitful period between the wars, which proved to be a time of experimentation similar to the early Victorian age, when different types of railways, such as Brunel's vacuum railway,[15] had been tried, most of which had failed. There were plenty of revolutionary ideas around in the interwar era, some sounder than others. This time both propeller and rocket-powered trains were developed in Germany, and on one test track a train invented by Fritz von Opel using a set of twenty-four rockets reached a speed of 157 miles per hour. Unfortunately, it left the tracks on the second run and crashed, putting a stop to further experiments since it had already attracted the opposition of General Motors, which bought a controlling share in the Opel car company in 1929. A propeller train, the *Rail Zeppelin*, invented by Franz Kruckeberg, was rather more successful, with a trial run on the Hamburg-Berlin main line averaging 160 miles per hour, but in truth that was too fast for the condition of the tracks or the needs of the railway, and running the train at slower speeds was inefficient, which made the project unviable. In France, Bugatti, a name normally associated with sports cars, developed a railcar in the 1930s running on a curious mixture of gasoline, alcohol, and benzol, and these were successfully introduced on several lines, including Paris-Lyon and Paris-Strasbourg, and survived until well after the war.

Although countries that had well-developed networks mostly stopped building new lines by the outbreak of World War I, with the exception of the United States, where the western states were still expanding their networks, there was still much growth elsewhere in the interwar period and several major projects were completed, particularly in less developed countries. As World War I was ending, a 640-mile-long meter-gauge line was completed between Bangkok and Singapore, linking several of the Malay states. Many other Asian countries, which had mostly been late starters, also experienced a major growth of new lines in this period, as did some well-established railways, such as those in India. The British colonial power showed a readiness to invest in the now largely nationalized Indian system, which continued to expand rapidly in the interwar period with 4,500 miles being completed in the 1920s, bringing the system up to 40,000 miles with a workforce of 800,000. In the 1920s, a program of electrification was begun with the conversion of the Bombay suburban lines, and it was soon extended to include the steep gradients of the Western Ghats described in Chapter 3.

In Australia, transport between New South Wales and Queensland was greatly improved with the construction of a new standard-gauge section that reduced journey times between Sydney and Brisbane from twenty-eight hours to under sixteen hours. It was not only Mussolini and Hitler, among the dictators of the period, who loved grand railway schemes. Stalin also saw railways as a great modernizing force. Under his regime Russia continued an old imperial project, the Turkestan-Siberia Railway (the Turksib), a 900-mile branch off the Trans-Siberian into what is now Turkmenistan that was completed in 1930. He also, incidentally, reopened the Trans-Siberian to western tourists and business travelers, who had been banned since the Communist takeover, and his government paid for the line to be double tracked all the way through to cater to increased traffic.

Iran, another late starter, built one of the most ambitious projects of the interwar years, the 865-mile Trans-Iranian Railway running from the Caspian Sea through Tehran to the Persian Gulf. This was yet another heroic railway where the builders encountered a new set of engineering challenges, and they attempted, but failed, to carve a tunnel through a salt dome and a hill made of pumice. The line, started in 1927 at the instigation of yet another dictator, Reza Shah, also adopted the technique of the Swiss railways in the Alps by using spirals carved in the mountains to limit the gradients, though they were still steep at 1 in 36. The railway opened in 1938, just in time to provide a vital supply route to the Soviet Union in World War II. In Africa, as mentioned in Chapter 7, the Benguela Railway in southwestern Africa was fully opened in 1932, and the Cape to Cairo reached its apogee.

In Belgium, which had the densest rail network in the world, there was still growth in the remarkable tramway system that had developed alongside virtually every major road. It was run by a separate company from the main Belgian railways, the Société Nationale des Chemins de fer Vicinaux, and by 1925 extended to nearly 2,500 miles of meter-gauge railways still largely operated by steam engines. The system had proved invaluable during World War I in carrying coal and was heavily used in World War II, as well, because gasoline and oil were in short supply.

Despite this growth and the modernization in various parts of the world, the interwar period was a difficult one for the railways. The sheer scale of the railways, with the enormous fixed investment tied up in track, signaling, stations, and rolling stock, made them inflexible and therefore slow to respond to change. Moreover, they were huge employers of labor, and in most countries the workforce became strongly unionized. This was hardly surprising, given that the workers were often poorly treated, having to work very long hours for little pay, and unionization was often won after a struggle with dyed-in-the-wool managements. Despite the managements' resistance, the nature of railways, in which disruptions were simple to cause and very expensive for the owners, meant that unionization was easier for the workers to achieve than in other industries. Britain had experienced its first attempts at railway labor organization in the 1860s and its first national strike in 1911, which succeeded in improving the workers' conditions, albeit after further disputes.

In 1912, Spain, too, had its first general railway strike, which resulted in a big increase in the workers' wages and pension entitlements. Workers were—crucially—winning the right to have shorter hours, and this greatly increased the costs of running railways, which previously had depended on exploiting workers by imposing unpaid overtime. In America, the first strike had been in 1877 on the pioneering Baltimore & Ohio and the Pennsylvania, and as mentioned in Chapter 4 there had been a particularly hard-fought struggle against Pullman in 1894. By the early twentieth century, the unions had established themselves as a powerful force in opposing management's attempts to impose pay cuts and other damaging changes to work conditions, though they were not always successful, and both Britain and the United States experienced railway strikes in the 1920s. The power of the unions, and the restrictive practices they adopted over such matters as manning, limited the ability of the railways to be flexible in the face of the growing competition from other modes of transportation.

Given this growth and the adoption of new technologies, which in many parts of the world offered the railways a new lease on life, it would be a mistake to date the beginning of the decline of the railways to the immediate aftermath of World War I. Of course, with the car and especially the freight truck beginning to make inroads, there was extra pressure on the iron roads, but these were still times of innovation, growth, expansion, and excellence in many parts of the world. As O. S. Nock put it, "It is remarkable that from the very depths of the great slump of the early 1930s there came some of the most enterprising and exciting developments that had been seen in the world of railways since the pioneer days of the nineteenth century."[16] As we have seen, steam locomotives were much improved, timings speeded up, and new technologies were being developed and introduced.

The railways faced another obstacle to progress, however, the traditional hostility against them from both politicians and people, the result of their long period as monopolists. As a result, the railways were often cruelly treated by the retention of burdensome regulation that had been imposed in a completely different economic climate. Unsympathetic politicians turned a deaf ear to their requests for the rules to be relaxed in the face of competition from other modes of transportation. New Zealand offers a neat example of this. Obliged, like many railways, as a "common carrier" to take all freight offered by customers, when the Great Depression of the 1930s started to bite and road transport became viable, the New Zealand railways started losing high-value traffic to freight trucks while being forced to keep on carrying the cheap freight. As Neill Atkinson, the historian of the New Zealand Railways, explained, the "farmers enjoyed generous rail subsidies on lime and fertilizer, but when it came to shipping their produce and other goods regarded New Zealand Railways as just another business competing for traffic."[17] To the government, the farmers appeared selfish and ungrateful, since, as Atkinson pointed out, "Hadn't railways opened up their lands and fostered their export industries?"[18] But expecting gratitude from the struggling farmers was hardly realistic. They, like everyone else, had long forgotten the crucial role the railways had played in creating their prosperity.

It would be the same everywhere else. The railways were no longer seen as an asset but rather in the same way that a halitosis-ridden guest at a party would be viewed, a bit of an embarrassment and, in the eyes of many, a pariah. In New Zealand, though, the railways were lucky, at least temporarily. There the government was willing to rescue them, passing legislation that controlled and licensed motor transport, and the railway managed to respond by flattening out its rates. It was a rare exception. For the most part, governments were not willing to bail out the rail industry; as a result, in the middle years of the twentieth century a combination of competition and hostility from governments hastened its decline. Memories of the big dominant monopolist companies outlasted the reality by several decades even though most major countries had by now nationalized their railways.

The ability of the railways to compete was weakening almost daily as first the automobile and later the airplane, especially in America, ate away at their income, and yet governments sat on their hands, refusing to help, either through deregulation or more directly through subsidies or modernization programs, until it was far too late. First, though, the railways had another opportunity to shine when World War II broke out, and again they passed the test remarkably well.

TWELVE

Decline But Not Fall

I N THE QUARTER OF A CENTURY BETWEEN THE TWO WORLD WARS, MOTOR vehicles established themselves as a convenient and flexible form of transportation, and consequently it might have been assumed that the railways were destined to play only a minor supportive role in World War II. Nothing could be further from the truth. While the railways could not function in the same way as in World War I, because, thanks to the development of airplanes and tanks, battles were far more mobile than in the 1914–1918 conflict, the railways were still crucial to all the combatants, including, most surprisingly, the United States.

The ability of the railways to deliver vast quantities of supplies and men using far less fuel than the equivalent road transport was a crucial aspect of their value. The military on both sides soon learned that in spite of technological developments in road transport, the railway was far more efficient both in terms of fuel and space, since a single wagon could carry over 100 troops or a load of more than 50 tons. As a War Ministry official wrote, in every war "there comes a time when locomotives are more important than guns."[1] The major combatants all rushed to produce new types of locomotives, paring down existing designs to create "austerity" engines that could be built as quickly and as cheaply as possible. The Germans, for example, produced the *Kriegslok*, which took only 8,000 man-hours to produce, just over a third of the time normally required, while hundreds of "MacArthurs" rolled off the U.S.

production lines and the British produced an Austerity locomotive based on an existing London, Midland & Scottish design.

The railways retained a vital role partly because they proved harder to destroy from the air than it was at first assumed would be possible. Although major stations and goods yards were a relatively easy target for bombers, and bridges were clearly vulnerable, that thin strip of track crossing vast swathes of the countryside was not so easy to hit, which is why sabotage by partisans eventually wreaked more permanent damage on the European rail networks than aircraft did.[2] In fact, random bombing proved more effective than targeted bombs, according to John Westwood, the historian of war and the railways, who wrote that "it is significant that the unaimed flying bomb of 1944 was as likely to cause serious railway damage as the aimed bomb of 1940."[3] Keeping the railways operational was vital, and engineers on all sides became incredibly adept at restoring damaged tracks with remarkable haste; consequently, the railways were rarely out of action for long periods. Even bridges could be replaced quickly. For example, on the night Coventry Cathedral was destroyed, the local rail lines were hit in no fewer than 122 places, but they were operational again within a week.

Just as in 1914, as soon as war broke out in August 1939 British railways were taken under government control. They faced two early tests: the evacuation that summer of children into the countryside in anticipation of the Blitz and then, the following year, much more urgently, the removal of troops from the Channel ports after the disaster of Dunkirk. They were extremely successful in both operations; the Dunkirk evacuation was carried out so effectively that Westwood suggested that perhaps far too much store was set by advance preparation. Railways, with experienced operators in charge, were far more flexible than expected: "The sudden, urgent and intense traffic which presented itself as British and Allied troops arrived at the Channel ports from Dunkirk was handled so admirably that it raises the question of whether long-term planning of military rail movements serves any real purpose."[4] So not only did all that planning in World War I make conflict more likely, but it was also a waste of time and effort!

Britain's railways continued to have a vital role throughout the war, carrying record numbers of people and unprecedented amounts of freight, including, for a time, guns on armored trains that patrolled the

Channel coast but were fired only once—in a test that caused mayhem locally by breaking all the windows in houses nearby. The railways were heavily used and kept running by skilled railway managers who worked closely with the army administrators. Injunctions to civilians to keep off the railways to make room for the military fell on deaf ears, however, as they rushed off to the seaside or up to London at every opportunity.

In contrast, the German war effort was hampered somewhat by the strained relations between the Third Reich's military managers in its Directorate of Transportation and their Reichsbahn equivalents. According to Geoffrey Freeman Allen, "railwaymen denigrated the military officials as interfering staff officers basically ignorant of transport and the Directorate's structure as rigidly bureaucratic and mistrustful of initiative."[5] Allen then added playfully that the railway managers "were not best pleased, either, to discover in the later years of the war that wagons labelled as high-priority war supplies not infrequently bore champagne or other luxuries for the officers' messes." This was not just a gratuitous piece of jingoism on Allen's part. The failure of senior Nazi officers to understand railway principles may well have had a deleterious effect on German logistics that affected the running of the war: "Railway operation, difficult in any case, was needlessly disturbed by the special trains in which Nazi dignitaries liked to cruise," wrote Westwood.[6]

Despite Hitler's support for the diesel Flyers, which stayed in sidings because of the shortage of fuel, he was far more enamored with roads and cars than trains, having developed the concept of motorways (*autobahnen*), in which he invested heavily. The railways, therefore, had been somewhat neglected, and this hampered the initial attack when the Germans again went through Belgium into France. As in 1914, the Belgians had sabotaged their mainline railways, and while the Germans rushed westward in their motorized transport, reinforcements were slow to arrive because of the logistical difficulties; only the rapid surrender of the French prevented the advance troops from being cut off.

The French made heavy use of their railways in this early period of the war to bring troops to the front and, after their surrender, to evacuate people fleeing from Paris. The Germans quickly took over the French railways, with the same personnel but supervised by a few German officers. They then exploited them to the full, both for military traffic and for the large quantities of building materials needed to defend the Channel and

Atlantic ports. The importance to the German occupiers of keeping the lines fully functioning gave the French railway workers ample opportunity for quiet sabotage. It was easy for paperwork to go astray, trucks to be sent to the wrong depot, coal to be thrown on the track, and so on. French railway workers were often used as couriers and spies by the Resistance. The most famous exploit was the regular concealment of messengers in the tender water tank of the locomotive used to haul the train of Pierre Laval, the collaborationist prime minister of "Vichy France," between Paris and Vichy. Later, of course, the French railwaymen would play a big part in helping the Resistance sabotage the system in anticipation of the Allied invasion, an episode featured in several films.

It was, however, on the eastern front where the Germans had the most trouble with the railways. The lack of railways in the west of the Soviet Union, which had hampered the Russians in World War I, would this time benefit them, becoming more of an obstacle for the invading Germans than a disadvantage to the defending Russians. The view that good railway transport tends to help the defender rather than the attacker, borne out by the experiences in 1914–1918, was reinforced by the Russians' defense of their homeland. The Soviets had cause to be grateful for the tsar's decision a century before to insist on a different gauge from the standard one used by all their European neighbors, precisely for the reason that proved prescient: protection against invaders. It worked. Having invaded Poland in 1939, the Germans were able to make use of the Polish network of more than 10,000 miles because it was standard gauge but once they tried to move into Russia, they found that the change in gauge was, indeed, a considerable barrier. When the Germans invaded Russia with an army of three million men, their efforts to maintain supplies into Russia were greatly hampered by the need to transship between trains at the frontier. They concentrated their efforts on motor transport, but the Russian roads were notoriously awful; according to Westwood, "in practice motor transport could do little better than horses."[7] The Russians left very little rolling stock behind and consequently the Germans found themselves short of transport, with line capacity around one-tenth of what was needed. This greatly slowed their advance toward Leningrad (St. Petersburg). The Germans changed the gauge of hundreds of miles of railway to help their invasion but only at a huge cost of manpower that further slowed their thrust into Russia. These problems delayed, possibly fatally, the German attack on Leningrad.

The Russians, for their part, used their railways to best effect, having learned from their failings in World War I, and now helped by having a strong centralized state. Hitler had expected the Russian railways to collapse, as they had in the previous war, but in fact they performed creditably throughout the conflict. Despite the paucity of lines, in the early part of the war the railways were used to evacuate factories and their workers from the west in the face of the German advance. The successful functioning of the railways was helped by the fact that, having moved everything eastward in anticipation of the German attack, the Russians had a far larger stock of locomotives available to them than the Germans, who were hindered by their own engines being less well adapted to the cold weather. The railways played a vital role in the defense of both Leningrad and Moscow. A newly built orbital railway around Moscow was crucial in allowing the defenders to move troops around the country without entering the capital, while in Leningrad—the site of the greatest siege of the war—the city was never entirely isolated from outside supplies thanks to the railway. Even though lines into the city were cut off, motor transport ran over the frozen Lake Ladoga to meet trains arriving on the opposite shore.

In the United States, the railroad companies, aware that there was a lot of money to be made in wartime, were keen to avoid being taken over by the government, as had happened in World War I. Consequently, they coordinated their efforts so well that there was no question of a similar move. With a shortage of car tires, and gas rationing in force, the number of passengers carried by the railways, which had been declining since the start of the Depression a decade before, soared, doubling from prewar levels. The railways were still the main way to travel long distance, accounting for all but 3 percent of travel by servicemen and women during the war. Even though the front was thousands of miles away, the adaptability of the railroads proved crucial. According to Geoffrey Freeman Allen, "when U-boats frustrated the normal oil supply of Eastern seaboard cities by tanker from the Gulf ports in the south, the railroads, until then negligibly involved in this traffic, somehow rustled up the tank wagons and improvised train working to cope."[8] In carrying the unprecedented amount of freight, the railroads were helped by the improvements in efficiency over the past quarter of a century and the readiness of their management to cooperate in order to ward off any threat of a federal government takeover. As a result, the railroads had a very profitable war, although as soon as it ended, the decline in passenger numbers, which had already been marked in the 1930s, resumed.

The railways are indelibly associated with two of the worst atrocities of World War II, most infamously, the transportation of millions of Jews and other "undesirables," such as Gypsies and homosexuals, to death camps in Poland and what is now Belarus. The victims were transported in unimaginably inhumane conditions in freight cars that were stifling in the summer and freezing in the winter, with many dying even before they reached the camps. The industrial scale of this ethnic cleansing would have been impossible without the railways, given the enormous numbers involved and the shortage of fuel for other forms of transport.

The other great railway-related war crime was committed by the Japanese using forced labor to build the Thailand-Burma Railway. The 260-mile railway was intended as a supply line for the movement of Japanese troops and materiel to the Burma front and for the planned invasion of India. A total of 250,000 men were forced to work on the railway, most of whom were local people, but there were also 60,000 British troops captured when the Japanese overran Singapore. The line linking the Thai and Burmese railway systems was built simultaneously from both ends, Thanbyuzayat in Burma and Nong Pladuk in Thailand, and was completed in just sixteen months. The appalling conditions, disease, starvation rations, lack of sanitation, and brutal behavior of the Japanese and Korean overseers took an enormous toll with more than 100,000 workers perishing, including more than 13,000 prisoners. The death rate was particularly high in the final three months as the Japanese were desperate to complete the work. The starving men were forced to do hard manual labor around the clock with reports of some being forced to work for sixty-two out of seventy-two hours. When opened, the line immediately formed a vital part of the Japanese war effort, as the Burma front became a critical supply line when the Japanese lost control of the South China Sea. After the war, the Burma railway soon fell into disuse and was abandoned, though a small section was re-opened in 1957 by the Thai government.

In modern conflicts the railways no longer have much of a role, since other forms of transport have taken over. After World War II, however, there was one last theater, the Korean War, in which the railways played a vital part. In August 1951, the Americans launched "Operation Strangle" against the North Korean railways, hoping to oblige them to use motor transport rather than trains to carry their supplies. Even by then, despite much technical improvement since 1945, the bombs were not sufficiently accurate, and little damage was caused despite huge efforts on the part of the United States Air Force. Moreover, when the track or bridges were hit, the North Koreans became so adept at repairing the damage that the operation was abandoned as futile. A second attempt, "Operation Saturate" launched the following year, met with a similar fate. The North Koreans kept their railways running throughout the war, but that was really the last time that railways played a significant role in wartime. By the outbreak of the Vietnam War, alternative transport systems were used and more sophisticated weaponry, able to better target railways and their structures, became available, changing the nature of how wars were fought.

World War II left many parts of the European railway network ruined almost beyond repair. Even in Britain, where damage was relatively slight, the railways had been so overused and undermaintained that they were in a parlous state. While railway nationalization was an idea first espoused by the political Left, the Depression of the 1930s had made public ownership of railways in Europe a pragmatic necessity. Most politicians accepted that they were a vital part of the nation's infrastructure and that they needed state support to survive. Britain's nationalization following the election of a Labour government in the aftermath of the war can be seen, therefore, not so much as a politically motivated whim but part of a Europe-wide trend. Government intervention was essential because only the state could afford the massive repair bill.

In Holland, for example, where all the major bridges had been blown up in advance of the German invasion, the rail system had quickly been repaired and was very heavily patronized throughout the

war. In an effort to hasten the end of the war, the Dutch railway workers went on strike; in revenge, when their defeat became inevitable, the
Germans wantonly destroyed most of the system, blowing up all the
major bridges again but this time ensuring that even the piers were destroyed. Stations, locomotives, and carriages were wrecked, although
oddly, the Germans left behind some of their own rolling stock, which
helped to restart services. Money obtained through the Marshall Plan,
the big American aid scheme to reconstruct Europe, enabled the Dutch
railways to recover relatively quickly. The Marshall Plan funds were
also used to reconstruct the damaged railways in other parts of Europe,
and several countries used the opportunity to embark on large-scale
electrification plans.

Britain did not use its Marshall Plan money for the railways, and it
was not until 1955 that the government adopted a modernization plan
promising an end to steam-locomotive production. Nothing expresses
the desperate attempt by Britain to retain steam locomotives in the face
of overwhelming evidence from abroad that they were in decline more
than the opening of the country's first national testing center for steam
locomotives in October 1948, barely twenty years before the last British
steam engine would be pensioned off.

Although on the Continent the pace of change was faster than in
Great Britain, electric or diesel services remained in the minority, and
traveling on European trains in the 1950s was largely a dismal experience. The railways had been patched up, and money spent on basics,
but there was little modernization. Bryan Morgan, touring in France,
West Germany, and Italy at the time, was distinctly unimpressed.[9] In
France, despite being one "*qui se passionne pour*" the French lines,
Morgan concluded that "for practical purposes they are the most inadequate of any important country." The "ruthlessly inflexible" train services were designed entirely around Paris, with expresses leaving at 8
A.M. and 10 P.M. aimed at allowing people a day trip to or from the capital, but offering little for other rail passengers, who were likely to end
up at junctions where "little moves save in relation to a once-daily
event." The planners, he concluded, had succeeded in "immobilizing
the country."

Morgan was far more impressed with Italy, which, he said, "not only
has trains in a big way; it believes in trains." While largely satisfied with

the service, Morgan was most dismayed by the "deplorable" habit of opening the booking offices only five minutes before a train was due, "which is all right at a sleepy Spanish halt but idiotic when handling several thousand passengers daily," especially as he always found himself behind the same woman, "a war widow, remarried to the victim of an industrial accident, having fifteen children and visibly expecting another, going on her annual holiday and with a brother in law working on the railways." Consequently she was "entitled to a 95 per cent discount but is holding out for 100 per cent and is prepared to argue the point until we have all missed the express." While praising the number of services, he had great difficulty in finding a timetable and was appalled at the huge quantities of people and luggage using the trains. West Germany, he reckoned, was the "finest of the major national railways of Europe," despite "its unpunctualities and its deplorable branch-line stock."

The railways, therefore, were fair game for the competition, and inevitably they lost out. The motor manufacturers, too, had suffered in the war, and for the rest of the 1940s they were not in a position to increase output rapidly enough to pose much of a threat to the railways. But as the economies of Europe and the United States recovered and moved away from a war footing, they unleashed a torrent of cars onto the market, which, together with massive road-building programs, undermined the economics of the railways and shrunk their market share. West Germany, for example, which had just 2.4 million motor vehicles in 1950, more than tripled that number in the ensuing decade and by 1970 had some 16.8 million. In 1950, Deutsche Bahn, the state-owned rail organization, had 37.5 percent of the total passenger transport market, more than either buses or private cars, but within twenty years that proportion had fallen to just 7.8 percent while the share of the automobile had risen to 81 percent.[10]

Though this was a more marked decline for the railway than was seen in neighboring countries—the result of the Germans having a particularly buoyant vehicle-manufacturing sector stimulated by a national passion for fast cars—all Western European railways suffered similarly during the period. Freight, too, disappeared from the railways at an alarming rate as cheap trucks, able to carry ever bigger loads, proliferated. The railways became competitive only for the carriage of heavy low-value items such as aggregates or coal, and industrial products like steel. In the East, behind the Iron Curtain, the picture was very different because the Communist authorities discouraged individual car use; there, the railways remained heavily used, though mostly starved of adequate investment for modernization. In East Germany, for example, as late as the 1970s, the railways were still carrying two-thirds of the country's long-distance freight because the government wanted to keep expensive oil imports bought with hard currency to a minimum.

With the loss of much of their traffic, it was inevitable that the Western European railways would start to close lines and drastically reduce staffing levels. Branches had started closing in many countries since the 1930s, but the pace accelerated greatly once cars started rolling onto the improved roads in the 1950s and 1960s in huge numbers. The inflexibility of the railways was proving a great burden, as stations and whole lines had to be shut. In Britain there was the infamous Beeching report published in 1963 which led to the closure of 4,000 miles of railway, a quarter of the total, and 3,000 stations, while in France most of the departmental lines built after the clamoring of local interests in the 1870s were closed in the 1940s and 1950s.

There were, however, some heroic attempts to resist the onslaught from cars, freight trucks, and airplanes. Gradually, after the chaos of the immediate aftermath of the war, railway administrations across Europe began to understand that the affluence that spawned the automobile also offered them opportunities to provide a new kind of service. Rather than expecting passengers to make do with the slow and dirty steam-hauled trains, which differed little from their prewar predecessors, the more forward-thinking railway managers realized that a better standard of service was expected by their now more affluent passengers.

In the mid-1950s, inspired by the creation of the European Coal and Steel Community, which was the precursor of the Common Market, the head of the Dutch railways, F. Q. den Hollander, suggested a network of Trans-Europ Expresses linking the major centers on the European mainland that would allow business travelers to reach their destinations fast enough to do a day's work and return the same evening. Fearing that competition from airlines would take away this lucrative business, seven railways, including those of France, Italy, and Germany, combined to create a series of international first-class-only trains intended to provide a genuine alternative to flying. Traveling on international trains had until then been a slow and cumbersome experience, requiring lengthy stops at borders for customs and immigration procedures as well as locomotive changes and new drivers. The restrictive practices of rail unions resulted in detailed "safety" checks, right down to the last lightbulb, before trains could proceed, making the services take far longer than necessary and thus preventing them from competing with the airlines.

Hollander's concept was to create a pan-European railway that would use the same fleet of luxurious trains across the continent and have its own dedicated staff to create a service far superior to that which was normally provided by the railways. This would have been a genuine alternative to the large European airline companies, especially because it would have offered a far wider range of journeys, but it was an idea too far ahead of its time. The grander aspects of the concept were thwarted as the individual railway administrations, all state-owned, were not ready to cede their independence, and the eventual scheme was a watered-down version of Hollander's dream. His plans to include meals in the ticket price, like airlines, and have special porters at every station were rejected, but the railways did agree on a common ticketing and reservation system that represented a radical improvement from previous arrangements, although even today booking cross-border journeys is not always possible on European railways.

A network of half a dozen Trans-Europ Expresses (TEEs) was launched in June 1957, connecting, for example, Paris with Amsterdam and Frankfurt with Zürich, and the concept quickly grew to encompass seventy-three European cities. The trains were initially all-diesel, though later some electric trains were included in the network, and several railways developed specially designed trains running at speeds of up to 124 miles per hour. All the trains were branded with traditional-sounding railway names, such as *Rheingold*, *Etoile du Nord*, and *Mediolanum*, and the lines were based around a series of hubs, such as at Frankfurt and Basle, rather like airlines. Although they did retain a significant number of business travelers who would otherwise have flown, they were fighting an uphill battle, particularly for journeys longer than a couple of hundred miles, at least until, as we shall see in the final chapter, the railways responded by creating high-speed trains on dedicated tracks, which greatly reduced journey times. Most Western European nations soon signed up to the TEE concept, and although the standard of service varied from one country to the next—the best was reputed to be the service on the Swiss trains—all provided meals and modern arm-chair-style seating. However, these early diesel trains were not as well sprung as the coaches hauled by steam locomotives, giving a somewhat bumpy ride. The concept nevertheless became so popular that it was ex-panded to include several domestic services. In the 1970s the TEE net-work started to be phased out and was replaced by Euro City trains offering both first- and second-class accommodation. Today, a similar concept, Railteam, has been developed for a Europe-wide network of high-speed train services (see next chapter).

The prestige trains in the United States continued running after the war, and were even improved upon with the addition of observation cars. As we have seen, they were not a new idea, having first been used by the Canadian Pacific in the 1890s, but they had subsequently been forgotten and were not resurrected until 1944, after a General Motors executive was so impressed by his view from the cab on a trip through the Rockies that he thought passengers would enjoy a similar experi-ence to while away the long journeys in the West. Since railroads in the United States mostly have far more generous height restrictions (some-times up to 20 feet) than railways elsewhere, there were no insuperable technical obstacles. Rival railroads rushed to add observation cars to their trains in an effort to retain their dwindling traffic, and one, the Baltimore & Ohio, was even daft enough to fit a searchlight onto its ob-servation cars in a misguided effort to give its passengers something to see through the night!

More successfully, several companies introduced "vista dome" trains, which had an upper deck for premium fare payers, often using them to serve meals, rather like those restaurants at the top of com-munication towers. The "domeliners," as they were called, were a valiant effort by the railroads to postpone the inevitable, and for a time it worked. The most famous of them was the *California Zephyr*, which ran between Chicago and San Francisco, and was carefully timed to ensure that it traversed the best scenery in daytime. This was no longer train travel designed merely to get people from point A to point B; rather, the journey was being sold as a land cruise, which clearly suggested that the railroads were beginning to lose the serious business traveler. The *Zephyr* flourished in the 1950s and early 1960s, often being full and booked well in advance, but by 1969 it was reported to be losing $2 million annually and the service was withdrawn—although the name was retained by Amtrak, the subsidized state-owned railway.

The collapse of passenger traffic on the U.S. railroads was faster and more sustained than anywhere else: In 1930, 75 percent of passenger traffic was by rail, but that figure had declined to just 7 percent in 1970, and most of those journeys were short commuting trips in the few densely built-up cities, such as Chicago and New York, that had retained their suburban rail networks. Although the obvious reason for the decline was growing competition from both cars and airplanes, there was a more fundamental one that goes to the soul of American society. There was almost a venal desire to rid the country of the railways, not least because their history of abuse and corruption was all too recent in the collective memory and the railroad companies were still intensely disliked. Moreover, the very collectivity of railways, and their dependence on fixed timetables limiting an individual's right to choose when to travel, were at odds with the American way of life.

The railroad companies, too, contributed to a lemming-like collective suicide of the industry by resisting the offers of state aid from government; in their American way, they were reluctant to relinquish their commercial freedom. They had, of course, forgotten that they mostly owed their very existence to subsidies from government, albeit hidden ones in the form of route surveys undertaken free by the army and land grants, and they barely put up a fight when the United States Post Office withdrew its lucrative contracts to send the mail by rail, which also had long been a hidden subsidy to the passenger railroads. Then there was the problem of high costs. Railway managers had over the years agreed to labor practices that were inflexible but that had been fine in the days of plenty, when railroads could meet the expense from their monopoly profits. Now these high costs were a millstone and impossible to negotiate away, especially as some were enshrined in regulation. By the 1960s, bankruptcies were becoming commonplace, while other railroads, including those fierce old rivals, the Pennsylvania and the New York Central, merged to stave off that fate.

Passenger trains in the United States only survived thanks to the creation of the government-supported Amtrak company in 1970, when the great free-enterprise nation of the world was forced to nationalize its passenger rail services. Amtrak has been propped up by huge amounts of government subsidy ever since then, largely because self-interested local politicians ensured the survival of long-distance trains that retained many of the illustrious traditional names, even though few people ever used them. As a result of this political feather-bedding, Amtrak has been a disaster for much of its existence and a barrier to the development of a sensible system of passenger railways. Instead of focusing on a few viable services, particularly in the northeast corridor, Amtrak has been obliged by the pork-barrel politics to run completely unviable services, usually once per week. These operate at very slow speeds and are frequently late because, in the long single-track sections, priority is given to freight trains, which are the mainstay of the U.S. railroads since they are highly profitable. If the railroads were relieved of their passenger services, they could concentrate on their freight operations, which are still vital to American industry. Freight loadings declined, too, in the 1970s, but, as we shall see in the next chapter, were revived in the 1980s following deregulation.

There was an odd beneficiary from the rapid decline of the American passenger trains. The now nationalized Mexican railways, which had suffered perennially from underinvestment, obtained large numbers of excellent American passenger coaches, many nearly new. This stock found its way onto the Mexican railways, which were able to launch a series of excellent passenger services from the capital with names such as *El Regiomontano* to Monterrey, the *Aguila Azteca* to Nuevo Laredo, and *El Tapatio* to Guadalajara, which was named after the local dancing girls.

It was not only in Europe and America that luxury trains were being developed at the same time that rail services generally were declining. In South Africa, the other *Blue Train*—not the one that goes between Calais and the Côte d'Azur—was launched in 1946. There had been a tradition of luxury service on the 1,000-mile journey between Cape Town and Pretoria stretching back to 1903. The *Union Limited* boasted shower baths, an observation lounge, and continuous valet service, but in 1946 it was transformed into the *Blue Train*, with air-conditioned train sets built by the British Metropolitan Cammell company. By the 1970s, with new rolling stock that included three baths, this was the last of the great luxury trains in the world, harking back to the old days when servants were cheap and railways offered unparalleled standards. The epitome was the sumptuous lead coach designed for no more than six passengers, the kind of accommodation provided elsewhere for royalty. It included a three-room suite for two, befitting first class on a cruise liner, and the rest of the accommodation on the sixteen-coach train was equally opulent, with a staff of twenty-eight ministering to just 107 passengers for the day-long journey. Geoffrey Freeman Allen, who devoted a whole chapter to the *Blue Train* in his book *Luxury Trains*—without ever mentioning that at the time in apartheid South Africa, it was only available to white people—did not doubt that it was "the most luxurious train in the modern age."[11] It survives today as a service aimed at tourists, taking twenty-seven hours, nearly four hours longer than in the 1970s, to cover the route, but still retaining many of its luxurious features.

In Australia there were belated attempts to resist the onslaught from aviation by improving long-distance train services. Thanks to the construction of a section of new line and gauge conversion, it became possible in 1970 to travel the 2,500 miles between Perth and Sydney in considerable luxury without changing trains. The *Indian Pacific* ran four days a week and took nearly three days, but passengers enjoyed excellent standards of comfort that almost matched those of the *Blue Train*. Most importantly, given that the train crossed the desert with, as mentioned previously, the world's longest section of straight track across the Nullarbor Plain, where temperatures could reach 60°C (140°F), it had the most modern form of air conditioning. Oddly enough, in other respects the Australian railways were slow to catch on to modern traveling needs. The tradition of meal stops for long-distance journeys, for example, continued into the 1970s on the Queensland Railway's services between Brisbane and Cairns, the last trains in the world to do so.

New Zealand was one of the countries where the railway lost its central place in the economy quite rapidly. Given how the railways had virtually created the nation, the sharpness of the decline was cruel for the railways. At the height of the war in 1944, when the railways in New Zealand were, as elsewhere, effectively the only way to travel throughout the country because of gasoline rationing, the number of long-distance railway passengers had reached a peak of 15 million annually; within a decade, the number fell to under 5 million. And that was before the airlines started seriously eating away at the remainder. In response to that threat, New Zealand Railways tried to retain its business passengers and attract tourists onto its long-distance services by introducing more comfortable trains with names such as *Blue Streak* and *Silver Fern* on the North Island main trunk route. The most illustrious was the *Silver Star* overnight service between Auckland and Wellington, launched in 1971 using Japanese-built trains, sporting bowtie-wearing stewards, and offering a shoeshine service. Successful at first, it only survived eight years, as sparsely populated New Zealand, with its relatively slow trains and long distances, proved easy meat for the airlines and their short journey times, despite fares that were three or four times higher than those of the railways. The country lost 1,000 miles of branch lines, nearly a third of the total mileage, and fewer than 100 of the 1,350 stations survived, although admittedly many of these had been little more than wooden shacks bearing the station name, designed to shelter the one family with a farm nearby. However, more positively, New Zealand Railways enjoyed growth on their suburban lines, a trend that was mirrored elsewhere as road congestion in built-up areas began to reach intolerable levels.

The decline in traffic on railways across the world coincided, too, with the need for investment. Since most lines had been built in the nineteenth century, the equipment was in urgent need of refurbishment or total renewal. Moreover, steam technology was becoming increasingly uneconomical—and unacceptable to passengers—and railways were consequently faced with the choice of modernizing or closing. Had railways been able to muddle through with aging engines and carriages, then many lines might have remained viable. However, that option was not open to railways, except in Eastern Europe, where passengers were presented with little choice, given that individual car ownership was not available to the masses. Elsewhere, dissatisfaction with standards of service only precipitated the rush to use the roads. Diesels were seen as the cheap alternative, but the haste to convert from steam locomotives contributed, in several countries, to the rapid decline of rail transport, and some railways scrapped their steam locomotives far too readily. Although diesel did offer savings on operating costs, they were expensive to buy, and many steam locomotives were written off unnecessarily early, especially as the cost of coal was far more stable than that of oil.

In the United States, conversion to diesel was remarkably fast, and the only major railroad still using steam in the mid-1950s was the coal-hauling Norfolk & Western Railroad. In Britain, after a slow start, a rush toward dieselization, stimulated by the Modernization Plan of 1955, saw the introduction of no fewer than fifty different types of diesel locomotives, many poorly designed, and some relatively new steam engines were scrapped. With its strong domestic market, the United States had become the world's leader in rail technology. Major companies had developed powerful and efficient diesel engines, and its export business flourished, thanks to aggressive marketing campaigns. However, as Westwood argued, "in many cases the diesel locomotive was oversold. Whereas in American conditions it usually had substantial operating and economic advantages over steam, this was not necessarily true everywhere."[12]

Therefore, oddly, a technology that should have been rail's salvation contributed, in several countries, to its decline. In the Third World, many railways, whose governments were induced by sales talks from General Motors and other U.S. locomotive manufacturers to buy diesels, found themselves with expensive machines that they did not have the skills or money to maintain, and many diesels were soon left mouldering in sidings. Latin American countries induced to buy U.S. diesels tended to convert their railways too quickly, while in contrast, France and Germany, which both made sure they had time to develop their own home-grown locomotives, took a more considered and, arguably, more sensible approach. South Africa and Australia were also among those who introduced diesels more gradually.

Electrification was potentially a better option than diesel, offering a cleaner and more efficient technology since it did not involve carrying heavy fuel on board. Certainly, as we see in the next chapter, its use on high-speed rail lines has contributed to their success. But again, when railways were faced with the need to modernize, electrification was an expensive option requiring considerable initial investment. Many railways were not courageous enough to make the necessary expenditure and lost out, but those that did invariably found that there was a "sparks effect," with faster, cleaner, and more reliable services attracting passengers back onto the railways. In the United States, the huge domestic diesel-locomotive industry and the availability of cheap oil left electrification confined to suburban services and a few lines on the eastern seaboard, apart from the odd failed experiment on longer routes. Instead, it was the French who became the world leaders in developing electric railway technology. Straight after the war, its state-owned railway, SNCF, began a rolling program of electrifying all its main lines using overhead wires and a 25,000-volt system, which is now widely regarded as the most suitable. The same system was used in India, which embarked on an extensive electrification program on its main lines, and even in England, where the first section of the West Coast Main Line was converted in 1966. Throughout Europe, however, a variety of types have been employed, and this makes operating across frontiers difficult as it either involves a change of locomotive or the installation of several systems, an expensive and wasteful process.

For a time in the 1960s and early 1970s, it seemed as if rail travel might have had its day. Who needed trains when they had cars that could go door to door at little cost in an age of seemingly endless cheap oil? Or when even buses could do the same journey more quickly for a lower fare, and trucks could take virtually any load anywhere? Certainly in the United States and Canada, that logic prevailed, and passenger services on the railways all but disappeared. Elsewhere, too, with the underinvested trains suffering from deteriorating performance at the same time as bus journeys were speeded up on newly opened expressways, the economics of the railways put their very survival at risk. For example, in the 1970s, it became possible to travel on a bus between Brazil's biggest cities, São Paulo and Rio de Janeiro, in six and a half hours with a ten-minute frequency, while the train, with just a handful of daily services, took nine hours, an hour longer than before because the track had deteriorated. In several countries in Latin America and Africa, such as Guatemala, Honduras, and Guinea, the passenger railways, often state owned and run unimaginatively, could not withstand the impact of the competition from road transportation and disappeared completely. Latin America, in particular, but to a lesser extent every other continent, is littered with redundant railways that could still serve a useful purpose if they had not been allowed to deteriorate and die out through lack of will and understanding during that period. Elsewhere it was war that shut down the railways or restricted their development. In several parts of Africa and the Middle East, such as Angola and Lebanon, the railways were closed as a result of a conflict, while in Iran, an ambitious plan for an expansion of the railways involving some 6,000 miles was brought to a halt by the overthrow of the Shah in 1979.

Even in these bad times, however, there were a few countries, principally Communist ones, that kept faith with the railways. China, in particular, where the system, as noted already, was far smaller than the size of the country warranted, maintained a continuous program of growth throughout the postwar years. China's railways had been badly damaged in the war, and with the upheaval that would lead to the Communist takeover in 1949, progress in rebuilding them had initially been slow. However, once Mao Tse-tung's regime came to power, the railways were seen as a transport priority for the People's Republic of China, which kept on building railways as if the automobile had never been invented. The mileage of track more than doubled in the twenty-five years after the war, reaching around 30,000 miles by 1980, despite the political and economic upheavals of the 1960s. Indeed, China was belatedly undergoing a railway revolution, and it was by far the postwar world leader in levels of investment, both for creating new lines and for modernizing its existing railways. These extensions to the system included the kind of heroic engineering feats that had been undertaken elsewhere in the world up to a century before, such as the crossing of the Yangtze River with a bridge that was more than a third of a mile long, a project completed in 1969 at Nanking, and the conquering of major mountain ranges in southeastern China with two new railways. Further north, the 1,200-mile Lanxin railway between Lanzhou and Urumqi was completed in 1962 and extended to the Kazakhstan border in 1982. Once a few connections across borders had been made, this was supposed to be part of what the Chinese have called the second Eurasian continental bridge, a revival of the old silk route, linking China with Rotterdam in the Netherlands by a rail line far shorter than the Trans-Siberian and much quicker than a sea voyage. Unfortunately, despite various announcements, the concept has never been properly developed, not least because of the customs and regulatory difficulties of transporting containers on the railways of half a dozen countries.

The Chinese investment program for their existing railways included electrifying several routes as well as doubling the track on many main lines to increase capacity. Despite these improvements, China retained steam longer than any other country in the world and built the last steam locomotive factory in the 1970s. Stations, too, were upgraded. Beijing obtained an entirely new station—built in just twelve months and with a dozen platforms—but there was a paucity of trains for such an enormous city. In 1978, from this magnificent new station, there were just fifty-eight daily departures, a tiny number for a population of 4 million people who did not have access to private cars and a railway with no competition from long-distance buses or domestic air services. No wonder there was always chaos at holiday times, when tickets were much sought after at peak times. The waiting hordes could, however, enjoy "The East is Red" chimes from a huge clock tower with a pagoda-style roof. Train travel was slow, too, with the fastest averaging barely 45 miles per hour, as the frequent freight trains took precedence over passenger services. Even the shift toward encouraging private cars, which began in the 1990s, has not stopped the Chinese from pouring money into their railways and opening up more lines, such as the Tibet railway, completed in 2006 (see next chapter), the most recent example of an imperial railway.

Russia, which had the world's largest unified system, invested heavily in its railways after they were heavily damaged in World War II. The ·Trans-Siberian, which had been spared, was electrified in stages, as were large sections of the rest of the Russian railways by the mid-1950s, since the Soviets were keen to promote a modernizing image for their railways. Indeed, the railways were used as a propaganda tool by the Russians, who opened a metro system in Leningrad (now St. Petersburg) in 1955 that was almost as grand as the one in Moscow (built between the wars) with its famous chandeliers. The busiest line, the newly electrified route between Moscow and Leningrad, boasted the *Aurora*, which took just five hours for the journey, an impressive average of 80 miles per hour. There were other relatively fast express services running to distant republics such as Azerbaijan and Kazakhstan, which were then part of the Soviet Union, but, as ever, away from the main lines services were pretty grim. Despite the Cold War, through services to other countries were made possible by gauge-change machinery at the borders starting in the 1960s. These ran from Moscow to Western capitals such as Paris and Vienna (called, rather light-heartedly, the *Chopin*) as well as to other Eastern-bloc countries such as Czechoslovakia and Romania. In 1963, a service was launched between Moscow and the Hook of Holland, a journey that took two and a half days for the 1,600 miles, which connected directly with the ferry to Harwich.

While these large Communist countries were building up their railways, the smallest, Albania, became the last significant European country to join the railway age, opening its first line in 1947. It was, though, not only in the Communist countries that railways were being opened— or indeed reopened. Freight lines continued to be built in several parts of the world, notably in Australia, Cameroon, and Brazil, mostly with the traditional function of linking the hinterland with a port. Despite this, in the 1960s, for the first time since the invention of the railways, the world's rail mileage went down, from 903,000 miles in 1960 to 886,000 in 1975.[13]

Yet trains were beginning to come back into fashion. The turning point came with the first oil price shock of 1973, which prompted the initial rethinking of the notion that the railway was an invention that had outlasted its usefulness. Transportation planners not only realized that the era of cheap oil might be finite but also learned that the car, which was causing endless traffic jams on the roads in towns and cities around the world, was not the panacea they had previously thought it was. As mass car ownership spread around the globe, the value of rail services was beginning to be recognized once more. Moreover, the railways had embarked on a project to regain some of that lost traffic, and it was beginning to bear fruit. The era of the high-speed line, a train service fit for the twenty-first century, had arrived.

THIRTEEN

Railway Renaissance

IT IS NO EXAGGERATION TO SAY THAT WHEN THE WORLD'S FIRST HIGH-speed line opened in the autumn of 1964 between Tokyo and Osaka, a new era of train travel began. The line had been long in gestation; construction of a new 100-mile-per-hour railway between the two cities had been started in the 1930s, but the scheme was abandoned at the start of the Pacific War in 1941. The impetus to create a new line after the war came from the severe overcrowding on the old route, which, despite electrification and modern signaling, could not take any extra trains. Japan is eminently suited for railway investment because its 125 million people are concentrated in large cities, typically 100 miles apart, on the coastal plains, making it a railway planners' dream. Postwar Japan, with American aid, was booming, and the number of passengers between the major cities was increasing rapidly. The old 320-mile railway between Tokyo and Osaka was the country's most heavily used line, accounting for a staggering quarter of both the passenger and freight traffic carried by Japanese National Railways, the state-owned railway, while it constituted just 3 percent of its mileage.

Japanese National Railways decided that rather than attempting to quadruple the existing tracks, it would build a new route that would allow faster trains to run on a completely separate alignment. Thus, the concept of "high-speed" lines was born, aimed at not only enabling passengers to travel faster but, crucially, also boosting capacity dramatically by providing an entirely new railway route, uncluttered by local services or freight trains, that would allow for more intensive usage. The number of trains a line can accommodate is greatly reduced if they run at different speeds, but on dedicated high-speed lines all the trains run at the same speed.

In order to ensure the line—dubbed *Shinkansen*, which means literally "new trunk line," but has now come to signify a high-speed line—was kept completely separate from the rest of the network, it was built to the standard 4-foot, 8.5-inch, gauge, in contrast to the 3-foot, 6-inch, gauge used elsewhere on the Japanese system. The Tokaido (Tokyo-Osaka) *Shinkansen* established the template for future high-speed-line projects: There were to be dedicated tracks, no sharp curves,[1] no level crossings, in-cab signaling (in other words, no signals outside the train for the driver to read), and a very limited number of stations. These lines are effectively the expressways of the rails, with few junctions and stops. Work on the Tokaido *Shinkansen*, funded by the World Bank, started in April 1959 and took five years to complete, with the first trains running in time for the Tokyo Olympics of 1964, which were held in October to avoid the full summer heat. Construction was a major engineering feat since, unlike the old nineteenth-century Japanese lines, the *Shinkansen* was built with as few curves as possible, thus requiring the construction of numerous bridges and viaducts, which accounted for a third of the mileage. Most of the intermediate stations, too, were put on loops off the main running line to allow trains to stop without delaying those behind them.

Inevitably, the project cost nearly double the original estimate, but it met the deadline of serving the Olympics. The trains operated initially at 125 miles per hour, slow for today's high-speed services, but nevertheless slashed the journey time between the two cities from 6 hours and 40 minutes to just 3 hours and 10 minutes. Although the trains ran smoothly and reliably, there were unexpected problems, notably the pain caused to people's ears by the high air pressure created when two trains crossed in one of the numerous tunnels, which comprised a total of 45 miles of the route. The powerful air currents generated in these tunnel crossings tended, too, to blow the water up from toilet bowls, much to the hapless user's embarrassment. As a result, the trains, which were already fitted with air conditioning and triple-glazed windows, had to be pressurized, an expensive and technically difficult requirement.

Earaches and misbehaving toilets did not deter the passengers. Within three months, 11 million people had traveled on the line, and it only took three years for the first 100 million to be carried. As a result, the *Shinkansen* was soon generating substantial profits, even taking into account interest payments on the cost of construction. This success led to the publication of a plan for a network of 4,500 miles of high-speed line to be completed by 1985. That proved far too optimistic. Work was started on two new lines in 1971, but the expansion plans became surprisingly controversial despite the success of the first *Shinkansen*. The proliferation of viaducts and bridges on the Tokaido line guaranteed that the noise from the frequent trains traveling along them at the speed of 2 miles per minute was spread far and wide, and the railway was obliged to make expensive infrastructure changes, pushing up the cost of construction. Concrete bridges rather than steel became the norm, and wherever the line passed through a built-up area, walls had to be built to contain the noise from the wheels. Even then opponents were not satisfied. This problem has been particularly acute in densely populated Japan, and other countries that have built high-speed lines have encountered similar opposition. In a world where expressways crisscross both urban and rural settings with impunity, new railways seem to generate far more opposition than appears warranted by their limited impact on the environment. It is as if the old suspicions and dislike of the railways are reignited by their expansion plans. In Britain, for example, two expressways were built through Kent in the 1960s,[2] but the Channel Tunnel Rail Link generated a far more vocal campaign by local residents.

Japan, too, was hit by the oil crisis of 1973–1974, which, while helping the railways by pushing up the cost of fuel, also had a negative effect as the consequent recession reduced passenger numbers. Plans for the high-speed network were scaled back; nevertheless, half a dozen lines, amounting to nearly 2,200 miles, have been built. In addition, train speeds have been accelerated, so that today's fastest *Shinkansen* trains travel at 186 miles per hour—the same as most European high-speed services—and there are plans for further speed increases. The *Shinkansen* has proved to be a tremendous boon for Japan's economy. In 1994, an economist bemoaning the loss of passenger services in the United States wrote: "High-speed rail has played an impressive part in reducing transportation costs in Japan and limiting the nation's oil imports. The International Institute for Applied Systems Analysis found the *Shinkansen* to be nearly three times more productive than aircraft serving the same route in terms of labor efficiency, five times more effective in terms of capital costs on equipment, and eight times more effective in terms of energy consumed."[3]

Yet, despite the success of the *Shinkansen*, it was over a decade before any other railway would rival Japan's achievement. There was widespread interest across the world in improving the speed of express trains from their ponderous average of 60–70 miles per hour, as this was seen as the only route to the salvation of the railways. At the time, the expressways were relatively congestion-free, and cars could easily average the same speed as the express trains of the day, with the added advantage of providing door-to-door service.

Several countries—including, interestingly, the United States—were considering high-speed rail projects, but building new railways is inevitably expensive, requiring government subsidy and long-term commitment. There was, too, still doubt about whether railways would remain a viable technology, given the ubiquity of motor vehicles; and, as we have seen, new rail projects were invariably controversial. Initially, too, European railways sought to improve timings on their existing lines through electrification and through the expansion of the Trans Europ Express network, rather than by building dedicated high-speed tracks, as the latter was such an expensive proposition. In Germany, for example, Deutsche Bahn introduced a daily train running between Munich and Augsburg that reached 125 miles per hour in 1965.

It was hardly surprising that it was the French who would be the first to commit themselves to the idea, as they had long been interested in high-speed trains and had even given the world the term "express." As far back as 1955, the French had captured the world speed record with an electric train that ran at 206 miles per hour, but that trial nearly resulted in disaster because the force of the train running so fast buckled the track. This experiment demonstrated that high-speed rail would require a far more sophisticated level of technology than conventional lines, or *lignes classiques*, as the French call them. Both track and trains need substantial improvement and investment for high-speed running. The French had the technology available and started to build a high-speed line between Paris and Lyon in 1976. As in Japan, the motivation was not only the desire to run at high speed but also the need to boost capacity. The old Paris-Lyon-Mediterranée line was full, despite electrification, and although it had four tracks in most places, the two-track sections would have been prohibitively expensive to double. Therefore, in late 1976 work started on the Paris–Sud Est Ligne à Grande Vitesse[4] (LGV), the first high-speed line in Europe.

The French Train à Grande Vitesse (TGV) was, though, more than that. It was a political project designed to boost the image of *l'Hexagone* to the outside world, as Lord Adonis, the British transportation minister, put it in 2009 when he said the TGV was "not just a train but a vision of the future and a means to bring it about, a force for national integration and regeneration and a source of intense national pride."[5] The concept was slightly different from the Japanese version. Whereas the *Shinkansen* was a totally separate line necessitated by its different gauge, the French trains ran into Paris Gare de Lyon on tracks they shared with conventional trains. Therefore the dedicated LGV veered off from the old tracks in the Parisian suburbs, greatly reducing the cost of construction (as the final stages of a new route into a major city are always the most expensive, while only marginally increasing journey times). The uniformity of gauge has the advantage of allowing the TGV to leave its dedicated tracks to serve many destinations that do not have big enough populations to warrant a high-speed line but that still benefit from the faster journey times. Work progressed relatively smoothly, and the Sud Est was fully open by September 1983.

The Sud Est line was designed to have a maximum running speed of 300 kilometers per hour (186 mph), although in practice the trains only operated at 270 kilometers per hour (168 mph). Nevertheless, that was fast enough to obliterate the market for air services between Paris and Lyon and to convince the French that they had hit upon a winner. The TGV, like the *Shinkansen*, has been both a technical and a commercial success, attracting millions of passengers onto the railways and helping France retain a major role in world railway technology through Alstom, the French designer and manufacturer of the TGV train sets. Over the next thirty years, lines radiating out of Paris on all four cardinal points[6] have been built, with the latest, Est, with a line speed of 320 kilometers per hour (200 mph), completed in 2007. It is now possible, for example, to travel almost the whole length of France, the 487 miles between Paris and Marseille, in three hours, making it the longest nonstop rail journey in the world. Services have proved so popular that on many routes there are duplex trains, with upper and lower decks, and there are even high-speed postal trains running at night for *La Poste*.

Other countries have followed suit. In Germany, there was again a rather different concept. Instead of having entirely separate routes, the Germans decided on a mix-and-match approach of building sections of new line interspersed with running on conventional track for its ICE (Inter City Express) trains, which can run at a maximum of 300 kilometers per hour (186 mph) on the dedicated sections. There is now an extensive network of ICE trains, many of which run into neighboring countries such as France, Denmark, and Holland. Italy, too, is joining the high-speed club, building a line running through the spine of the country that links Turin and Milan with Rome and Naples—and, uniquely, is allowing private operators to run on it.[7] The services, like those in France, will run for most of the journey on dedicated high-speed track. They will parallel Italy's main A1 expressway for long stretches, but the trains will run at more than twice the speed of the cars, a permanent demonstration of the superiority of rail.

In contrast, Spain, whose conventional trains operate on a 5-foot, 6-inch, gauge, decided to run its high-speed AVE (Alta Velocidad Española) trains on standard gauge, partly because the technology was imported from France and Germany and this will allow the same components to be used as in those countries. The first Spanish high-speed line was built between Madrid and Seville to coincide with Expo 92, and its success stimulated the adoption of a hugely ambitious plan for the biggest high-speed network in Europe. Already by the end of 2008 Spain was close to matching France, with just under 1,000 miles of high-speed line, and it was on the way to having a 10,000-kilometer (6,250-mile) network by 2020, which would ensure that 90 percent of Spaniards living within 50 kilometers (31 miles) of a station would be served by high-speed trains.

China, though, dwarfs even Spain in ambition. In 2008, China, which opened its first high-speed line in 2003, only had 345 miles in operation, but some ten times that mileage was either under construction or definitely planned, and China is expected to overtake Japan as the world leader in the mid-2010s. Elsewhere in Asia, Taiwan and South Korea already have high-speed services running on dedicated lines, using Japanese and French technology, respectively, and South Korea is already constructing an extension.

Overall, the world had over 6,500 miles of high-speed line by the end of 2009, and there were about 3,000 more miles under construction. It represents a new railway boom, not quite on the scale of the first in terms of its impact, but a fantastic affirmation of the iron road. Wherever they have been introduced, high-speed lines have proved popular, though not always meeting the optimistic expectations put forward by over-eager promoters of the projects. Mostly, too, they have proved profitable, though not necessarily when interest payments are taken into account—Eurostar, for example, linking London with Paris and Brussels through the Channel Tunnel, has a particularly high cost base, and therefore has not reached anywhere near the break-even point—but, crucially, as is always the case with railways, they deliver major economic benefits that cannot be captured through ticket sales. On the cost side, there have been considerable overruns on several projects as technical difficulties have emerged. The new South high-speed line in the Netherlands, for example, has been beset with difficulties and was more than two years late in opening.

Incompatibility between the differing European systems has also added to the costs of projects, but overall the investment in high-speed rail has demonstrated that rail can be a popular alternative to traveling either by air or by road while offering environmental benefits. As climate change rises inexorably up the political agenda, the environmentally friendly aspects of rail travel have been emphasized by governments and railway organizations, something they did not always do sufficiently in the past. Indeed, until the 1990s, railways did not pay sufficient attention to environmental considerations and were rather profligate with energy, introducing ever heavier trains with little regard for fuel consumption. Now, with the advent of regenerative braking (which puts the power used in braking back into the electricity grid) and far more efficient engines, the rail industry is able to be much more environmentally friendly, which has become a political imperative. The Eurostar service, which is mostly powered by nuclear-generated electricity, boasts of zero carbon emissions, a claim that is not entirely sustainable but at least demonstrates purposeful intent.

One of the great achievements of high-speed train travel is its superlative safety record. At the time this edition went to press in 2010, there have been no passenger fatalities on high-speed services running on dedicated lines,[8] though they have carried billions of passengers. Extremely high standards of safety have been designed in as an integral part of these railways. The Japanese lines have tried to ensure that their high-speed lines can resist the impact of an earthquake, and despite several serious tremors, there has been only one derailment related to them throughout the *Shinkansen*'s history. The most serious incident, in France, involved a derailment caused by subsidence resulting from World War I trenches, but there were no serious injuries.

Predictably, it is the two bastions of privately built railways, the United States and the United Kingdom, where the state has had the least involvement, that have missed out on the high-speed revolution. That is no coincidence. High-speed lines require a major commitment from the state, as the private sector is unable to fund such major schemes without subsidy and government support. Both countries have embarked on high-speed projects. In North America, the private railroads—and later Amtrak—tried to speed up their trains using high-speed technology, but the various trials, both in the United States and Canada, all ended in failure or only lasted for brief periods. In some cases the projects were dogged by technical problems, while in others, politicians were unwilling to allow the government to pay for the necessary investments.

Even today, the United States has failed to build a high-speed line, despite operating its flagship Acela trains in the northeastern corridor. These high-speed tilting trains are capable of running at 200 miles per hour but are in fact restricted to 135 miles per hour, and even much less than that, on most of the route. Acela provides a patchy service hindered by level crossings, the lack of a dedicated high-speed track, and various speed restrictions, which means that the trains take three and a half hours between New York and Boston, averaging only 86 miles per hour. This has proved good enough to grab a substantial part of the airline market but not to wipe it out, as has effectively happened between

Paris and Lyon, Paris and Brussels, or Madrid and Seville, which are served by efficient high-speed lines.

Of major developed economies,[9] only in the United States has intercity passenger rail travel become entirely marginal. Today, Amtrak still serves five hundred stations, but it boasts a mere 28 million passengers per year, barely ten days' worth of the passengers traveling by rail in far smaller countries such as Britain or France. However, this might change in the next decade. On election day in November 2008, California voters voted yes to a proposition to build a 650-mile high-speed network running initially between Los Angeles and San Francisco at speeds of up to 220 miles per hour, and later stretching up to Sacramento and down to San Diego, but there is no definite timetable for construction. Similar schemes in Texas and Florida have previously stalled. Barack Obama set about changing the climate for investment in rail with the launch, soon after his arrival in the White House, of an $8 billion stimulus package for high speed rail. There could be major investment in the U.S. rail system over the next few years for the first time in a generation.

America, though, does have a buoyant freight sector, which has enjoyed a sustained renaissance since it was deregulated in 1980 by the Staggers Act. Legislation stretching back to 1887 that had prevented the railways from setting their own prices, forcing them to offer the same rates to every shipper, was abolished. The Staggers Act, which actually should have been passed several decades before, gave the railroads commercial freedom but also ensured that they could not unreasonably block access to key junctions for rival companies. Over the next decade, rates and costs were halved, reversing the long-term decline of the U.S. railfreight business: New lines have opened and defunct ones reopened, allowing the United States to maintain its position as the world's biggest railway network, with over 155,000 miles of track. The act also helped improve the economics of what are called short lines, or "Mom and Pop railroads," which are mostly branch lines that carry freight for relatively short distances. Freed from any regulatory controls and expensive labor agreements, and often supported by the local state government, five hundred such short lines have flourished. They are a uniquely American phenomenon, as successful branch lines that thrive in countries whose governments are traditionally supportive of railways, such as Switzerland

and Germany, tend to provide passenger services subsidized by local or regional government but have little freight traffic.

Britain, too, has lost out by not choosing to adopt high-speed technology on dedicated lines. Instead, British Railways went down a different route in the early 1980s by attempting to develop the Advanced Passenger Train, an electric train designed to go faster on conventional lines by tilting at curves, rather in the same way that a motorcyclist leans into bends. The tilting enabled the train to go 50 percent faster on curves, and theoretically it was designed to reach 155 miles per hour, though any speeds above 125 miles per hour would have required a different method of signaling. Unfortunately, the combination of trying to introduce the train into passenger service too early and hostile press coverage caused the abandonment of the scheme, probably just at the point when it might have overcome the technical difficulties. Ironically, the Swiss and the Italians perfected the tilting concept, and today trains using that technology, called "Pendolino" after their Italian origins, run on Britain's West Coast line. But in the 1980s, British Rail chose the confusingly named but highly successful diesel "125 High Speed Trains," which run at 125 miles per hour. These trains have to share the tracks with other services on what is, effectively, an updated Victorian railway. The only dedicated high-speed line in Britain is the Channel Tunnel Rail Link, renamed High Speed One, running between London St. Pancras and the Channel Tunnel. This 67-mile line used for Eurostar and domestic Kent commuter services will, for the foreseeable future, be the only such line in the United Kingdom, as there are no definite plans to build any others.[10]

Elsewhere, high-speed trains are spreading around the world. Along with numerous extensions being constructed in countries with existing high-speed lines, such as France, Spain, and South Korea, plans are being discussed or built in places as far afield as Argentina, Ukraine, and Turkey. Having a high-speed line is becoming emblematic for nations in the same way that owning an airline used to be. France boasts of its TGV, and no Japanese tourist brochure would be complete without a picture of the *Shinkansen* running past Mount Fuji; it is one of the world's most used images. Expressways never receive such good press!

It is not only high-speed lines that are being built. New railways, mostly either for freight or suburban services, are being promoted in dozens of countries. It is China, already the postwar star in railway development, that is currently making the most investment in new rails. In October 2008, the Chinese government approved a $330 billion railway investment plan that forecast a 50 percent increase in the country's network, to 75,000 miles, by 2020. China is also building metro systems or suburban lines in several cities, and rail development is seen as a key part of the country's economic strategy—though it must be said that thousands of miles of expressway are also on the stocks.

Indeed, metro systems are a key component of railway growth. There are already well over a hundred cities with metro systems around the world, and new ones, as well as extensions and extra lines, are being built on every inhabited continent. Metro systems, both underground and elevated, are popping up in the most unexpected places, ranging from relatively small cities with populations of fewer than 1 million people, such as Lausanne in Switzerland or Bielefeld in Germany, to big megalopolises such as Mexico City or Shanghai, to countless little-known medium-sized cities such as Brescia (Italy), Maracaibo (Venezuela), and Bursa (Turkey). Dubai, a place that has become gridlocked by the availability of cheap gasoline and is synonymous with the car, opened a new metro system in September 2009.

New railway lines are also popping up in strange locations. Today there is probably more investment in railways than at any time in their history. For example, in a project that had been debated for decades, Gabon in West Africa opened its first line of 420 miles in 1987. The line is used principally to transport ore, but also carries passengers. In Brazil, the 250-mile Ferronorte Railway, completed in 2001, hauls massive trains with ninety wagons, each holding 120 tons of grain. Further plans for expansion are expected to be made. Large projects, such as the Ghan, the transcontinental north-south railway in Australia intended for freight, and the remarkable line up to Tibet, have been completed in the twenty-first century. Elsewhere railways are being extended, modernized, electrified, and doubled or quadrupled. In Saudi Arabia a massive project has been launched to build a "landbridge," a 900-mile railway cutting across the whole country from the Gulf to the Red Sea, to avoid having to ship freight around the coast and to carry pilgrims. Together with a couple of other proposed projects, the landbridge will double Saudi Arabia's existing railway network. Further ahead, a project to build a railway through Iran to link Europe with the Indian subcontinent is also being planned. Not surprisingly, as a result of all this activity the railways are big business and getting bigger. A railway trade exhibition, Innotrans, held in Berlin in 2008, was spread over fifteen halls and had over 1,600 exhibitors, nearly ten times more than in 1996.

The railways have turned the corner from the dark days of the 1950s and 1960s when it was thought they would lose any relevance to the modern world. What the Anglo-French railway writer Clive Lamming called *ferropessimisme*[11]—expressing the notion that the decline and marginalization of the world's railway systems were inevitable—is dead. Old alignments are being considered for reopening, and countless articles have been written bemoaning the short-sightedness of those who closed down railways in the postwar bonfire of lines. The notion that railways had to be speeded up in order to survive no longer holds, as congestion has slowed road traffic, and the comforts of relaxed rail travel have become more widely appreciated. In a way, a slow rail journey can be a better experience than a fast one. The great rail journeys of the world, highlighted in a series of books of that name,[12] are experienced on mountain railways, tortuous branch lines, and rickety old routes that have somehow survived into the twenty-first century.

The railways have proved far more durable than many other contemporary inventions. They have seen rival technologies, such as atmospheric railways and monorails—which never fulfilled their early promise—and the development of newer ideas, such as maglev. It was always likely that the railways would survive into the twenty-first century, as even in their darkest days they were carrying billions of passengers annually, but there were doubters about their ability to make it. Richard Hope, the veteran British writer on railways, recalled that "back in 1957, while doing my National Service in the Royal Air Force, one of my fellow officers announced that the passenger train was doomed. I took issue and bet him £10 that thirty years hence there would still be two daily trains between London and York."[13] He certainly won his bet: In fact there are more than thirty direct daily services in each direction today.

Railways have survived and flourished in unlikely places. It is not only in densely populated areas that they are still a vital part of the infrastructure of nations. Many of the Andean railways, for example, still cater to either heavy freight flows or tourists, with the latter enjoying excursions such as the magnificent line up to the Machu Picchu ruins in Peru. Other remote lines remain a lifeline for remote communities, such as the far less famous but still remarkable 100-mile Fianarantsoa Côte Est line in Madagascar, running up from the coast to the island's high plateau. The journey takes eight to ten hours, as the trains barely ever reach 20 miles per hour, but, given the lack of roads in the area, the line provides not only a fantastic scenic view for tourists but also a lifeline for the local community.

The French railways have shown, too, that, with investment, lines that are off the beaten track can flourish. The Mont Blanc railway, which celebrated its hundredth anniversary in 2008,[14] is a fabulous little railway, with perilous viaducts over torrents and steep gradients—which explains why it was one of the early electrified lines. In the past such branches would have been allowed to wither away and die, but now, with investment both from SNCF, the French state-owned railway company, and local government, it has flourished, now boasting new trains and improved track. As a result, numbers of passengers have soared, relieving local roads of congestion. Of course, all of this requires public money, as such railways cannot make a financial return on the investment, but the expenditure pays for itself in many other ways.

The railways are rarely accorded the credit they deserve. They get an unfairly hostile press, as passengers seem to demand the impossible of them, and yet these same people tolerate all the faults of road transport or aviation. They are an Aunt Sally for newspaper editors, criticized for the slightest delays and always accused of inefficiency. The smallest accident is blown up out of all proportion, even though—or perhaps because—they are the safest form of transportation, and their safety record has improved immeasurably in recent times thanks to more advanced technology and better management.[15]

This bad press has left the railways vulnerable to unfair treatment by politicians whose whims have frequently led to long-term damage, such as the overenthusiastic cutting back of the railways in Britain in the 1960s. More recently, the fashion for privatization has wreaked havoc on various railways across the world, resulting in some countries, such as Argentina, virtually abandoning passenger services. In Europe, a basic failure to understand the workings of railways has led the European

Union to enforce rules separating the infrastructure from the operations in order to allow "open access" to the tracks. These rules were designed to free up the market for rail transport and to break up the state-owned monopolies, but the laudable aims have created a huge bureaucracy, a bonanza for expensive consultants, lawyers, and regulators, which has complicated working arrangements and greatly added to operational costs.[16] Certainly, the politicians have done the railways no favors with their tinkering, but as this book has shown, railways throughout their history have been subject to politicians' whims.

The railways may have a diminished role as a transport provider, but they are still a vital component of the world's economy. They have survived partly because of the way they have so fundamentally changed the way we live. They helped to create the cities and the working patterns that now depend on their continued existence. The car and even the airplane may have taken away a large chunk of their market, but without the railways many cities would grind to a halt. The days of the general-purpose railway carrying everything have long gone, but by concentrating on the things they do well, the railways have found a new purpose, especially with their burgeoning high-speed rail networks, heavy freight lines, and metros. With the new emphasis on the environment, and a recognition of the comforts of rail travel—contrast, for example, the ability of trams to get people out of their cars compared with the failure of buses to do so—rail's role and importance will continue to grow. Their success is based on a fundamental scientific fact outlined by Geoffrey Freeman Allen: "Steel wheel rolling on steel rail achieves the most frictionless movement—short of levitation—yet devised; as a result, a train can move more tonnage per unit of traction, per unit of fuel consumed and per staff involved in the transit than any other medium."[17]

Rail travel at its best is unbeatable. The advantages of train travel in modern trains were set out by the German author Erwin Berghaus, in his characteristic eccentric way, and although he was describing the trans-European trains of the 1950s, his comments are still very much applicable today:

When one sees a streamlined diesel set tearing along the track, one can easily imagine wings beneath the long unbroken expanse of windows, and the general impression is similar to that of an ultra-modern jet plane. When one is seated in the comfort of a modern express, one may be excused for comparing the relative merits of road and rail travel. What does the man at the wheel of his car see, his eyes, ears, hands and feel all occupied in the driving. His holiday begins when he reaches his destination. For the railway passenger it begins on the platform.[18]

Sadly, the railways have not always not made sufficient effort to exploit these advantages. The railways partly became the architects of their own demise by not responding to change or pursuing innovations fast enough. International travel in Europe has never been sufficiently developed, despite the Trans Europe Expresses, as cross-border tickets have been difficult to obtain and technical differences have made frontier crossings inconvenient. For example, it is still impossible for many trains to run from one country to another without a change of locomotive and personnel. Since the 1990s, the European Union has been trying to remedy this situation through the adoption of shared technical standards, but it is an extremely lengthy process. If it were possible for a single locomotive to travel through Eastern Europe between Sweden and Italy, it would require over a dozen aerials to deal with the various safety and signaling systems. Only recently have the European railways created a joint pan-European agency, Railteam, to do what the airlines have done for decades, coordinating their marketing effort to create seamless travel across frontiers even when a change of train is required; even now, however, Railteam only deals with high-speed trains, not conventional services. Further afield, a train traveling on the longest possible continuous journey by rail, the 10,600 miles between Algeciras in southern Spain and Ho Chi Minh City in Vietnam, would, according to Clive Lamming,[19] require four changes of gauge.

It would be comforting to be able to say that the railways brought only good to the world. Certainly, they were the most important invention of the second millennium, transporting the Industrial Revolution from a few hot spots to large parts of the world. They were a democratizing force, too, allowing people to travel in an unprecedented way and opening up their eyes to the world, both literally and figuratively. They relieved much tedious and hard labor and spread economic development everywhere.

But, as we have seen, there were also negatives. In many parts of the world, they contributed to environmental degradation. Their tracks wrecked many pristine landscapes, even if, today, we view these same lines rather romantically. In particular, their use for military purposes was quickly exploited by governments, both to put down rebellions and to wage war. As John Westwood put it, "It was rail transport that sustained the mass armies and the mass participation of modern warfare. Total war was a product of the railway age, and without the railways would have been impossible."[20] Without them, too, the most evil crime of the twentieth century, the industrialized murder of millions of people in the Holocaust, would not have been possible. On balance, however, there is no doubt that the railways deserve to be celebrated with enthusiasm. They spread civilization around the world, creating the opportunities for unprecedented growth. Without them, we might be a hundred years further back in terms of economic wealth and industrial development.

There are numerous reasons why railways have not only survived, but flourished. In most countries with a sophisticated and modern rail network, passenger numbers are increasingly driven by both pull and push factors. Rail travel in modern trains is more attractive and pleasant than any other means of travel. With access to the Internet becoming possible on many services, train journeys can be seen as productive time in a way that other forms of travel can never be. With the Internet, too, buying tickets has become easier and information on services far more readily available. Meanwhile, congestion in the sky and at airports, and endless traffic jams in towns and on expressways, have combined to make people look to rail as a better way of traveling. In economic terms, too, as fuel prices rise, pushing up the cost of both road transport and aviation, trains will attract even more customers. More recently, a crucial extra reason has come into the equation, ensuring that the railways will have a rosy future. Trains are far more environmentally friendly than other modes of transportation and will become more so as railway organizations realize that this is a crucial part of the argument to convince governments to invest in them.

Railways may have lost out to the car and the truck, and in America and other big countries, to the airplane. But the fact that they survived and now thrive shows their resilience and flexibility. Trains may be of the past, but they are still the future. They will improve, not just on high-speed lines, but elsewhere too as technology makes them more efficient, more comfortable, and faster. And there is the rather delicious prospect that they might conceivably outlive the car. It may be a fanciful idea, but then not even Stephenson realized quite what an impact his Liverpool & Manchester Railway would have. Although in most places today, rail's modal share of travel is tiny compared with that of road transport, that situation could easily change. All this personalized mobility has not necessarily delivered any overall benefit to society. Are the Chinese better off now with their traffic jams and beltways than they were twenty years ago when bicycles and buses were the dominant form of urban transport, and trains took them between cities? Would it have been better if transport technology had atrophied at the turn of the century and the car had never come to dominate the world?

With every town or village within a few miles of a station or a tram stop, and buses for shorter journeys, a far more rational system of transport and land use would have been developed. Imagine a world without parking lots, expressways, or service stations. Sure, there might have been eight- or ten-track railways connecting major cities, with huge termini and massive bus stations at each end, but it is an alternative vision that has many attractions. Think of all those delightful towns and cities not blighted by the permanent gridlock that affects them today. And all those horrible housing estates, accessible only by car, that would never have been built. We all know that the oil will run out at some point, and as it starts to become too expensive and governments recognize that it should be carefully rationed, trains may regain their place at the center of the transportation system. Now there's a prospect to warm the heart.

NOTES

Preface

1. O. S. Nock, *World Atlas of Railways* (London: Mitchell Beazley, 1978), 8.
2. For the most part I have used the term "American" to denote the United States, although I am aware that the term is sometimes confusing and contentious.

Chapter 1

1. He was Dutch despite his curiously English-sounding name.
2. This is also the name of the national railway company in Spain today, but the modern company was not created until the 1940s.
3. The Spanish did not choose 5 feet, 6 inches; rather, they chose 6 Castilian feet, which is precisely 5 feet, 5.81 inches.
4. See, for example, my previous book, *Fire & Steam: A New History of the Railways in Britain* (London: Atlantic Books, 2007), from which some of this early history of the British railway has been taken.
5. Often spelled "waggonways."
6. Maurice W. Kirby, *The Origins of Railway Enterprise* (Cambridge: Cambridge University Press, 1993), 9.
7. A. F. Garnett, *Steel Wheels* (Waldenbury, East Sussex, U.K.: Cannwood Press, 2005), 6.
8. Ibid., 16.
9. As with several elements of this brief story, there are differing accounts, and some suggest that this engine never ran on rails.
10. They would only become cities later in the nineteenth century.
11. See Wolmar, *Fire & Steam*, 87ff.

12. There are some suggestions that Stephenson had come to the same conclusion about tubular steam boilers and that his proposal was better.

13. Whereas the British use the term "railway," Americans usually prefer "railroad." In this book, I have used the terms interchangeably, but I have tried to use "railroad" when referring specifically to a U.S. line.

14. Quoted in O. S. Nock, *World Atlas of Railways* (London: Mitchell Beazley, 1978), 13.

15. As suggested by Michael Robbins in *The Railway Age* (London: Penguin Books, 1962).

Chapter 2

1. Sir Arthur Helps, *Life and Labours of Mr. Brassey* (London: Evelyn, Adams and Mackay, 1872; reprinted 1969), 72.

2. Most European railways conform to the Bern gauge, which is considerably wider and higher than the British standard.

3. As seen in Chapter 4, this was notable in the United States, where the loading gauge allowed for particularly large rolling stock.

4. Michael Robbins, *The Railway Age* (London: Penguin Books, 1962), 119.

5. Quoted in the seminal history by Henri Pirenne, *L'histoire de Belgique*, vol. 6 (Brussels: Lamartin, 1931).

6. This is very similar to the story of the catering company in Swindon, which insisted that all trains stop there until its rights were bought out by the Great Western. See Christian Wolmar, *Fire & Steam: A New History of the Railways in Britain* (London: Atlantic Books, 2007), 71.

7. Nicholas Faith, *The World the Railways Made* (London: Bodley Head, 1990), 27.

8. Quoted in Erwin Berghaus, *The History of Railways* (London: Barrie and Rockliff, 1964), 48.

9. Ibid., 49.

10. Quoted in Faith, *The World the Railways Made*, 58.

11. Quoted in A. F. Garnett, *Steel Wheels* (Waldenbury, East Sussex, U.K.: Cannwood Press, 2005).

12. Nord, Est, Ouest, Paris-Lyon-Mediterranée, Paris-Orléans, and Midi.

13. Robbins, *The Railway Age*, 124.

14. The small German states imposed customs duties on goods crossing their territories. Someone traveling south from Hamburg to Austria would be required to pay customs duties in ten different states, using different currencies and at different rates.

15. Berghaus, *The History of Railways*, 26.

16. Quoted in Margaret Esther Hirst, *The Life of Friedrich List and Selections from His Writings* (New York: A. M. Kelley, 1965), 38.

17. Ibid.

18. See the excellent and comprehensive history of the line by John Lace at http://easyweb.easynet.co.uk/~jjlace/index.html.

19. Robbins, *The Railway Age*, 119.

20. Lace, http://easyweb.easynet.co.uk/~jjlace/index.html.

21. See Faith, *The World the Railways Made*, 134.

22. P. M. Kalla Bishop, *Italian Railroads* (New York: Drake, 1972), 16.

23. David Taylor, "Naples in the Avant Guard," http://faculty.ed.umuc.edu/ ~jmatthew/naples/train.html.

24. Bishop, *Italian Railroads*, 25.

25. See Albert Schram, *Railways and the Formation of the Italian State in the Nineteenth Century* (Cambridge: Cambridge University Press, 1997), 29.

26. Ibid., 3.

27. Portugal actually chose 5 Portuguese feet, or 5 feet, 5.5 inches, which was later assessed as 1,664 millimeters, 8 millimeters smaller than the Spanish gauge, but this did not prevent through traffic, suggesting that gauge was not that accurate. Both eventually compromised, using 1,668 millimeters.

28. In fact, it is slightly smaller than that, at 4 feet, 11.875 inches, but is usable by 5-foot rolling stock.

29. Kevin Fink, "The Beginnings of Railways in Russia," http://www.fink .com/papers/russia.html.

30. See Berghaus, *The History of Railways*, 61.

31. Robbins, *The Railway Age*, 109.

Chapter 3

1. Michael Robbins, *The Railway Age* (London: Penguin Books, 1962), 106.

2. Thanks to the development of many cheap narrow gauge "tramways," which were economically never viable, the Irish railways reached a peak of 3,442 route miles in 1920, a very dense concentration, given the rural nature of the country, and more than double today's number, representing a proportionately much greater reduction than in the United Kingdom over the past century.

3. Hugh Casserley, *Outline of Railway History* (Newton Abbot, Devon, U.K.: David and Charles, 1974), 22.

4. Ibid., 20.

5. Bombay is now called Mumbai. I have retained names that were in use in the nineteenth century, such as Calcutta and Madras.

6. K. R. Vaidyanathan, *150 Glorious Years of Indian Railways* (Mumbai: English Edition, 2003), 2.

7. Rahul Mehrotra and Sharada Dwivedi, *Anchoring a City Line* (Mumbai: Eminence Designs, 2000), 9.

8. Letter reproduced in Roopa Srinivasan, Manish Tiwari, and Sandeep Silas, eds., *Our Indian Railway: Themes in India's Railway History* (Delhi: Manas Saikia for Foundation Books, 2006), 18.

9. Now called Kolkata.

10. Letter reproduced in Srinivasan et al., *Our Indian Railway*, 16.

11. Quoted in Ian Kerr, *Engines of Change: The Railroads That Made India* (Santa Barbara, Calif.: Praeger, 2007), 17.

12. This story is told by Tim Fischer in *Transcontinental Train Odyssey* (St. Leonards, NSW, Australia: Allen and Unwin, 2004), 7, but seems rather incredible given that Dalhousie was relatively expert in railway technology and not given to such bizarre, unscientific notions.

13. In today's parlance, a memorandum, reproduced in full in Srinivasan et al., *Our Indian Railway*, 23–39.

14. Kerr, *Engines of Change*, 18.

15. Dalhousie memo, in Srinivasan et al., *Our Indian Railway*.

16. *Illustrated London News*, June 4, 1853.

17. Quoted Kerr, *Engines of Change*, 6.

18. From Srinivasan et al., *Our Indian Railway*, 23–39.

19. This was in fact optimistic for the 1840s, when most engineers reckoned that 1:200 was the most that could be coped with on a working railway.

20. Kerr, *Engines of Change*, 40.

21. Quoted in Anthony Burton, *On the Rails: Two Centuries of Railways* (London: Aurum, 2004), 135.

22. Ibid.

23. See Terry Coleman, *The Railway Navvies* (London: Hutchinson, 1965).

24. Estimated in Kerr, *Engines of Change*, 37.

25. Quoted in ibid., 45.

26. Quoted in Burton, *On the Rails*, 135.

27. Kerr, *Engines of Change*, 45.

28. Ibid., 47.

29. Ian Kerr, *Building the Railways of the Raj, 1850–1900* (Oxford: Oxford University Press, 1995), 13.

30. Figures from Kerr, *Engines of Change*, 49.

31. Oscar Browning, *Impressions of Indian Travel* (1903); quoted in Vaidyanathan, *150 Glorious Years of Indian Railways*, 15.

32. Margaret MacMillan, *Women of the Raj* (London: Thames and Hudson, 1988), 74.

33. Technically, at the time, the various states of Australia were independent from each other.

34. Quoted in Patsy Adam Smith, *The Romance of Australian Railways* (Adelaide, South Australia: Rigby, 1973), 15.

35. Quoted in ibid., 25.

36. Ibid.

37. David Burn, writing in the *Tasmanian Journal of Natural Sciences* in 1842, quoted in Smith, *Romance of Australian Railways*, 26.

38. Smith, *Romance of Australian Railways*, 25.

39. Ibid., 33.

40. Now called Woodville.

41. C. C. Singleton and David Burke, *The Railways of Australia* (Melbourne: Angus and Robertson, 1963), 28.

42. Taken from a contemporary report and quoted in Smith, *Romance of Australian Railways*, 33.

43. This would prove to be almost a tradition, as at least two other openings resulted in the similar fatality of a dignitary.

44. Arthur Helps, *Life & Labours of Mr. Brassey* (London: Evelyn, Adams and Mackay, 1969 [1872]).

Chapter 4

1. The second biggest network at the time was the Soviet Union's, which had 86,000 miles of rails at its peak.

2. A. F. Garnett, *Steel Wheels* (Waldenbury, East Sussex, U.K.: Cannwood Press, 2005), 103.

3. Jim Harter, *World Railways of the Nineteenth Century: A Pictorial History in Victorian Engravings* (Baltimore: Johns Hopkins University Press, 2005), 248.

4. This was called a "running light," in railway parlance.

5. See Christian Wolmar, *Fire & Steam: A New History of the Railways in Britain* (London: Atlantic Books, 2007), 52–53.

6. Quoted in John F. Stover, *American Railroads* (Chicago: University of Chicago Press, 1961), 32.

7. These were referred to as "hot boxes" and still are an occasional problem today, particularly on freight trains.

8. Quoted in Albro Martin, *Railroads Triumphant: The Growth, Rejection and Rebirth of a Vital American Force* (Oxford: Oxford University Press, 1992), 51.

9. Stewart H. Holbrook, *The Story of American Railroads* (New York: Bonanza Books, 1947), 36.

10. This happened in England, too, on the Stockton & Darlington Railway. See Wolmar, *Fire & Steam*, 18.

11. Daniel J. Boorstin, in the preface to Stover, *American Railroads*, v.

12. Holbrook, *Story of American Railroads*, 3.

13. Highlighted in Chapter 1 of Wolmar, *Fire & Steam*.

14. Stover, *American Railroads*, vi.

15. Ibid., 31.

16. Quoted in ibid., 17.

17. Ibid.

18. Ibid., 29.

19. Terry Gourvish, *Mark Huish and the London & North Western Railway* (Leicester, U.K.: Leicester University Press, 1972), 32.

20. A 4-4-0 wheel arrangement predominated on the American railroads for the first half-century of their existence.

21. Holbrook, *Story of American Railroads*, 88.

22. Ibid., 8. Indeed, that tradition continues in some places. I remember helping the little train on the Darjeeling Himalayan Railway back onto the tracks in 2000.

23. A railway term for drivers who travel on trains as passengers.

24. Holbrook, *Story of American Railroads*, 39.

25. Ibid., 40.

26. Ibid., 41.

27. Martin, *Railroads Triumphant*, 15.

28. Ibid.

29. Nicholas Faith, *The World the Railways Made* (London: Bodley Head, 1990), 264.

30. Stover, *American Railroads*, 20.

31. Estimated in ibid., 20.

32. Arkansas, Missouri, Tennessee, and Vermont.

33. The report was by the Windom Committee of the U.S. Congress.

34. He was supported by a fellow senator, William R. King of Alabama, as the act also covered land grants for the Mobile & Ohio in Alabama and Mississippi.

35. An earlier act, in 1835, had allowed the Tallahassee Railroad Company the right of way 30 feet either side of the line, the free use of timber from public lands within 100 feet, and land for a terminal, but this was a far more modest measure that made little difference to the railroad's ultimate profitability.

36. Indeed, that may be the case, but they remained a danger to railroads, causing the Big Bayou Canot train disaster as recently as September 1993. In this incident, a barge hit a bridge, pushing the rails out of alignment, and a train plunged into the abyss, with the loss of forty-seven lives.

37. Martin, *Railroads Triumphant*, 46.

38. Ibid., 367.

39. One version was the first film I ever saw in a cinema, in the late 1950s.

40. Faith, *The World the Railways Made*, 317.

41. Allan Nevins, quoted in Faith, *The World the Railways Made*, 318.

42. Stover, *American Railroads*, 61.

43. Quoted in Faith, *The World the Railways Made*, 322.

44. Quoted in ibid., 320.

Chapter 5

1. Albert Schram, *Railways and the Formation of the Italian State in the Nineteenth Century* (Cambridge: Cambridge University Press, 1997), 14.

2. In practice, several countries, such as France and Italy, allowed slight variation, up to 1,465 millimeters, while the track in Germany was always precisely 1,435 millimeters. The Bern Convention of 1886 agreed that while 1,435 millimeters should be the standard, a gauge of up to 1,465 millimeters would be allowed on tight curves.

3. Allan Mitchell, *The Great Train Race: Railways and the Franco-German Rivalry* (Oxford, U.K.: Berghahn Books, 2000), 106.

4. The locomotive was also named after von Engerth, though he designed it in cooperation with another engineer, Fischer von Röslerstamm.

5. Then part of the Austrian Empire but now in Italy.

6. Today it is a UNESCO world heritage site.

7. Parts of it were actually in operation by 1981.

8. Clive Lamming, *Larousse des trains et des chemins de fer* (Paris: Larousse, 2005), 45 (author's translation).

9. Ibid., 47.

10. His name, sadly, is not reported in contemporary accounts.

11. P. M. Kalla-Bishop, *Italian Railroads* (New York: Drake, 1972), 28.

12. Confusingly, this tunnel is known by the French-sounding name Fréjus in Italy but is usually referred to as the Mont Cenis Tunnel in France.

13. Susan A. Ashley, *Making Liberalism Work: The Italian Experience, 1860–1914* (Santa Barbara, Calif: Greenwood, 2003), 40.

14. Ibid., 44.

15. Indeed, a Conservative MP, Robert Adley, was one of the fiercest opponents of rail privatization during these debates.

16. See Christian Wolmar, *Fire & Steam: A New History of the Railways in Britain* (London: Atlantic Books, 2007), 97ff.

17. This was a safety initiative that was resisted strenuously in the United Kingdom until the Armagh accident in 1886, which killed scores of schoolchildren. See Wolmar, *Fire & Steam*, 166ff.

18. Mitchell, *The Great Train Race*, 136.

19. Ibid., 149.

20. Ibid., 153.

21. Ibid., 171.

22. Augustus J. Veenendaal Jr., *Railways in the Netherlands: A Brief History, 1834–1994* (Palo Alto, Calif.: Stanford University Press, 2002), 30.

23. From Schram, *Railways and the Formation of the Italian State*, 69. The exact figures for France, Germany, and Great Britain at the turn of the century were, respectively, 0.98, 0.92, and 0.93. These comparisons are, by necessity, fairly arbitrary. Different countries count their railways in different ways. For example, some include sidings or narrow-gauge railways, while others do not. However, these statistics do give a broad idea of the levels of development of each country's system.

24. The figures in 1900 for Italy, Spain, and Austria-Hungary were, respectively, 0.51, 0.71, and 0.75. Schram, *Railways and the Formation of the Italian State*, 69.

25. In terms of kilometers of railway per square kilometer.

26. Maxime Hélène, "Louis Favre, Constructor of the St Gothard Tunnel," *Scientific American*, Supplement No. 365, December 30, 1882.

27. Mitchell, *The Great Train Race*, 146.

28. Ibid., 44.

Chapter 6

1. Then called New Granada.

2. In terms of cost per mile.

3. The canal, incidentally, is reckoned to have claimed 27,500, but it did take nineteen years to build. Work was started by the French in 1881 but ended in 1889, and was restarted in 1904 by the Americans.

4. "History of the Panama Railroad," http://www.trainweb.org/panama/history1.html.

5. Calculated as amount of freight carried per kilometer of line.

6. Albro Martin, *Railroads Triumphant: The Growth, Rejection and Rebirth of a Vital American Force* (Oxford: Oxford University Press, 1992), 29.

7. Anthony Burton, *On the Rails* (London: Aurum Press, 2004), 94.

8. Oscar Lewis, *The Big Four* (New York: Alfred A. Knopf, 1938), 3.

9. Ibid., 23.

10. Ibid.

11. Quoted in many places, including John F. Stover, *American Railroads* (Chicago: University of Chicago Press, 1961), 71.

12. Lewis, *The Big Four*, 62.

13. John Hoyt Williams, *A Great and Shining Road: The Epic Story of the Transcontinental* (Lincoln: University of Nebraska Press, 1996), 183.

14. Stover, *American Railroads*, 75.

15. Stewart H. Holbrook, *The Story of American Railroads* (New York: Bonanza Books, 1947), 171.

16. Leland Stanford Jr., who died of typhoid in Italy in 1884.

17. Quoted in Stephen E. Ambrose, *Nothing Like It in the World* (New York: Pocket Books, 2005), 130.

18. Dailey, Elliott, Joyce, Kennedy, Killeen, McNamara, Shay, and Sullivan— which demonstrates that not all the Central Pacific workers were Chinese.

19. Variously called Promontory Summit or Promontory Mountain, it is 6 miles west of Ogden, but the main line no longer goes through that spot.

20. Holbrook, *The Story of American Railroads*, 170.

21. Seymour Dunbar, *A History of Travel in America*, vol. 1, reprint (Santa Barbara, Calif.: Greenwood Press, 1968 [1937]), 265.

22. There were earlier sleeping compartments, but Pullman made his with the intention of attracting affluent travelers with higher standards.

23. Stover, *American Railroads*, 76.

24. He was actually born in Canada.

25. Which was not yet a state.

26. Quoted in A.F. Garnett, *Steel Wheels* (Waldenbury, East Sussex, U.K.: Cannwood Press, 2005), 120.

27. Nicholas Faith, *The World the Railways Made* (London: Bodley Head, 1990), 66.

28. David Cruise and Alison Griffiths, *Lords of the Line: The Men Who Built the CPR* (New York: Viking, 1988), 138.

29. Quoted in Nick Mika and Helma Mika, *The Railways of Canada: A Pictorial History* (New York: McGraw-Hill Ryerson, 1972), 100.

30. Ibid., 128.

31. An anonymous writer quoted in Faith, *The World the Railways Made*, 67.

32. Brian Fawcett, *Railways of the Andes* (East Harling, Norfolk, U.K.: Plateway Press, 1963), 18.

Chapter 7

1. Steven G. Marks, *Road to Power: The Trans Siberian Railroad and the Colonization of Asian Russia, 1850–1917* (Ithaca, N.Y.: Cornell University Press, 1991), xii.

2. Mile of railway per square mile of territory.

3. Quoted in Marks, *Road to Power*, 124.

4. There was, of course, no Pakistan at the time, and India shared a long border with Afghanistan.

5. In today's currency that represents around £4 billion (or $6.5 billion), but even that huge figure gives a misleading impression because the cost in terms of the gross domestic product (GDP) of Russia at the time was enormous, second only to its expenditure on World War I. Moreover, estimates vary because of corruption and bad accounting; the cost may have been nearly double, as much as 1,400 million rubles (say, around £140 million or $700 million in money of the day, or approximately £7 billion or $11 billion in today's currency).

6. Alexander Polunov, quoted at http://forum.warfare.ru/index. php?show-topic=1251.

7. Marks, *Road to Power*, 126.

8. Ibid., 135.

9. There were occasional exceptions, which at times led to food riots, but for the most part food standards were surprisingly good given the inevitable supply difficulties.

10. Quoted in Marks, *Road to Power*, 191.

11. Ibid., 166.

12. Ibid., 169.

13. Ibid., 225.

14. *Railroad Gazette*, November 12, 1897, 798.

15. Bryan Morgan, ed., *The Great Trains* (New York: Crown, 1973), 149.

16. The color of Britain's colonies on maps of the day.

17. George Tabor, *The Cape to Cairo Railway and River Routes* (London: Genta, 2003), 3.

18. Quoted in ibid., 11.

19. Ibid., 83.

20. These friendly-looking herbivore hippos are, in fact, the biggest killers in Africa today. They are incredibly fierce if they feel threatened.

21. Quoted in Tabor, *Cape to Cairo Railway*, 87.

22. Ibid., 85.

23. Ibid.

24. Now Mutane.

25. Tabor, *Cape to Cairo Railway*, 95.

26. Now Zimbabwe.

27. Technically, the second Boer War, as there had been a brief one in 1880–1881 when the Boers successfully resisted a British attempt to take over the Transvaal. The second war lasted until 1902 and ultimately resulted in the creation of the Union of South Africa.

28. Tabor, *Cape to Cairo Railway*, 150.

29. From his journal, quoted in Tabor, *Cape to Cairo Railway*, 175.

30. Now in the Democratic Republic of Congo.

31. Now Ilebo.

32. Now Shaba.

33. Then Portuguese East Africa.

34. Although it was not until a battle further south in November 1899 at Umm Diwaykarat that the Mahdi's forces were finally defeated.

35. About which he wrote a book, *The River War: An Historical Account of the Reconquest of the Sudan* (London: Longmans, Green, 1899).

36. Tabor, *Cape to Cairo Railway*, 237.

37. M. F. Hill, *Permanent Way*, vol. 2, *The Story of the Tanganyika Railways* (Nairobi: East African Railways and Harbours, 1957), 83.

38. Remarkably, he had not lost a single man to disease or hostile action.

39. The distance varies in different sources, ranging from 297 to 309 miles.

40. 3 feet, 6 inches; 4 feet, 8.5 inches; and 5 feet, 3 inches.

Chapter 8

1. These figures are taken from O. S. Nock, *World Atlas of Railways* (London: Mitchell Beazley, 1978). This figure refers to route miles rather than the length of track, which, of course, would be far greater given that some railways were double tracked and huge sidings were already being built in goods yards.

2. Ignoring tiny states like Gibraltar or Andorra.

3. Albania did not exist as a separate state until 1912 and only built its first railway in 1947.

4. As with many stories of early pioneering railways, this version of events has been challenged, but no coherent explanation has been given for such an odd choice of gauge.

5. Not including Alaska.

6. Now called states.

7. From the *Correio Mercantil*, quoted in Pedro C. da Silva Telles, *A History of Brazilian Railways*, Part 1: *The First Railways*, translated by Dr. Paul E. Waters (Bromley, Kent, U.K.: P. E. Waters and Associates, 1987), 55.

8. Just as with the Argentinian story relating to its adoption of the 5-foot, 6-inch, gauge, this tale has also been challenged. Historians of the Brazilian railway confess to being uncertain of the reason for using a gauge of 5 feet, 3 inches.

9. Later called the Santos a Jundiaí Railway.

10. Telles, *The First Railways*, 42.

11. Indeed, a bizarre campaign has been waged by a few eccentrics in Britain for many years advocating that the railways be paved over and used by buses and trucks, an utterly unworkable concept.

12. Telles, *The First Railways*, 56.

13. Oscar Zanetti and Alejandro García, *Sugar and Railroads: A Cuban History, 1837–1959* (Chapel Hill: University of North Carolina Press, 1998), 80.

14. Manuel Moreno Fraginals, quoted in Zanetti and García, *Sugar and Railroads*, 99.

15. The list is contained in John Marshall, *The Guinness Book of Rail Facts and Figures* (Guinness, 1975), 71. The exceptions are Pike's Peak on the Manitou and Pike's Peak Railway in Colorado (14,109 feet), and the Climax Spur on the Colorado & Southern (11,465 feet).

16. Brian Fawcett, *Railways of the Andes* (East Harling, Norfolk, U.K.: Plateway Press, 1997 [1963]), 27.

17. A maximum of 4.4 percent was specified in the contract, but in truth there were several stretches of 4.9 percent.

18. It stayed British for nearly a century until it was taken over by a Chilean company in 1980.

19. D. Trevor Rowe, *The Railways of South America* (Cornwall, U.K.: Locomotives International, 2000), 46.

20. P. M. Kalla-Bishop, *Mediterranean Island Railways* (Newton Abbot, Devon, U.K.: David and Charles, 1970), 97.

21. Ibid., 19.

22. The towns are 268 miles apart on today's highway.

23. Quoted from *The Railway Magazine*, March 1898, in O. S. Nock, *Railways of Asia and the Far East* (London: Adam and Charles Black, 1998), 148.

24. Then sited further inland than today.

25. Quoted in Ralph William Huenemann, *The Dragon and the Iron Horse: The Economics of Railroads in China, 1876–1937* (Cambridge: Harvard University Press, 1984), 38.

26. Ibid., 5.

27. Ibid., 38.

28. Nicholas Faith, *The World the Railways Made* (London: Bodley Head, 1990), 159.

29. Neill Atkinson, *Trainland: How Railways Made New Zealand* (New York: Random House, 2007), 39.

30. Quoted in ibid., 30.

31. Ibid., 13.

32. Ibid., 57.

Chapter 9

1. Ernest de Selincourt, ed., *Wordsworth's Guide to the Lakes*, 5th ed. (Oxford: Oxford University Press, 1970 [1835]), 156.

2. Albro Martin, *Railroads Triumphant: The Growth, Rejection and Rebirth of a Vital American Force* (Oxford: Oxford University Press, 1992), 13.

3. Neill Atkinson, *Trainland: How the Railways Made New Zealand* (New York: Random House, 2007), 15.

4. This story is told in more detail in my previous book, *Fire & Steam: A New History of the Railways in Britain* (London: Atlantic Books, 2007), 141.

5. Nicholas Faith, *Locomotion* (London: BBC Books, 1993), 153.

6. Teresa Van Hoy, *A Social History of Mexico's Railroads* (Lanham, Md.: Rowman and Littlefield, 2008), 209.

7. Frank Norris, *The Octopus* (New York: Doubleday, 1901).

8. John F. Stover, *American Railroads* (Chicago: University of Chicago Press, 1961), 107.

9. Ibid., 108.

10. Ibid., 110.

11. Quoted in Terry Gourvish, *Railways and the British Economy, 1830–1914* (London: Macmillan, 1980), 48.

12. Quoted in Nicholas Faith, *The World the Railways Made* (London: Bodley Head, 1990), 59.

13. Ibid., 60.

14. Faith, *Locomotion*, 148.

15. At the outbreak of World War I, the British railways, for example, employed 625,559 people to provide just under 20,000 miles of track. Crudely extrapolating those figures would suggest that across the world at that stage, at 30 workers per mile and 780,000 miles of railway, the industry employed some 23 million people.

16. Allan Mitchell, *The Great Train Race: Railways and the Franco-German Rivalry* (Oxford, U.K.: Berghahn Books, 2000), 57.

17. Harold Edmonson, "Across America," in *Trains Around the World* (London: Octopus Books, 1972), 28.

18. Gourvish, *Railways and the British Economy*, 21.

19. Van Hoy, *A Social History of Mexico's Railroads*, 210.

20. Quoted in Faith, *The World the Railways Made*, 114.

21. A. W. Currie, *The Grand Trunk Railway of Canada* (Toronto: University of Toronto Press, 1957), 4.

22. Quoted in Faith, *The World the Railways Made*, 116.

23. Principally by a group of economists in the 1960s who used "new economic history" to analyze the effect of social trends such as slavery or the development of the railways. Despite their claims to rigor, their work has largely been discredited.

24. Mitchell, *The Great Train Race*, 57.

25. Faith, *The World the Railways Made*, 134.

26. Walden Pond, Henry David Thoreau's beloved home.

27. Quoted in Faith, *The World the Railways Made*, 136. The name of the yard is not specified but it is presumably Willesden.

28. Ibid., 142.

29. Ibid., 136.

30. Michael Robbins, *The Railway Age* (London: Penguin Books, 1962), 157.

31. Daniel J. Boorstin, "Preface," in Stover, *American Railroads*, vi.

32. Martin, *Railroads Triumphant*, 93.

33. Stewart H. Holbrook, *The Story of American Railroads* (New York: Bonanza Books, 1947), 354.

34. There were at first four zones across the United States; several states had not yet been admitted to the union, including Alaska and Hawaii, which added two more zones.

35. Quoted in ibid., p. 354.

36. Quoted (in French) in Bryan Morgan, ed., *Great Trains* (New York: Crown, 1973), 127.

37. Oddly for an American rail company, the term "railway" rather than "railroad" was used.

38. Faith, *Locomotion*, 163.

39. Jack Simmons, *The Victorian Railway* (London: Thames and Hudson, 1991), 330.

40. See Wolmar, *Fire & Steam*, 133–136.

41. Basil Cooper, *A Century of Trains* (Colchester, Essex, U.K.: Brian Trodd, 1988), 64.

42. Martin, *Railroads Triumphant*, 119.

43. Jim Harter, *World Railways of the Nineteenth Century: A Pictorial History in Victorian Engravings* (Baltimore: Johns Hopkins University Press, 2005), 417.

44. Ibid., 420.

Chapter 10

1. T. T. Bury, in 1831, quoted in P. J. G. Ransom, *Locomotion: Two Centuries of Train Travel* (Stroud, Gloucestershire, U.K.: Sutton, 2001), 22.

2. September 13, 1845.

3. Wolfgang Schivelbusch, *Railway Journey: The Industrialization of Time and Space in the 19th Century* (Oxford, U.K.: Berg, 1977), 67.

4. All these quotes are taken from Ian Kerr, *Engines of Change: The Railroads That Made India* (Westport, Conn.: Praeger, 2007), 95–96.

5. Ibid., 96.

6. Quoted in ibid., 98.

7. Quoted in Bryan Morgan, ed., *Great Trains* (New York: Crown, 1973), 32.

8. Geoffrey Freeman Allen, *Luxury Trains of the World* (London: Bison Books, 1979), 9.

9. Quoted in Albro Martin, *Railroads Triumphant: The Growth, Rejection and Rebirth of a Vital American Force* (Oxford: Oxford University Press, 1992), 51.

10. Allen, *Luxury Trains of the World*, 13.

11. The last operated until 2004 when GNER's *Tyne Tees Pullman* and *Yorkshire Pullman* lost their names.

12. W. M. Acworth, *Railways of England*, 5th ed. (London: John Murray, 1900; London: Ian Allan, 1965), 281.

13. This account is from Morgan, ed., *Great Trains*, 123.

14. Neill Atkinson, *Trainland: How Railways Made New Zealand* (New York: Random House, 2007), 28.

15. Ibid., 139.

16. Ibid., 140.

17. Except that some "express" trains were reserved for only first- and second-class passengers.

18. Published in E. Foxwell and T. C. Farrer, *Express Trains, English, French and Foreign: Being a Statistical Account of All the Express Trains of the World* (London: Smith, Elder, 1889).

19. Augustus J. Veenendaal, *Railways in the Netherlands: A Brief History 1834–1994* (Palo Alto, Calif.: Stanford University Press, 1998), 59.

20. Acworth, *Railways of England*, 280, 461.

21. Allan Mitchell, *The Great Train Race: Railways and the Franco-German Rivalry* (Oxford, U.K.: Berghahn Books, 2000), 160.

22. Acworth, *Railways of England*, 282.

23. Ibid., 469.

24. Ibid., 470.

25. Martin, *Railroads Triumphant*, 62.

26. Reproduced in Ransom, *Locomotion*, 55–60.

27. Ibid., 62–65.

28. Reproduced in Morgan, ed., *Great Trains*, 168.

29. O. S. Nock, *Railways Then and Now: A World History* (London: Paul Elek, 1975), 124.

30. Jim Harter, *World Railways of the Nineteenth Century: A Pictorial History in Victorian Engravings* (Baltimore: Johns Hopkins University Press, 2005), 277.

31. Demolished in the 1960s.

32. Dr. Dionysus Lardner, quoted in Steven Parissien, *Station to Station* (London: Phaidon, 1997), 8.

33. Harter, *World Railways of the Nineteenth Century*, 278.
34. Ibid., 275.

Chapter 11

1. This concept is fully aired in L.F.C. Turner's book *The Origins of the First World War* (London: Edward Arnold, 1970).
2. John Westwood, *Railways at War* (London: Osprey, 1980), 130.
3. Ibid., 131.
4. Ibid., 146.
5. T. E. Lawrence.
6. The term "Arabia" was used at the time to cover what is now mostly Saudi Arabia.
7. Hitler commandeered 2419 when he conquered France and in June 1940 forced the French to sign their surrender in it. However, the SS blew it up before it could be used again as the scene of further humiliation.
8. Geoffrey Freeman Allen, *Railways of the Twentieth Century* (New York: W. W. Norton, 1983), 9.
9. Albro Martin, *Railroads Triumphant: The Growth, Rejection and Rebirth of an American Force* (Oxford: Oxford University Press, 1992), 120.
10. Figures from Martin, *Railroads Triumphant*, 124. Interestingly, around the same number are carried annually on Britain's rail network today, though that figure includes commuters.
11. Quoted in Allen, *Railways of the Twentieth Century*, 70.
12. The term "Pacific" refers to a common type of locomotive in the United States and elsewhere. It has a wheel formation of 4-6-2, four bogie wheels, six large driving wheels, and two trailing wheels at the back.
13. In fact, the distance between Chicago and St. Paul was 410 miles. According to a contemporary *Time* magazine report (January 14, 1935), the train took 420 minutes, thus averaging 58.8 miles per hour.
14. Part of the Brighton route was originally electrified using overhead line, but it was soon converted to the third rail system.
15. An attempt to power railways with stationary engines that created a vacuum in a tube between the rails, and thus involved a piston attached to the train. The method proved inefficient and ultimately was unsuccessful because of the difficulty in sustaining the vacuum.
16. O. S. Nock, *Railways Then and Now: A History of the World's Railways* (London: Paul Elek, 1975), 161.
17. Neill Atkinson, *Trainland: How Railways Made New Zealand* (New York: Random House, 2007), 120.
18. Ibid.

Chapter 12

1. Quoted in Michael Robbins, *The Railway Age* (London: Penguin Books, 1962), 154.

2. Toward the end of the war, the British developed the Barnes Wallis Tallboy bomb, made famous in the Dambuster story, which could destroy railway viaducts even without hitting them directly, but these were in limited supply.

3. John Westwood, *Railways at War* (London: Osprey, 1980), 189.

4. Ibid., 188.

5. Geoffrey Freeman Allen, *Railways of the Twentieth Century* (New York: W. W. Norton, 1983), 124.

6. Westwood, *Railways at War*, 201.

7. Ibid., 195.

8. Allen, *Railways of the Twentieth Century*, 123.

9. Bryan Morgan, "Comments on Three Nations," in Bryan Morgan, ed., *The Railway-Lovers Companion* (London: Eyre and Spottiswoode, 1963), 429–437.

10. Figures from Allen, *Railways of the Twentieth Century*, 131.

11. Geoffrey Freeman Allen, *Luxury Trains of the World* (London: Bison Books, 1979), 149.

12. John Westwood, *The Pictorial History of Railways* (London: Bison Books, 1988), 161.

13. According to O. S. Nock, *World Atlas of Railways* (London: Mitchell Beazley, 1978), 8.

Chapter 13

1. Actually defined as no curves with a radius of less than 5 kilometers (about 3 miles).

2. The M20 was actually completed in the 1970s and 1980s.

3. Mark Reutter, "The Lost Promise of the American Passenger Train," *Wilson Quarterly* (Winter 1994).

4. Ligne à Grande Vitesse means "high-speed line"—the trains are called Trains à Grande Vitesse, a term often wrongly used to refer to the route.

5. Lord Adonis, Chancellor's Lecture, University of Kent, January 30, 2009.

6. There are various other sections of line, such as the *interconnection*, the detour around the east side of Paris, and several lines split, but there are four main LGVs emanating from Paris.

7. Beginning in 2010, all high-speed lines in Europe will be open to rival operators as a result of EU rules.

8. There was a major accident involving a high-speed train traveling at 125 miles per hour at Eschede in Germany in 1998, which caused 101 deaths, but the disaster took place on conventional track.

9. Of emerging economies, Brazil is probably the least well served by passenger railways.

10. The situation is quite fluid, however. The Conservative opposition announced in 2008 that it would support the construction of a north-south high-speed line that, provided the Tories win the 2010 election, would be completed in the mid-2020s, but no new money was promised for the project. Labour, which had long refrained from making any such commitment, announced late in 2008 that it would commission an assessment of the viability of a high-speed line between London and Birmingham by a newly created government company called HS2. This report will be published early in 2010.

11. Clive Lamming, *Larousse des Trains et des Chemins de Fer* (Paris: Larousse, 2005), 85.

12. See, for example, *Great Railway Journeys of Europe* (London: Insight Guides, 2002); Tom Savio and Anthony Lambert, *Extraordinary Rail Journeys* (London: Ted Smart, 2004); and Max Wade-Matthews, *Classic Railway Journeys of the West* (London: Anness, 2001.

13. Richard Hope, "Foreword," in Murray Hughes, *Rail 300: The World High Speed Train Race* (Newton Abbot, Devon, U.K.: David and Charles, 1988), 7.

14. I covered this in the column I write for *Rail* magazine in issue 596—obtainable via my website, www.christianwolmar.co.uk.

15. As ever, the issue is more complex than might be anticipated. In terms of accident per trip, rail is safer than aviation, but in terms of accident per million miles, flying is around seven times safer, according to statistics in the 1990s.

16. See, for example, Carlo Pfund, *Separation Philosophy of the European Union—Blessing or Curse* (Bern: LITRA, 2001); Christian Wolmar, *On the Wrong Line—How Ideology and Incompetence Wrecked Britain's Railways* (London: Aurum, 2005).

17. Geoffrey Freeman Allen, *Railways of the Twentieth Century* (New York: W. W. Norton, 1983), 123.

18. Erwin Berghaus, *The History of Railways* (London: Barrie and Rockliff, 1964), 195.

19. Lamming, *Larousse des Trains et des Chemins de Fer*, 123.

20. John Westwood, *Railways at War* (London: Osprey, 1980), 219.

BIBLIOGRAPHY

One of my reasons for writing this book was that there is so little literature that focuses on the social history of the railways. There are a few countries that are well covered by a book or two that convey the importance of the railways to their history, but there are huge gaps. For example, Cuba, New Zealand, India, and Mexico are all blessed with particularly good studies, whereas I struggled to find an equivalent on France, Germany, or Spain. Although I may have missed them, I suspect they do not exist. Therefore this bibliography is by definition patchy, and in a way I am hoping to inspire academics and railway writers to fill the gaps.

There is, too, the problem that many writers on the railways focus on narrow technical aspects of the story and other details that are geared toward rail enthusiasts rather than general readers. I have mentioned some of these works here not only because otherwise it would be a particularly short list, but also because they do often contain sections of wider interest.

This bibliography is, of necessity, highly selective, barely scratching the surface of the huge amount of railway literature (although, sadly, that is rather a flattering description for much of it). I have included those books in the coffee-table style that contain sufficient text to be enlightening, but I have omitted books that cover solely the British railways, as there is a thorough bibliography on that subject in my previous book *Fire & Steam: A New History of the Railways in Britain* (London: Atlantic Books, 2007). Apart from a couple of French publications, I have not included books only available in foreign languages. Some of the books mentioned below are obviously old and out of print, but I have included them anyway because they contain useful information; they can often be obtained online through secondhand book sites or in libraries (though my copies of most of them are ones that have been discarded by libraries).

General

There are surprisingly few general histories of the world's railways, apart from various coffee-table books. Michael Robbins's short but classic *The Railway Age* (London: Penguin Books, 1965) is a great starting point. An entertaining and quirky early one is *The History of Railways* by Erwin Berghaus (London: Barrie and Rockliff, 1964).

The coffee-table-sized books that I found most useful include: *The World's Railways* by Christopher Chant (Kent, U.K.: Grange, 2002); *On the Rails* by Anthony Burton (London: Aurum, 2004); *Locomotion* by Nicholas Faith (London: BBC Books, 1993); O. S. Nock, *World Atlas of Railways* (London: Mitchell Beazley, 1978), which has very useful maps and a good overview; Bryan Morgan, ed., *Great Trains* (New York: Crown, 1973); O. S. Nock, *Railways Then and Now: A World History* (London: Paul Elek, 1975); O. S. Nock, ed., *Encyclopaedia of Railways* (London: Book Club Associates, 1977); P.J.G. Ransom, *Locomotion: Two Centuries of Train Travel* (Stroud, Gloucestershire, U.K.: Sutton, 2001); John Westwood, *The Pictorial History of Railways* (London: Bison Books, 2008); Geoffrey Freeman Allen, *Railways: Past, Present and Future* (London: Orbis, 1982); and Basil Cooper, *A Century of Train* (Colchester, Essex, U.K.: Brian Trodd, 1988). One of the most comprehensive books on world railways is *Larousse des Trains et des Chemins de Fer* by Clive Lamming (Paris: Larousse, 2005), which, despite the author's name, is only available in French and is laid out in dictionary format.

The best book on how the railways changed the world is the wonderfully titled *The World the Railways Made* by Nicholas Faith (London: Bodley Head, 1990). A rare philosophical book on the impact of the railways is *Railway Journey: The Industrialization of Time and Space in the Nineteenth Century* by Wolfgang Schivelbusch (Oxford, U.K.: Berg, 1986). An excellent volume on the British influence on railways across the world is Anthony Burton, *Railway Empire* (London: John Murray, 1994). Two books by Geoffrey Freeman Allen, one of the great railway authors, were particularly helpful: *Railways of the Twentieth Century* (New York: W. W. Norton, 1983) and the lavishly illustrated *Luxury Trains of the World* (London: Bison Books, 1979).

Railways at War by John Westwood (London: Osprey, 1980) covers rather unevenly the role of the railways in wars from the American Civil War to the Korean War. An absolutely sumptuous book on the nineteenth century is Jim Harter, *World Railways of the Nineteenth Century: A Pictorial History in Victorian Engravings* (Baltimore: Johns Hopkins University Press, 2005), which has an enormous number of beautiful engravings.

On stations, there is the lavish *Station to Station* by Steve Parissien (London: Phaidon, 1997), and on art, the book specially produced for a 2008 exhibition in Liverpool, *The Railway Art in the Age of Steam*, by Ian Kennedy and Julian Treuherz (New Haven, Conn.: Yale University Press, 2008).

Europe

O. S. Nock's *Railways of Western Europe* (London: Adam and Charles Black, 1977) is, like many of his books, rather patchy, being a mixture of history and personal experience, but is nevertheless useful. On France, the best basic history is written in French by four authors, *Histoire du Réseau Ferroviaire Français*, by Patricia Laederich, Marc Gayda, Pierre Laederich, and André Jacquot (Valignat, France: Editions de l'Ormet, 1996). Brian Perren's comprehensive *TGV Handbook* (London: Capital Transport, 1998) tells the story of the development of high-speed lines in France. *ICE: High-Tech on Wheels*, by various authors (Hamburg: Hestra-Verlag, 1991), covers the German high-speed concept, and *Rail 300* by Murray Hughes (Newton Abbot, Devon, U.K.: David and Charles, 1988) gives an early overview of high speed trains. Few books compare different rail networks, but a rare example is the superb examination of the role of the railways of France and Germany in the three wars between the two nations, *The Great Train Race: Railways and the Franco-German Rivalry*, by Allan Mitchell (Oxford, U.K.: Berghahn, 2000).

Oddly, the relatively small Dutch network is blessed with a superb straightforward history, *Railways in the Netherlands: A Brief History, 1834–1994* by Arthur J. Veenendaal Jr. (Palo Alto, Calif.: Stanford University Press, 2001). H. C. Casserley's *Outline of Irish History* (Newton Abbot, Devon, U.K.: David and Charles, 1974) focuses on technology rather than social impact. *Railways and the Formation of the Italian State in the Nineteenth Century* by Albert Schram (Cambridge: Cambridge University Press, 1977) covers exactly what its title suggests.

P. M. Kalla-Bishop wrote several books in a series entitled "Railway Histories of the World" that are rather too focused on technical aspects but do include some background to various railways, including *Mediterranean Island Railways* (Newton Abbot, Devon, U.K.: David and Charles, 1970), *Italian Railroads* (New York: Drake, 1972), and *Hungarian Railroads* (New York: Drake, 1973).

Americas

The United States is obviously very well covered. The most useful general histories I came across—though I am sure there are many others—were: Albro Martin, *Railroads Triumphant: The Growth, Rejection and Rebirth of an American Force* (Oxford: Oxford University Press, 1992); John F. Stover,

American Railroads (Chicago: University of Chicago Press, 1961); Stewart H. Holbrook, *The Story of American Railroads* (New York: Bonanza Books, 1947). Stephen E. Ambrose, *Nothing Like It in the World* (New York: Pocket Books, 2005), is a terribly choppy but thorough account of the building of the first transcontinental, and Oscar Lewis's *The Big Four* (New York: Alfred A. Knopf, 1938) gives the background to its four main protagonists.

Canada, too, has a good literature, including general histories such as O. S. Nock, *Railways of Canada* (London: Adam and Charles Black, 1973), and Nick and Helma Mika, *The Railways of Canada: A Pictorial History* (New York: McGraw-Hill Ryerson, 1972). On the construction of the transcontinentals, there is David Cruise and Alison Griffiths, *Lords of the Line: The Men Who Built the Canadian Pacific Railway* (New York: Viking, 1988).

David Rollinson's *Railways of the Caribbean* (London: Macmillan, 2001) provides an excellent outline of the many obscure and mostly sadly defunct rail systems on the islands. *Sugar and Railroads: A Cuban History, 1837–1959*, by Oscar Zanetti and Alejandro García (Chapel Hill: University of North Carolina Press, 1998), and Teresa Van Hoy's *A Social History of Mexico's Railroads* (Lanham, Md.: Rowman and Littlefield, 2008) are two examples of those rare books that provide highly detailed context to railway development.

The classic work on South American railroads is D. Trevor Rowe, *The Railways of South America* (Cornwall, U.K.: Locomotives International, 2000), which was reprinted relatively recently, and Brian Fawcett, *Railways of the Andes* (East Harling, Norfolk, U.K.: Plateway Press, 1997), covers these magical railways in great detail.

Asia

The ever prolific O. S. Nock wrote a general overview, *Railways of Asia and the Far East* (London: Adam and Charles Black, 1978). India is well covered not least thanks to the writings of Ian J. Kerr with *Engines of Change: The Railroads That Made India* (Santa Barbara, Calif.: Praeger, 2007) and *Building the Railways of the Raj, 1850–1900* (Oxford: Oxford University Press, 1995), as well as various Indian authors, such as K. R. Vaidyanathan, *150 Glorious Years of Indian Railways* (Mumbai: English Edition, 2003), and Roopa Srinivasan, Manish Tiwari, and Sandeep Silas, *Our Indian Railway: Themes in India's Railway History* (New Delhi: Foundation Books, 2006).

On China, there is little literature, perhaps because it was a latecomer to major railway development, but the book by Ralph William Huenemann, *The Dragon and the Iron Horse: The Economics of Railroads in China, 1876–1937* (Cambridge: Harvard University Press, 1984), gives a good account of the railway's origins with considerable stress on the economics. There are fabulous pictures in *China: The World's Last Steam Railway*, by John Tickner, Gordon Edgar, and Adrian Freeman (London: Artists' and Photographers' Press, 2008).

Japan, in contrast, has a large literature, some of it translated into English, including Shoji Sumita, *Success Story, The Privatisation of Japanese National Railways* (London: Profile Books, 2000), which provides a basic history, and *High Speed in Japan* by Peter Semmens (South Yorkshire, U.K.: Platform 5, 2000), the story of the *Shinkansen*. There are several books on the Trans-Siberian, including Steven G. Marks, *Road to Power, The Trans-Siberian Railroad and the Colonization of Asian Russia, 1850–1917* (Ithaca, N.Y.: Cornell University Press, 1991).

Africa

Africa is not well served with railway books. George Tabor does his best to give coherence to the highly complex story of the *Cape to Cairo Railway and River Routes* (London: Genta, 2003). I happened to come across M. F. Hill, *The Permanent Way*, vol. 2, *The Story of the Tanganyika Railways* (Nairobi: East African Railways and Harbours, 1957), but not its earlier companion volume that covers the Kenya and Uganda railways, and it is full of fascinating tales of the early years of railways in East Africa. John Day's *Railways of South Africa* (London: Arthur Barker, 1963) provides basic information on the southern African railways, and he wrote a similar volume on North African railways.

Australia

Australia has a number of histories, including *Romance of Australian Railways*, which is by the prolific Patsy Adam Smith (Adelaide, South Australia: Rigby, 1973), and *Railways of Australia* by C. C. Singleton and David Burke (Melbourne: Angus and Robertson, 1963). Tim Fischer, a former deputy prime minister of Australia, covers the railways of both his own country and others in his entertaining *Transcontinental Train Odyssey* (St. Leonards, NSW, Australia: Allen and Unwin, 2004). I will end on the best social history of any railway that I came across: *Trainland: How Railways Made New Zealand* by Neill Atkinson (New York: Random House, 2007). It not only puts the railways in context but gives a thorough account of their history and impact, and is also lavishly illustrated with fantastic photographs and publicity posters. If only other countries had similar histories, writing this book would have been far easier.

INDEX

Photo credit: Paul Bigland

Christian Wolmar is a writer and broadcaster specializing in the social history of railways and transport. He has written for major British newspapers for many years and has contributed to many other publications, including the *New York Times* and *Newsday*. He frequently appears on TV and radio as an expert commentator. His most recent books are the widely acclaimed *The Subterranean Railway*, a history of the London Underground, the world's oldest system; *Fire & Steam*, the story of Britain's railways; and *Engines of War*.

www.christianwolmar.co.uk.